1st d
20 —

ADIRONDACK UPLAND FLORA:
An Ecological Perspective

Cinnamon fern (*Osmunda cinnamomea*)
This fern's fronds are dimorphic. The vernal, fertile, spore-producing, orange-brown ones wither by July, while the sterile, photosynthetic ones remain green until autumn.
© Michael Kudish 1992.

ADIRONDACK UPLAND FLORA:
An Ecological Perspective

Michael Kudish

The Chauncy Press
Saranac, New York
1992

ADIRONDACK UPLAND FLORA:
AN ECOLOGICAL PERSPECTIVE
Copyright © 1992 by Michael Kudish

All Rights Reserved No part of this book may be reproduced without written permission from the publisher; nor may any part of this book be stored in a retrieval system, or transmitted in any form or by means electronic, mechanical, photocopying, recording, or other without written permission from the publisher.

Library of Congress Cataloging-in-Publication Data

Kudish, Michael, 1943–
    Adirondack upland flora: an ecological perspective / Michael Kudish
      p.   cm.
    Includes bibliographical references and index.
    ISBN: 0-918517-16-8 (hard cover) : $45.00
1. Botany—New York (State)—Adirondack Mountains Region.
2. Botany—New York (State)—Adirondack Mountains Region—Ecology. I. Title.
QK177.K83   1992
581.9757′5-dc20   91-44851   CIP

Printed in Hong Kong
Produced by Blaze Int'l Productions, Inc.

THE CHAUNCY PRESS

Dedication

I dedicate this work
to
My Parents,

Aaron and Esther Kudish,

in recognition of their
inspiration and encouragement

Marsh marigold (*Caltha palustris*) The five yellow structures are sepals, not petals. Because many species in the Buttercup family lack petals, the sepals have taken over the responsibility of insect attraction. © Jim Kraus 1992.

FOREWORD

The *Adirondack Upland Flora* is the first of its kind to cover a major region of the Adirondack Mountains, specifically, the Adirondack Upland. It focuses on those vascular plants, both native and naturalized, growing in this geographic region. The plants themselves are characterized by special cells and tissues whose prime function is to support and conduct. They are, in fact, plants with plumbing, and include all ferns, clubmosses, horsetails, conifers and flowering plants. Native species, those which have migrated into the region without human assistance, number 456. Naturalized species, those introduced by people into the region but which have reproduced without human assistance, number 107. A comprehensive list of plants within the Flora area is included in Chapter 5. Chapter 1 includes a map which locates the thirty-mile radius circle showing that portion of the Adirondack Upland covered by this Flora, and its upper and lower elevational limits.

Discussions of those factors which influence plant invasions and migrations follow. These include relationships between postglacial conditions, people and plants; those between plants and factors which affect them from above, i.e. climate; and those between plants and factors which affect them from below, i.e. soils.

Protected species categorized by the New York State Department of Environmental Conservation as either endangered, threatened, rare, or (in current jargon) exploitably vulnerable are covered in Chapter 6.

All information about the plants included in *Adirondack Upland Flora* was compiled from several of the author's resources:

Herbarium specimens, collected since 1965, provided location, and if applicable, phenological information.

Field notes, also collected since 1965, provided both phenological and geographical information. Many were compiled by the author while hiking abandoned railroad rights-of-way when researching for *Where Did The Tracks Go* (Kudish, 1985).

35-mm slide transparencies, dated and identifying location, provided information on geography and phenology. Some of the photographs in the text are from approximately 2000 slides taken in and around the Adirondacks since 1965. Professor James Kraus of Paul Smith's College provided the others.

Notes compiled in preparation for ground cover, wildflower, edible, and poisonous plant walks, plus lecture notes specifically designed for students at Paul Smith's College and for members of various organizations provided resources for topics such as plant physiology, plant strategies, changes in vegetation with increased elevation on Algonquin Peak, relations between green plants, dendrology (tree identification), and forest soils.

Nearly 600 soil pits, dug by the author and Paul Smith's College classes since 1973, have contributed to the explanation of the distribution of many species. Some of the pits provided pH data, some textural data, others profile data, and still others vegetational data. Many exhibited combinations of these kinds of data.

Appendices provide information on tree ages, plant reproductive strategies, and a place-name latitude-longitude table.

* * *

Readers wishing visual identification of each species may find it helpful to use an identification manual along with this Flora. References, including identification manuals, are provided in the bibliography for readers who desire far more information than can be provided here. For a list of introduced, but not naturalized, species, readers may wish to consult a volume on cultivated plants. Other floras available on the Upland but covering only small areas are Peck's *Plants of North Elba* (1899), and Heady's *Annotated List of the Huntington Forest* in Newcomb, now the Adirondack Ecological Center, (1940), and Kudish's *Paul Smith's Flora I* (1975) and *II* (1981). All species listed in this Flora have been observed and confirmed by the author.

ACKNOWLEDGEMENTS

For reviewing the initial draft of the manuscript and making many helpful suggestions for improving the text: Dr. Edwin H. Ketchledge, Daniel Spada, Dr. Paul Jamieson and Dr. Donald Cox. Dr. Ketchledge's ideas on defining the geographic limits of the flora area and arranging the plant families according to Mitchell were especially welcome.

For most carefully and thoroughly editing the manuscript in each of its phases to improve the clarity of writing and the succession of ideas: Madge Heller, Editor, The Chauncy Press.

For valiantly typing almost indecipherable pages of field notes, frequent redrafts of the text, reams of edited prose, and the final manuscript, Teresa Bordeau, Secretary to the Social Sciences/Humanities Division of North Country Community College.

For generously contributing some of his rare and striking photographs for inclusion in the text, Professor James Kraus of Paul Smith's College.

For permission to reproduce his Labrador tea (*Ledum groenlandicum*) on the dust jacket, former Paul Smith's College Student, John Wood.

For intelligent and skillful reproductions of my maps, Susan Kautz of the Adirondack Park Agency.

For reporting on species of plants which the author would otherwise not have observed and confirmed: Lang Elliott, Dr. Mildred Faust, Professor Theodore Mack, Ruth Schottman, and Dr. Alfred E. Schuyler.

For occasional discussions on the regional flora: Professor John Brown, Peter Fry, Dr. Stephen T. Jackson, Lewis Staats, and Lois Wells.

For providing courier service between the author and his typist, Mrs. Ruth Woodward.

* * *

To note all those others: students, colleagues, teachers, and friends who have contributed to this work is not possible here. Nonetheless, I wish to acknowledge, also, their contributions to my research, teaching and writing.

American Beech (*Fagus grandifolia*) Beech possesses a most unusual phyllotaxy. The leaves and buds are arranged alternately, yet the bud scales are opposite. The leaf scars are rotated nearly 45 degrees from the lateral buds which they subtend. © Jim Kraus 1992.

TABLE OF CONTENTS

Foreword	ix
Acknowledgements	xi
Table of Photographs	xv
Table of Maps	xvi
Chapter 1—The Adirondack Upland	1
Chapter 2—Plants on the Move	5
The Fossil Pollen Record	7
The Macrofossil Record	7
The Replacement of Arctic Tundra by Boreal Forest	8
The Replacement of Boreal Forest by Northern Hardwoods Forest	11
The Replacement of Northern Hardwoods Forest by Southern Hardwoods Forest	11
Plants on the Move: Floating Across a Pond	15
Independence of Species	16
An Invasion of Humans	17
An Invasion of Weeds	18
Chapter 3—Plants and Climate	21
Plant Phenology of the Adirondack Upland	23
Growing Season, Elevation Change, and Phenology	30
Growing Season, Latitude Change, and Phenology	34
Introductions	36
Precipitation	36
Wind	40
Light Intensity and Daylength	42
Winter Severity	43
Lightning	44
Aspect and Vegetation	44
Chapter 4—Plants and Soils	47
Plants and Soil Minerals	49
Plants and Soil Texture	50
Plants and Soil Depth	53
Plants and Soil Water	54
Plants and Soil Nutrients	54
Plants, Humus, and Soil pH	55
Herb Dependence on Particular Tree Species	60
From "The Silt Trip" to "The Subtle Pioneers"	61
References	64

Chapter 5—The Adirondack Upland Flora	65
Sequence of Families	67
Information on each Species	71
Information on Maps	72
Species Descriptions	77
Chapter 6—Plants in Jeopardy	241
Appendices:	
I. Ages and Diameters of Native Trees	248
II. Plant Strategies: Reproduction	249
III. Table of Latitudes and Longitudes	252
Bibliography	259
Index	285

TABLE OF PHOTOGRAPHS

Red Maple	*Acer rubrum*	James Kraus	Front Cover
Cinnamon Fern	*Osmunda cinnamomea*	Michael Kudish	iv
Marsh Marigold	*Caltha palustris*	James Kraus	viii
American Beech	*Fagus grandifolia*	James Kraus	xii
Spring Beauty	*Claytonia caroliniana*	James Kraus	4
Pitcher Plant	*Sarracenia purpurea*	James Kraus	24
Sundew	*Drosera rotundifolia*	James Kraus	45
Bog Laurel	*Kalmia polifolia*	James Kraus	48
Blueberry	*Vaccinium angustifolium*	James Kraus	66
Black Cherry	*Prunus serotina*	Michael Kudish	75
Bunchberry	*Cornus canadensis*	James Kraus	236
Sugar Maple	*Acer saccharum*	James Kraus	240
Milkweed	*Asclepias syriaca*	James Kraus	243
Cardinal Flower	*Lobelia cardinalis*	James Kraus	247
Bluets	*Hedyotis caerulea*	James Kraus	250
Wild Raisin	*Viburnum cassinoides*	Michael Kudish	254
Witchhobble	*Viburnum lantanoides*	James Kraus	274
Daisy	*Leucanthemum vulgare*	James Kraus	313
False Hellebore	*Veratrum viride*	James Kraus	316
Labrador Tea	*Ledum groenlandicum*	John Wood	Back Cover

TABLE OF MAPS

Chapter 1
    The Adirondack Upland,   2

Chapter 2
    Major Post-Glacial Migration Routes Into New York by Plant Species, (Long Island migration routes had little or no impact on those of northern New York).   6

Chapter 5
    Species Distribution Maps

Balsam Fir	(*Abies balsamea*)	89
Eastern Larch	(*Larix laricina*)	91
White Spruce	(*Picea glauca*)	93
Black Spruce	(*Picea mariana*)	95
Red Spruce	(*Picea rubens*)	97
Red Pine	(*Pinus resinosa*)	99
Pitch Pine	(*Pinus rigida*)	101
Eastern Red Cedar	(*Juniperus virginiana*)	105
Witch Hazel	(*Hamamelis virginiana*)	113
Butternut	(*Juglans cinerea*)	117
Northern Red Oak	(*Quercus rubra*)	119
Mountain (Green) Alder	(*Alnus viridis* ssp. *crispa*)	123
Balsam Poplar	(*Populus balsamifera*)	137
Three-toothed Cinquefoil	(*Potentilla tridentata*)	155
Silver Maple	(*Acer saccharinum*)	171
Poison Ivy	(*Toxicodendron radicans*)	175
Maple-leaved Viburnum	(*Viburnum acerifolium*)	195
Heart-leaved Aster	(*Aster cordifolius*)	199

Chapter 1

THE ADIRONDACK UPLAND

The Adirondack Upland has a characteristically distinct flora.

Kudish

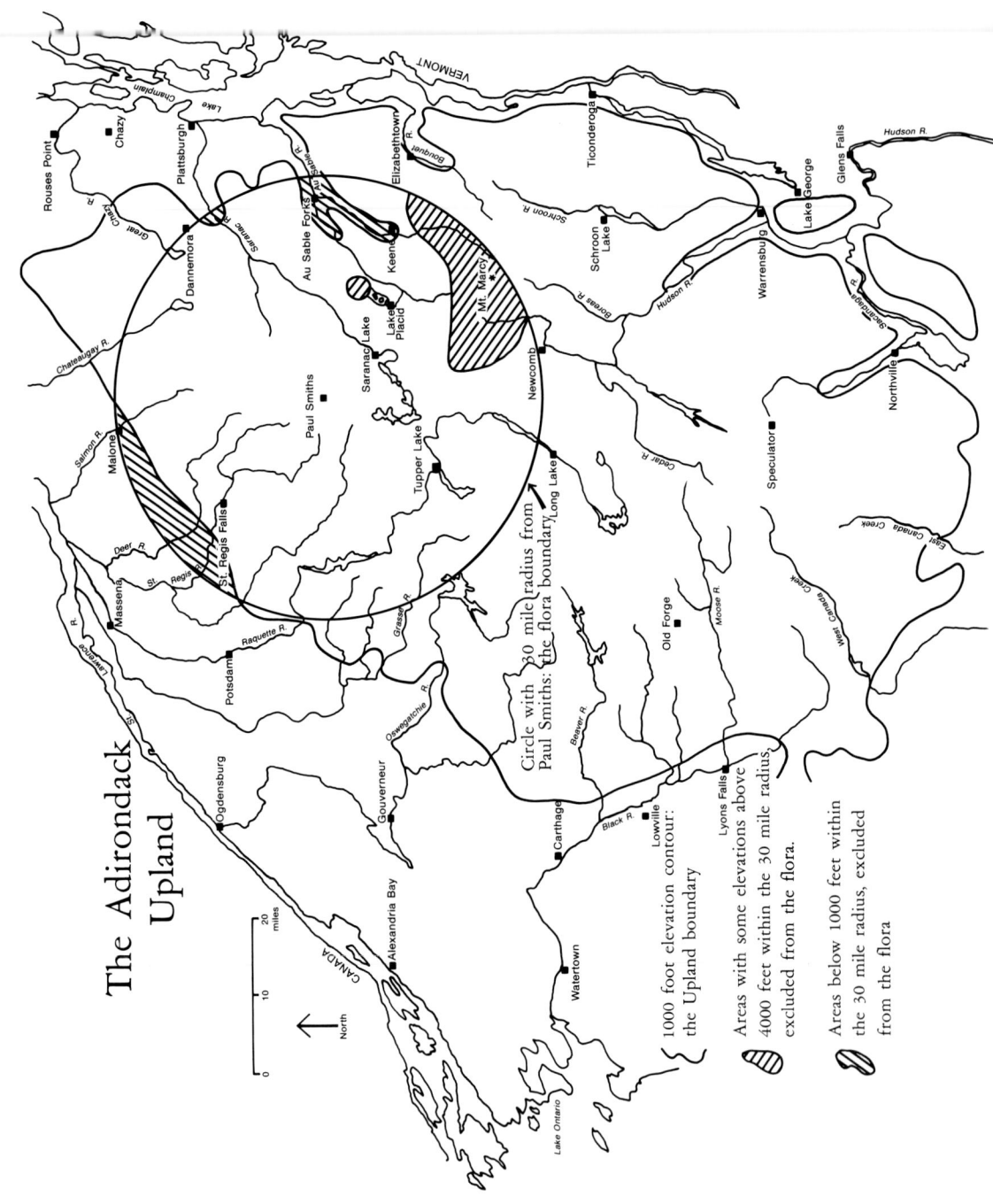

The geographic region covered in the Flora is a circle 30 miles, about 48 km, in radius, centered at the Paul Smith's College Cubley Library at Paul Smiths, New York, latitude 44°26' north, longitude 74°15' west. This 30-mile radius barely enters the Saint Lawrence Valley on the northwest and north, and falls just short of the Champlain Valley on the east. The area of 2827 square miles, about 7350 square kilometers, is almost wholly restricted to the Adirondack Upland. The elevation limits are from 1000 to 4000 feet, 305 to 1219 meters, within this circle. The circumference, sweeping clockwise from the north, passes through Malone, Lower Chateaugay Lake, Chazy Lake, Saranac, Ausable Forks, Keene Valley, Tahawus, north of Long Lake, Horseshoe Lake, just east of Cranberry Lake, Five Falls Reservoir on the Raquette River, Parishville, Fort Jackson, and Moira. Most of the circle's area is within the Adirondack Park, except for a small segment between Parishville and Lower Chateaugay Lake.

The lower elevational limit of 1000 feet is coincidental with the perimeter established by the New York State Department of Environmental Conservation which has found it useful as a boundary between forest types. Most of the southern affinity species are confined to the Lake George, Champlain, and Saint Lawrence Valleys and do not climb above 1000 feet. These valleys have a very different flora because of the lower elevations, longer growing season, and longer period of human disturbance history. Many species of southern affinities reach their northern limits in these valleys and do not enter the Adirondack Upland. The only areas below 1000 feet within the 30-mile radius are between Parishville to just east of Malone, the Saranac River Valley around Moffitsville, and the Ausable River Valleys below Wilmington and Keene.

The 4000 foot elevation upper limit marks another major floristic change. A number of peaks above 4000 feet exhibit the Arctic-alpine, or Hudsonian zone, another world biologically. The typical Adirondack Upland forest ceases to exist here. (For discussions of those plants growing on those high peaks within the 30-mile radius and above 4000 feet see entries in the bibliography.)

The Adirondack Upland has a characteristically distinct flora, and it is this quite uniform flora which this volume describes. Indeed, the plants, soils, and climate are quite uniform throughout the Adirondacks within the Upland elevational belt—the flora of the Inlet-Eagle Bay area, for instance, is almost identical to that of Paul Smiths. Therefore, much of the data in this Flora will apply to portions of the Adirondack Upland between 1000 and 4000 feet outside the 30-mile radius. (Examples of areas in this category include the region from Cranberry Lake west to Fine, from Long Lake southwestward through Old Forge to Forestport, and from Indian Lake southward past Speculator.)

Spring beauty (*Claytonia caroliniana*)
Each flower has five sepals, five white petals with pink stripes, five stamens and an ovary consisting of three fused leaves known as carpels. © Jim Kraus 1992.

Chapter 2

PLANTS ON THE MOVE: SPECIES MIGRATIONS AND INVASIONS

Every stand of trees and plot of ground has its own history. And that which we see today is the latest stage in a long process of adjustment to past treatments and continuing processes.

Ketchledge

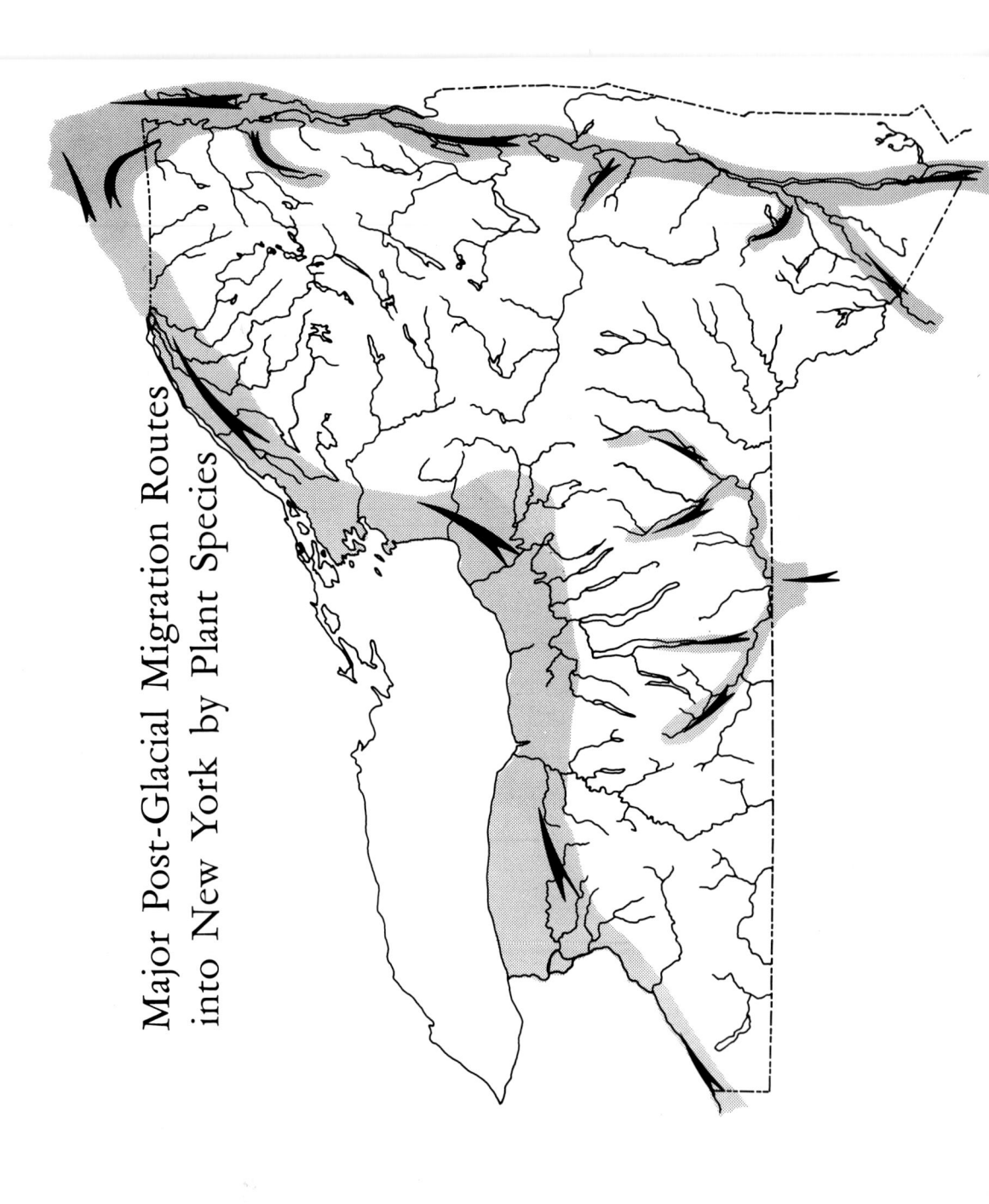

Major Post-Glacial Migration Routes into New York by Plant Species

What was the original Adirondack forest like before the advent of disturbance on a grand scale by people of European ancestry in the nineteenth century? How did vegetation develop following deglaciation over approximately 13,000 years? Our ideas on this original vegetation are sketchy at best since no human records of the arrival of the plants were made. Evidence used to reconstruct the development comes from several sources: (1) fossil pollen; (2) macrofossils; (3) studies of the present vegetation well to the north in Canada and also in the Pacific Northwest and Alaska; (4) examinations of modern Adirondack soils and determinations of what plant species these soils can support; and (5) maps of the present distribution of species whose northern limits are the southern Adirondacks.

THE FOSSIL POLLEN RECORD

Pollen grains preserved in peat in bogs can be used to reconstruct past vegetation, but there are problems with this method. First, pollen can be carried by wind over long distances, miles to tens of miles and more, so that some of the pollen extracted from a bog might have come in from afar where the vegetation was different. Second, those plants which are wind-pollinated produce many times more pollen grains than those plants which are animal (largely insect) pollinated; thus, wind-pollinated plants always appear disproportionately abundant in the fossil pollen record, and animal-pollinated plants appear conversely almost always disproportionately scarce.

Pollen studies from Cape Cod across Rhode Island and Connecticut, across southeastern New York and into New Jersey and northern Pennsylvania show generally (a) spruce-fir forest with some arctic tundra plants soon following deglaciation, then (b) northern hardwoods: beech, sugar maple, yellow birch and (c) most recently, what I call southern hardwoods: a mixture of oaks, hickories, tulip tree, chestnut, black birch, flowering dogwood, black gum, eastern red cedar and pitch pine. The vegetation in the region today is predominately type (c), types (a) and (b) having long gone. In contrast, in the Adirondacks, all the original tree species which soon followed deglaciation are still here: fir, spruces, larch, paper birch, balsam poplar, pin cherry, aspens, and mountain ash so that the flora has not really changed that much on many Adirondack sites since deglaciation.

Examples of pollen studies in and around the Adirondacks are:

Cox (1959) at Chestertown, Perch Lake (north of Watertown), and Consaulus Bog (west of Ballston Lake).

VanLeeuwen (1977) at Roakdale Bog east of Onchiota.

Whitehead, Charles, Jackson, Reed and Sheehan (1986) at Heart Lake, Lake Arnold, and Wallface Ponds in the High Peaks.

THE MACROFOSSIL RECORD

Recent evidence derived from macrofossils by Whitehead et al. (1986) from the three High Peaks ponds listed above has given us absolute proof that certain species existed around these ponds in the past and in accurate proportions. A summary of these findings is presented by Ketchledge (1989). Twigs, leaves, needles, fruits, cones, and most seeds cannot travel far in the wind. Therefore, they fall within several hundred feet of the plant. The macrofossil record reveals that hemlock, yellow birch, and white pine were higher on the slopes of Adirondack peaks in the past than at present. All three are at Heart Lake today, but Lake Arnold and Wallface Pond no longer harbor them. If one cause of the lowering of maximum elevation for these species is the shortening of the growing season due to the lowering of mean annual temperature, we can easily estimate how much the Adirondacks have cooled off since the warmer period, the Hypsithermal Period, about 9000 to 3000 years ago. My calculations show that it was only three to five degrees Fahrenheit warmer then—no more. Nonetheless, a minor change for people in mean temperature of a few degrees Fahrenheit is a major change for plants. (See Chapter 3 for detail on the relationship between growing season and the elevational range of plant species.)

Like pollen, the oldest macrofossils are found in the bottom of the peat bog or lake sediments and the youngest on top. Hence, by its vertical position in the peat or sediments, the relative age of a fossil can be determined. Using the Carbon-14 (radiocarbon) dating method, the botanist can estimate the absolute age of the fossil.

THE REPLACEMENT OF ARCTIC TUNDRA BY BOREAL FOREST

Because Canada is the Adirondack's past vegetationally, examination of modern vegetation well to the north may provide some insight into what vegetation looked like in the Adirondacks soon after deglaciation. Even so, it may not produce a completely accurate picture. What has happened in Canada may or may not have happened in northern New York. Vegetation works on a very local level.

Examination of the Pacific Northwest states Washington and Montana, the Canadian Rockies, and Alaska in the 20th century where glaciers (smaller mountain glaciers, not continental) are retreating, reveals that vegetation often very quickly invades fresh mineral deposits usually within a few years. Thus, after a century a forest has developed. Therefore, it seems reasonable to assume that vegetation similarly quickly colonized Adirondack tills and outwash within several years or several tens of years at most.

Can we assume that the Adirondack Upland at elevation 1500 to 2000 feet supported pure arctic tundra for a long period of time and then rapidly became totally forested by the boreal spruce-fir? I doubt it. My recent studies in Newfoundland and Labrador revealed that arctic and boreal forest vegetation forms a "patchwork quilt," a mosaic, not determined so much by elevation as by wind exposure, soil depth, and random distribution of seeds and spores. Arctic tundra plants on shallow-soiled ridges exposed fully to winds occur at or near sea level, while on sites up to 1000, or even 1500 feet above sea level where the soils are deep and there is protection from wind, boreal forest consists of fir, paper birch, and black and white spruce trees up to 50-feet tall. My hypothesis is that the Adirondack Upland looked like this: not pure tundra and then suddenly boreal forest, but rather a long period of "patchwork quilt" of perhaps centuries to a millennium before continuous spruce-fir forest moved in. Because arctic tundra species cannot tolerate shade they were eliminated on sites where the forest invaded. All plant invasions, arctic tundra, boreal forest, northern hardwoods, and southern hardwoods must have come in from the south—perhaps also from the southeast and/or southwest as the continental ice sheet wasted away to the north.

Today, the further north one travels in Canada, the more abundant the arctic-alpine species become and the closer to sea level they approach. North of the tree line, this vegetation type, also known as tundra, is the only vegetation type. The term "arctic" refers to these small species which dominate the low elevations of northern Canada. Further south, in the Adirondacks and in northern New England, these same plants become and are called "alpine" because they occur largely on summits of the highest peaks. The term "arctic-alpine" includes these plants in both geographic areas. All Adirondack and northern New England alpine species occur commonly in the Arctic as well, but there are some arctic species which no longer exist in the United States. A trip to relatively nearby summits of the Shickshock Mountains of the Gaspé in Québec and to Newfoundland will confirm this—one need not travel to the northern limit of forest in Canada.

For a time following deglaciation, tundra vegetation covered all of New York State and New England as low as elevations approaching sea level. Today the only tundra vegetation left, as remnants, in New York is on the highest Adirondack summits. This vegetation, along with alpine areas in the Green Mountains of Vermont and the White Mountains of New Hampshire, marks the southern limit of tundra in eastern North America. In the Catskills and other more southerly portions of the Appalachians no tundra exists. In both, the mountains are forested to their summits.

More specifically, tundra species today grow on the summits of seventeen of the highest Adirondack peaks, generally above 4700 feet, as well as on a few extraordinarily exposed lower knobs and bluffs as low as about 4000 feet. These species persist where a combination of environmental factors are, and have been since deglaciation, too harsh to support a forest. Growing season is very short, about three months, and winds are excessive, causing dehydration and breakage to plant parts exposed above the snowpack (see Chapter 3 on climate for more detail). Soils are very thin, usually consisting of an organic mat of accumulated dead plant debris, often less than a foot thick, sitting directly on bedrock (see Chapter 4 on soils). The environmental conditions on the alpine summits are very much like those in the tundra of northern Canada. Such an alpine zone reveals that these plants have been here since deglaciation and that the summits have never been forested over the last 12,000 years.

(Because this Flora does not deal in detail with elevations above 4000 feet, specifically the Alpine summits, and because the arctic-alpine species are not typical of Adirondack Upland vegetation, most of these are excluded from Chapter 5. They present a biologically different world from that found in typically forested Adirondacks and should be treated in a separate volume.) References already available on Adirondack alpine summits include:

Adams et al. (1920)	Houghton (1948)	Phelps (1964; 1970)
Cate (1979)	Ketchledge (1982; 1984)	Riebesell (1981)

DiNunzio (1972) McMartin (1984) Woodin (1950)

Two species of the arctic-alpine flora are included in Chapter 5 because they occur also under 4000 feet. One is *Vaccinium uliginosum,* the alpine bilberry, which occurs on McKenzie Mountain at 3861 feet and on Catamount Mountain near Silver Lake in Clinton County at 3100 feet. The other is *Potentilla tridentata,* the three-toothed cinquefoil, which is widely scattered on the Upland from Paul Smiths and Bloomingdale eastward at elevations as low as 1550 feet. (Consult Chapter 5 for detail on these two species. A map accompanies the cinquefoil.)

Some twelve of the other typical arctic-alpine species observed in the Adirondacks occurring only above 4000 feet are listed below with their New York State rarity codes. (Consult the reference sources listed above for specifics on additional, uncommon, alpine-summit species).

Because these alpine species are at the southern limits of their natural ranges and confined to several summits, they are invariably rare in New York State, despite the fact they become abundant further north in Canada. These species are included in Forest Preserve lands and are thus protected by New York State law from collectors and those who disturb vegetation for economic gain. This is not enough, however, because the greater danger is their being trampled by hikers swarming the alpine summits each summer. Public education programs and attempts at revegetating the eroded and trampled sites have been in existence since the 1960s—see Ketchledge (1982; 1984).

Rarity codes for the arctic-alpine species follow. The three sources are Mitchell (1986), Birmingham (1988) and Clemants (1989). (See Chapter 6 for an explanation of these codes.) A dash indicates that no rarity code is given for this species.

TABLE 1
ARCTIC-ALPINE SPECIES: RARITY CODES

SPECIES	RARITY CODES		
	Mitchell	Birmingham	Clemants
Betula glandulosa, dwarf birch.	Rare E-3	Endangered	G4G5, S1
Carex bigelowii, Bigelow's sedge.	—	Rare	G5, S2
Diapensia lapponica, mountain bride.	Rare T-3	Threatened	G5, S2
Empetrum nigrum, black crowberry.	Rare	Rare	G5TU, S2
Juncus trifidus, Arctic rush.	Rare T-3	Threatened	G5, S1
Lycopodium selago, alpine clubmoss.	—	Exploit. vul.	—
Minuartia groenlandica, mt. sandwort.	Rare	Rare	—
Rhododendron lapponicum, Lapland rosebay.	Rare T-3	Threatened	G5, S2
Salix uva-ursi, bearberry willow.	Rare T-3	Rare	G5, S2
Scirpus caespitosus, deer's hair.	Rare T-3	Threatened	G5, S2
Solidago cutleri, Cutler's goldenrod.	—	—	G4, S2
Vaccinium boreale, northern blueberry.	Rare	—	G3, S2

THE REPLACEMENT OF BOREAL FOREST BY NORTHERN HARDWOODS FOREST

The replacement of boreal forest (spruces, fir, paper birch, and mountain ash) by northern hardwoods (sugar maple, beech, and yellow birch) with red maple, black cherry, white and red pines, and hemlock was probably largely attributable to soil types, since all these tree species can occur on the Adirondack Upland at the same elevation and with the same climate. Simply put, northern hardwoods, with sugar maple the most selective or most demanding species, moved in on the better sites. Sugar maple will not grow well where the sites are (1) too sandy (less than 5% silt-plus-clay, as in many outwash areas), (2) too wet (swamps, marshes, bogs, and fens having water tables within two feet of the surface except where water is flowing downslope as in well-oxygenated springs), and (3) too shallow to bedrock (generally less than 18 to 24 inches of soil). Any one of these three factors alone can limit sugar maple and good northern hardwoods growth; two or three of these factors combined will rapidly eliminate sugar maple saplings and prevent northern hardwoods development on the site. Such sandy and/or wet and/or shallow soils still harbor remnant boreal forest, and have done so since the replacement of tundra species because invasion by northern hardwoods has proven difficult to impossible.

The fossil record shows that northern hardwoods invaded the better sites in the Adirondacks within about 3000 years after deglaciation so that they are not truly newcomers to the region. The northern hardwood forests must have also invaded from the south, following the boreal forest by several millennia, and several hundred miles behind the spruces and fir.

The species of trees (see remnant population maps in Chapter 5) the three spruces, balsam fir, and larch were most likely all present and common at one time in the Saint Lawrence, Champlain, Mohawk, and Black River Lowlands. Today, they are uncommon, rare, or absent altogether. A few remnant pockets do exist, largely in swamps where northern hardwoods have not been able to replace them during the last 12,000 years or so. (See the species maps for larch, the least rare of these remnants, and for balsam fir with a station at Brasher Center and several more near Chazy.) Who can find remnants of spruce stands below elevation 700 feet in the Lowlands today? The Adirondacks remain as an island refuge for these five boreal conifers, surrounded by Lowlands from which they began to be displaced as early as 10,000 years ago.

[Further reading on the post-glacial development of New York State vegetation can be found in Bray (1915); Cox (1959); Storey (1977); VanLeeuwen (1977); and Whitehead, Charles, Jackson, Reed, and Sheehan (1986). Further reading on the post-glacial development of vegetation throughout the eastern United States can be found in Delcourt and Delcourt (1984) and in Sears (1942 and 1948).]

THE REPLACEMENT OF NORTHERN HARDWOODS FOREST BY SOUTHERN HARDWOODS FOREST

We can learn more about how some plant species migrate by mapping the contemporary distributions of species absent on the Adirondack Upland but present in the Saint Lawrence, Mohawk, Hudson, and Champlain-Lake George Lowlands, and in the lower reaches of the Ausable, Schroon, and Boquet Valleys. In these Lowlands and lower river valleys, certain species whose main populations are well to the south are near or at their northern limits. I refer to these as southern hardwood forest species although they occur as natives in northern New York.

Many of these southern species are uncommon in northern New York because they are at or near the northern ends of their natural ranges. Further south, even as close by as central and southern New York, they may be abundant. This is the reverse situation from that of the arctic-alpines which reach their southern limits of their natural ranges in the Adirondacks and which are also rare.

The Connecticut, Hudson-Champlain, Great Lakes-Saint Lawrence, and, to a lesser extent, the Susquehanna Lowlands must have provided major plant migration routes north (see map on page 6). The Lowlands, because of their elevation, largely below 500 feet, have and must have had a higher mean annual temperature; a higher mean annual temperature creates a longer growing season similar to that of southern Pennsylvania, New Jersey, southeastern New York, and coastal New England from Connecticut through Cape Cod. Hence, some southerly species could and do survive in northern New York, Vermont, and New Hampshire well to the north of the main portions of their natural ranges in, and only in, these lowlands.

Explaining how southern hardwood forest species replaced northern hardwood forest species is more difficult than explaining how the boreal forests replaced tundra and how, in turn, northern hardwoods replaced boreal forest. Certainly, the longer growing season in the lowlands favors the survival of species most abundant in the warmer climates further south. Many of the southern hardwood forest species, especially the oaks, hickories, black birch, and pitch pine follow major disturbances. This is because these tree species generally sprout better after burns or cutting and are less tolerant of shade than northern hardwoods. They can invade disturbed areas more rapidly than a northern hardwood forest could in most cases.

Most areas in New York State dominated by oaks and hickories, black birch, and pitch pine, in contrast to areas where these trees are present in small numbers in a northern hardwoods forest, have been burned over and/or cleared repeatedly for centuries and for perhaps millennia. The fires and clearings originated not only from people of European ancestry in the last three-and-one-half centuries, but also from Amerindians prior to that. Amerindians removed forest for agriculture, blueberries, better hunting, defense, better travel, etc., largely by fire. But they did so only in certain areas; not everywhere. In the eastern Adirondacks, Europeans nearly clearcut many areas today dominated by oaks, in order to obtain charcoal for the iron industry in the nineteenth and early twentieth centuries.

In the Hudson-Champlain Lowland, sassafras, tulip tree, rose bay rhododendron, flowering dogwood, poison sumac, and highbush blueberry reach their northern limits in the Albany area. Other species have migrated northward further along the Lowland but have not reached Québec. Marie-Victorin (1935 and 1964) who compiled a superb flora of the Laurentians and Montréal area does not include the following ten species noted below with their northern limits. (These are not included in Chapter 5 of this Flora because they occur below the 1000–foot minimum elevation.)

Asclepias amplexicaulis, curly leaf milkweed, below Ausable Chasm.
Castanea dentata, American chestnut, at Thurman.
Cimicifuga racemosa, black cohosh, along Boquet River above Wadhams.
Geranium maculatum, cranesbill, at Fort Ticonderoga.
Phaseolus polystachios, wild bean, at Point au Roche.
Platanus occidentalis, sycamore, at Ausable River Delta.
Quercus ilicifolia, scrub oak, at Prospect Mountain above Lake George.
Quercus prinus, chestnut oak, on Coot Hill above Crown Point.
Quercus velutina, black oak, below Ausable Chasm.
Vaccinium pallidum (vacillans), blueberry, also below Ausable Chasm.

Many of the species present in the Hudson-Champlain and Saint Lawrence Lowland, but absent in this Upland Flora, today range as far north as Québec. Marie-Victorin includes the following 38 species in his flora:

Agalanis (Gerardia) purpurea, purple gerardia.
Alisma plantago-aquatica, water plantain.
Anemone riparia, anemone.
Apocynum cannabinum, Indian hemp.
Asarum canadense, wild ginger.
Betula lenta, black birch.
Carya cordiformis, bitternut hickory.
Carya ovata, shagbark hickory.
Ceanothus americanus, New Jersey tea.
Celtis occidenatlis, hackberry.
Conopholis americana, squaw-root.
Draba arabisans, whitlow-grass.
Echinochloa muricata, barnyard grass.
Elymus canadensis, wild rye.
Eupatorium perfoliatum, boneset.
Hieracium paniculatum, panicled hawkweed.
Lilium philadelphicum, wood lily.
Phryma leptostachya, lopseed.
Polygonum lapathifolium, pale smartweed.
Polygonum pensylvanicum, pinkweed.
Populus deltoides, cottonwood.
Quercus alba, white oak.
Quercus bicolor, swamp white oak.
Ribes americanum, wild black currant.
Sanguinaria canadensis, bloodroot.
Shepherdia canadensis, buffaloberry.
Smilacina stellata, starry false Solomon's seal.
Stachys palustris, woundwort.
Symplocarpus foetidus, skunk cabbage.

Teucrium canadense, American germander.
Trichostema dichotomum, blue curls.
Trillium grandiflorum, white trillium.
Typha angustifolia, narrow-leaved cattail.
Ulmus rubra, slippery elm.
Ulmus thomasii, cork elm.
Uvularia perfoliata, bellwort.
Zanthoxylum americanum, prickly ash.
Zizia aurea, golden alexanders.

(These are not included in Chapter 5 of this Flora because they do not occur above the 1000-foot elevation minimum. They are not typical of the Adirondack Upland vegetation presenting as they do another biologically different world. A whole, separate volume is needed to do them justice.)

The Great Lakes-Saint Lawrence Lowland has provided another major access route for plants into northern New York, but in this instance it is for species from the midwest. These entered New York from Pennsylvania along the Erie and Ontario Lake plains, and stay generally well below 1000 feet elevation. Cucumber tree magnolia reaches the Syracuse-Utica area today, while red (green) ash and bur (mossycup) oak have reached all the way down the Saint Lawrence Lowland into Québec and swept around southward up the Champlain Lowland to around Essex and Westport.

Bitternut and shagbark hickories have completely encircled the Adirondacks, but have not yet invaded them. The mountains are thus an island of absence of these two species. Populations moved north through the Champlain Lowland and, by another route, northeast through the Saint Lawrence Lowland, the two populations meeting near Mooers and Rouses Point. Crash? Maybe not, as arrival times may have been centuries apart.

Plant migrations in the lowlands surrounding the Adirondacks, especially those from the east, are different from those in the major valleys which penetrate the interior of the Adirondacks. One remarkable example of plant migration is in the Ausable Valley, especially the East Branch which maintains a relatively low elevation deep into the mountains (Keene Valley is barely over 1000 feet). A number of southern species have apparently migrated up the East Branch from north to south as far as Keene Valley. These reversed their direction of invasion. They made a "U" turn as they left the Champlain Lowland moving north, turned first west up the River to Ausable Forks, and then finally headed south. Migration into Keene Valley from the south via the Schroon, upper Boquet Valley and Chapel Pond is unlikely as these species are absent on the divide southeast of Chapel Pond. I call this population of southern species the Ausable Lobe because when viewed on a map it resembles a branch off the Champlain Lowland. (In the Catskills, the Esopus Valley creates a similar vegetational lobe off the Hudson Lowland at Kingston.)

Species which occur in this Flora area only at Keene Valley, barely above the minimum 1000-foot elevation threshold, include:

Aster cordifolius, heart-leaved aster.
Aster divaricatus, white woodland aster.

Carpinus caroliniana, American hornbeam or musclewood.
Circaea lutetiana, enchanter's nightshade.
Cornus amomum, silky dogwood.
Cornus rugosa, round-leaved dogwood.
Hamamelis virginiana, witch hazel.
Juglans cinerea, butternut.
Juniperus virginiana, eastern red cedar.
Pedicularis canadensis, wood betony.
Viburnum lentago, nannyberry.

White oak (*Quercus alba*) appears no higher than 900 feet at Palmer Hill north of Ausable Forks and therefore misses the Flora limits by 100 feet.

Other species seem to have migrated westward out of the Champlain Lowland directly up onto the Adirondack Upland without taking the major river valleys' routes. They have climbed moderately high, 2000 feet or so, on the slopes of the eastern Adirondacks, and one species, the mountain alder (*Alnus viridis*), reaches into the alpine zone. None of these species occurs farther west from a line drawn from Lake Placid through Whiteface Mountain to Catamount Mountain in Clinton County at any elevation on the Upland. Included in this group in addition to the mountain alder are sweet fern (*Comptonia peregrina*) with one exceptional site near Bloomingdale, maple-leaved viburnum (*Viburnum acerifolium*), waterleaf (*Hydrophyllum virginianum*), and white snakeroot (*Eupatorium rugosum*).

Some species have not yet invaded the Adirondack Upland. The reason may be neither too short a growing season nor too poor soils, but simply that they have not had a chance to migrate in yet. Plants appear to move slowly because they do not limit themselves to short human life spans. Give them time, several thousand more years perhaps, and have patience. Many are still migrating north, and haven't reached their limits yet, their attempts greatly complicated by human activities. Currently, a number of species, each independently of the other, has reached particular northern and upper elevational limits, all different. Look again "tomorrow" and see that they all have moved; each in its own way and each at its own rate.

Clear indications of how slowly plants move are found in Johnson and Adkisson (1986) who showed that oaks migrated northward about 10,000 years ago in the central and eastern United States at an average rate of 380 yards per year. This is equivalent to about five years for one mile. Spruces were slower, averaging 275 yards per year or 6.4 years to move one mile. The rate of glacial retreat in the Paul Smith's area, estimated at least 0.6 mile per year, is quite rapid compared with tree migration rates.

PLANTS ON THE MOVE: FLOATING ACROSS A POND

Plants can migrate by using a floating vegetation mat. On May 14, 1985 some faculty called my attention to such a mat out in the middle of Jones Pond, three and a half miles east-northeast of the Paul Smith's College Campus. The mat had not been there the day before. Winds were out of the southwest and the only possible source was the marsh at the southwest outlet of the Pond. The mat, perhaps fifty feet across, had apparently broken away from the marsh. Carried northeastward by the wind across the open water, it soon became lodged against a point along the west shore of the Pond.

On May 15, 1988 others noted the same phenomenon. Another mat had appeared in the middle of Jones Pond where it had become moored until September when it moved to the north shore. An examination of the north and northwest shores of the Pond revealed another half-dozen mats ranging in diameter from perhaps twenty-five to seventy-five feet. These had apparently accumulated over the years.

The dominant plant on each mat is the cattail, and with it a number of other species all typical of marshes: touch-me-not, calla lily, sweet gale, marsh Saint John's-Wort, sensitive fern, marsh fern, and even a white pine seedling. [See Hogg and Wien (1988).]

This type of plant migration is limited, of course, to within a body of water. But if the body is large, plant-bearing mats could travel for miles. Perhaps a study of them on such bodies of water as Long Lake, Indian Lake, the Sacandaga Reservoir, or Lake Champlain would reveal that they do in fact travel extensively.

INDEPENDENCE OF SPECIES

The preceding discussions of plant species migrating into and around the Adirondacks may suggest that these plants moved in definite species groups or teams, often called communities, plant associations, forest types, or ecosystems. The terms arctic-alpine, boreal, northern hardwoods, and southern hardwoods are intended only for introductory explanation and are thus oversimplifications. In actuality, green plant species do not migrate in groups (floating across a pond may be a rare exception). Each species migrates independently of all other species, at its own rate, with its own seed and spore dispersal agents, and to its own kinds of sites where it will grow and reproduce. Because green plants are autotrophs, i.e. self-feeders, they do not require the presence of any other species of green plants to thrive, reproduce, and migrate.

Plant species which often grow together require, by sheer coincidence, similar sites and conditions; not each other. Beech and sugar maple often grow together; so do red spruce and balsam fir, black spruce and larch, alders and willows, yellow birch and red maple, oaks and hickories. But each of these species will grow just as well without the other.

One exception to this rule of independence occurs when one heterotrophic plant is partly or totally dependent upon another plant or totally dependent on a fungus. Examples are Indian pipe, pinesap, coral root orchid, and beechdrops which are parasitic on fungi. Wood betony, also called lousewort, is partly parasitic on other plants.

Vegetation invading a new site, such as fresh till, outwash, or a glacial lake bottom, is like a huge unruly mob of people bursting into a store at the moment a great sale begins. No one knows the others, and there is no teamwork. Inside the store, the mob pushes, shoves, pokes, and competes so that randomly and chaotically individuals grab whatever merchandise they can. Similarly, individual plant species seize available plant growth sites.

AN INVASION OF HUMANS

Just as vegetation invaded the Adirondacks from the Mohawk, Champlain-Hudson, Saint Lawrence, and Black River Lowlands thousands of years ago, people of European ancestry invaded the Adirondacks during the last two to two and one-half centuries. The lowlands, with their longer growing season, more fertile and less stony limestone and shale soils, and less steep slopes were settled first in the late 17th through early 19th centuries. Lands were cleared for farm and pastureland with the forest so devastated that little original remnants remain. Gradually, settlers invaded the Adirondacks by largely following the major river valleys upstream in the early to mid-Nineteenth century since the lower reaches of these valleys were most similar to the lowlands and were more readily cleared for agriculture. These settlers migrated up the same river valleys that many plant species had done millennia earlier: the Hudson and the Schroon; the Boquet, Ausable, Saranac, and Chazy from Lake Champlain; the Chateaugay, Salmon, Saint Regis, Raquette, Grass, and Oswegatchie from the Saint Lawrence Lowland; etc. The interior portion of the Adirondack Upland and the inaccessible high peaks were settled last, in the mid-19th century. A great portion of the forest in the interior was never cleared for agriculture, but of this forest, much was logged and some burned. As a result little still remains of original old growth.

In the Paul Smiths area, first settled in the 1850s and 1860s, there was inadequate time for all the lands to be cleared and/or logged and/or burned by the time that the Forest Preserve came into being in 1885. Hence, a majority of these lands in the original (1885) Preserve remain in old growth. Many people consider the value of old growth forests to be spiritual, but there is another equally if not more compelling reason to preserve the few we have left. These forests are relatively simple since only natural factors have been at work in them. Once such human interventions as agriculture, forestry, mining, residences, and industry occur, the factors which have shaped the forest are more complex—a combination of natural and human ones. In studying an old growth forest, we can eliminate the human factors and concentrate on the natural ones. This value of old growth, then, lies in that we can appreciate/understand the natural processes of forest evolution free of man's intervention.

As one travels from the central interior of the Adirondack Upland: Paul Smiths, Saranac Lake, Tupper Lake, Long Lake, Blue Mountain Lake, Indian Lake, and Raquette Lake outward toward the lowlands, one passes through increasingly more disturbed forests. These include not only those cleared for agriculture, but also those cleared for charcoal to support the vast iron refining industry above the Champlain and St. Lawrence Lowlands. One way of quickly learning the geographic distribution of various kinds and intensities of disturbances is to examine the railroads, described in Kudish (1985), which served all the major industries in the area: forestry, agriculture, mining, and tourism.

Recorded history suggests that Amerindian people lived in the lowlands and ventured into the Adirondack Upland mainly during the warmer months to hunt, fish, and gather edible and medicinal plants. Their impact on the native vegetation of the Adirondacks was minimal compared with that of people of European ancestry. Certain exceptions are lands that had been cleared to raise maize, squash and beans, or burned to increase deer and/or blueberry crops. But such lands were not extensive within the Adirondacks.

AN INVASION OF WEEDS

Along with people, their crops, their hay, and their domesticated animals came the weeds. An army of European species, largely unwanted, invaded North America in the 17th through 20th centuries. Some were deliberately brought in from Europe for gardens, food, fiber, and ornament; these escaped from cultivation, reproducing on their own. Others, brought over inadvertently in hay in the hulls of ships, also reproduced on their own. All these species, wanted or not, have become naturalized. Most are shade-intolerant. That is, most require full sun and grow only in such disturbed areas as untended gardens and lawns, roadsides, abandoned fields, abandoned quarries and gravel pits, railroad grades, powerlines; etc. They thus leave native shade-tolerant woodland plants alone. Of the 563 species listed in Chapter 5 of this Flora, 107, about 19%, are naturalized. A list of the more common and familiar European weeds includes:

Achillea millefolium, yarrow or milfoil.
Agrostis tenuis, red top grass.
Chenopodium album, lamb's quarters or white goosefoot.
Cichorium intybus, chicory.
Dactylis glomerata, orchard grass.
Daucus carota, Queen Anne's lace or wild carrot.
Hieracium aurantiacum, orange hawkweed or devil's paintbrush.
Hieracium caespitosum, yellow hawkweed or king devil.
Hypericum perforatum, common St. John's-wort.
Leucanthemum vulgare, daisy.
Oxalis corniculata, lady's sorrel.
Phleum pratense, timothy.
Plantago lanceolata, narrow-leaved plantain.
Plantago major, common plantain.
Polygonum aviculare, knotweed or knotgrass.
Polygonum persicaria, lady's thumb.
Potentillia argentea, silvery cinquefoil.
Potentillia recta, rough-fruited cinquefoil.
Prunella vulgaris, self-heal or heal-all.
Ranunculus acris, tall meadow buttercup.
Rumex acetosella, sheep sorrel.
Silene latifolia, evening lychnis.
Silene vulgaris, bladder campion.
Stellaria media, chickweed.

Taraxacum officinale, dandelion.
Trifolium spp., all species of clovers.
Verbascum thapsus, mullein.
Vicia cracca, vetch.

If they enter the woods, European weeds follow old tote roads, logging skid trails, fire truck trails, and driveways to private camps. The seeds travel on vehicles, on shoes, on boots, on pant legs, and on dog fur, often penetrating miles into the forest. In the woods off the trail, all the plants, with perhaps one or two exceptions discussed below, are native. As trails radiate out into the woods and penetrate the forest like fingers of an open human hand, so do the Europeans penetrate native woods along these long and narrow trails. I have seen mullein on the Tongue, a tract of Paul Smith's College land west of Campus, at least two miles from the nearest stand of European weeds along a highway. More mulleins were growing on a log road at Hidden Pond (Lot 15) nearly three miles from the nearest public highway (Route 30 at Mountain Pond) and their cousins. Other Europeans such as plantains which grow along logging roads can tolerate trampling.

There are a very few exceptions to the shade-intolerance rule for Europeans in North America. *Epipactis helleborine,* a greenish-flowered somewhat inconspicuous orchid, is as shade-tolerant as many native woodland plants and has become at home in northern hardwoods and mixed hardwood-conifer stands, often miles from the nearest road or building.

One plant which moves exceptionally fast is *Hieracium lachenalii.* This moderately shade-tolerant hawkweed can survive in lightly-shaded, semi-open stands of northern hardwoods and mixed woods, suggesting also a native origin. I first noticed it in 1981 near Bartlett Carry between Upper and Middle Saranac Lake. By 1985, it had spread to Wawbeek across the Upper Lake. The invasion moved northward following Route 30 to Heron Lane and near the Paul Smith's College Sugarbush east of Osgood Pond in 1986, covering a distance of 15 air-miles in five years. By 1986, this hawkweed had also spread eastward and had taken over the southeast slopes of Mount Pisgah in Saranac Lake Village. In 1987, I found it on Raquette Lake at Camp Huntington (Pine Knot). This far southwest, however, the species could have been established well before the Bartlett Carry invasion. Has this plant been spreading over the whole Adirondack region from southwest to northeast? The fruits of hawkweeds travel largely by wind. I wonder if people and their vehicles have aided in accelerating the distribution rate.

In 1985–1986, I was asked to do a botanical survey of the Four Brothers Islands, off Willsboro Point in Lake Champlain. (Each island is lettered rather than named.) My first impression of these islands, because of the large number of naturalized European weed species, was that of one huge barnyard. Over-grazing by domesticated animals, heavy trampling, and phenomenal vent (voiding) activity of nesting birds must have played an important role in the history of these small islands. I tallied the number of naturalized species and compared this with the total native and naturalized number of species. My conclusion was that the percentage of naturalized species is directly proportional to the degree of disturbance.

Other examples show that plants are constantly on the move today just as they were in post-glacial times. To explain just exactly how these plants migrate is next to impossible. Many are flown in by birds or by wind-borne seeds, and can come from virtually anywhere. The European weed, Hawk's-beard (*Crepis*), suddenly appeared in abundance in 1981 along the roadside of State Highway 86 between Paul Smith's College and Easy Street, a distance of about a mile. It disappeared in 1982, 1984, 1986, 1987, and 1989. It reappeared in 1985, 1988, and 1990. The native weedy species, Cudweed (*Gnaphalium uliginosum*), appeared on the Paul Smiths College Campus in 1976 and 1977 almost overnight, but hasn't been seen since. Figwort (*Scrophularia*) did likewise in 1986, probably arriving with some landfill materials behind the Maintenance Building of the College. Similarly, the native pondweed (*Potamogeton epihydrus*) nearly filled the west end of Cooler Pond in 1977, but was gone the following summer.

Trying to keep track of about 500 migrating plant species is like trying to watch over 500 children moving independently around a playing field. Neither the locations of the plants nor those of the children are the same from one moment to another.

Chapter 3

PLANTS AND CLIMATE

Plants have different strategies for adapting to various environmental factors.

Kudish

Temperature, precipitation, wind, and light, those phenomena which determine climate, further compound the impact of post-glacial conditions on plants. Inevitably temperature and length of growing season have significant effects over time on plant phenomenology (these are documented in Table 2 which provides a listing of botanical and climatic events occurring throughout the year, as well as in Table 3 which provides a list of several selected species and times during the growing season when mapping them proves most expeditious). Both elevation and latitude impact on temperature and, in turn, on growing season length. They have less influence on the phenomenology of plants introduced to the region.

The Adirondacks are fairly uniformly wet everywhere so that precipitation: rain, snow, and related forms play only a minor role in plant distribution. Only when there are rare and extreme droughts or floods are the plants affected. (For discussions of acid deposition occurring in all precipitation consult publications listed in the bibliography.)

In addition, wind affects plants, surprisingly not so much by breakage as by dehydration.

Further, light intensity and duration affect plant species with different shade tolerances.

PLANT PHENOLOGY OF THE ADIRONDACK UPLAND

Phenology, a contraction of phenomenology, is a term used to describe the study of appearances, occurrences, or events. In biology particularly, one studies when living organisms perform activities, especially activities that are cyclical (daily, monthly, yearly, etc.). In the case of plant activity, it describes such events as leafing, flowering, fruiting, and leaf fall.

The following table is arranged according to time of year rather than by species. Readers botanizing in the Adirondack Upland at a specific time of the year may find it a helpful reference.

The table has five vertical columns. The first lists the date for any year. The second lists the mean temperature in degrees Fahrenheit for any date on the Upland, most accurate for the 1500 to 2000 foot elevation range. Because this is the mean, daily minimums average about ten degrees cooler and daily maximums average about ten degrees warmer. The third column lists day length in hours and minutes from sunrise to sunset. Daylength is uniform over the whole Upland for any given day, varying only in summer and winter a few minutes latitudinally from the Canadian border to the Mohawk Valley. The fourth column records special astronomical and climatic, but non-biological events such as the longest day, shortest day, coldest day, warmest day; etc.

The fifth column lists biological events, specifically plant phenology. Although some of the events involve leafing, fruiting, and leaf fall, most listed are median flowering dates. The species included are natives from which such medians are available. Earliest and latest flowering dates for many species will be found in Chapter 5.

For winter, early spring, and late fall, when plants are inactive, one date per week for each month is listed: the 1st, 8th, 15th and 22nd. If an additional date features some special climatological or astronomical event, that date also is listed. Interpolation is necessary to determine the mean daily temperature and daylength for dates not included in the Table. During the late spring, summer, and early fall when plants are active, temperature and daylength (plus special climatological and astronomical event dates) are still listed for the four dates per month; additional dates are listed for phenological events of the plants.

For example, on June 1, the mean daily temperature for the Upland is 55°F, and the daylength is 15 hours and 25 minutes. There are no special climatic nor astronomical events that day, the last trees are leafing out, and this is the median flowering date for Jack-in-the-pulpit, Chokecherry, and Solomon's plumes.

TABLE 2
PLANT PHENOLOGY FOR THE ADIRONDACK UPLAND

Date	Mean Daily Temp. °F	Day Length h:m	Climatic Phenology (long-term averages)	Plant Phenology
January				
1st	16	8:56		
4th	15	8:59	Latest sunrise, 7:38 AM EST	
8th	15	9:03		
15th	15	9:12		
22nd	15	9:25	Coldest days of the year, January 21st through 23rd.	
February				
1st	15	9:46		
8th	15	10:02		
10th	16	10:08	Warm-up begins.	
15th	17	10:22		
22nd	18	10:44		
March				
1st	20	11:05		
8th	22	11:27		
15th	24	11:49		
22nd	27	12:13	Vernal equinox: March 20th to 23rd; spring begins; daylength increasing at greatest rate.	
April				
1st	30	12:45	First Sunday in April: Eastern Daylight Savings Time begins: set clocks one hour ahead.	

Pitcher plant (*Sarracenia purpurea*)
The leaves are highly modified into "pitchers". An inner slippery surface in one portion, and downward-pointing hairs in another make escape almost impossible for an insect. © Jim Kraus 1992.

TABLE 2 (cont.)

Date	Temp	Time	Notes	Flowering
April				
5th	32	12:57	Mean daily temperature begins to rise above freezing.	
8th	33	13:07		
15th	36	13:29	Most snow changes to rain.	
20th	38	13:40	Median date of ice-out on Lower Saint Regis Lake.	
22nd	39	13:51	Days warming up at the greatest rate: April 20th through 23rd.	
May				The following are MEDIAN flowering dates:
1st	43	14:16		
2nd				Trout lily & Red maple
3rd				Spring beauty
4th				Trailing arbutus
5th				Dutchman's breeches and Squirrel corn
6th				Round-leaved violet
7th				Sweet gale
8th	46	14:35		Sweet white violet
9th				Painted trillium
10th				Strawberry
11th				Bartram's serviceberry
12th				Purple trillium
14th				Leatherleaf
15th	49	14:52		Blue violet, Early saxifrage, Canada honeysuckle, and Kidneyleaf buttercup. Most hardwoods begin to leaf out rapidly.
17th				Marsh marigold & Sugar maple
19th				Bellwort
20th				Goldthread & Toothwort
21st				Canada violet & Serviceberry
22nd	52	15:07	Median date of the last frost except in pockets where it can freeze even in summer.	Witchhobblebush, Downy yellow violet, & Red cherry. Most hardwoods in full leaf.
23rd				Dwarf ginseng
25th				Red elderberry
30th				Foamflower
31st				Pussy toes
June				
1st	55	15:25		Last trees to leaf out: Big-tooth aspen, White & Black

TABLE 2 (cont.)

June				
				ashes. Jack-in-the pulpit, Chokecherry, & Solomon's plumes.
2nd				Low sweet blueberry
3rd				Wild sarsaparilla
5th				Clintonia and Bog rosemary
6th				Bunchberry & White baneberry
7th				Canada mayflower & Bog laurel
8th	57	15:34		
9th				Starflower & Black chokeberry
10th				Fringed polygala, Three-eaved false Solomon's seal & Nannyberry.
12th				Labrador tea, Indian cucumber, & Small cranberry.
13th				Pink lady's slipper
14th	59	15:39	Earliest sunrise: 4:13 AM EST or 5:13 AM DST	Hooked buttercup
15th	59	15:40		Blue-eyed grass
17th				Wild raisin & Common cinquefoil
20th				Sheep laurel
22nd	61	15:41	Longest days, June 20–23: 15h 41m. Summer solstice.	
26th				Bush honeysuckle and Tall meadow rue.
27th				Wild iris
28th	62	15:40	Latest sunset: 7:50 PM EST or 8:50 PM DST.	
29th				Wood sorrel
July				
1st	63	15:38		Blackberry
4th				Large roundleaf orchid
5th				Yellow pond lily & Leafy white orchis.
8th	64	15:32		
12th				Swamp rose and Ragged fringed orchid.
13th				Grass pink
15th	65	15:22		
16th				Robbin's ragwort
18th				Shinleaf, False violet & Rose pogonia.

26

TABLE 2 (cont.)

Date				
July 19th				Dogbane, Virgin's bower and Cowwheat.
21st				Bluebell
22nd	65	15:09	Warmest days of the year, July 20th through 23rd; mid-point of growing season.	Plants at all elevations phenologically equal: Higher elevation plants catch up with lower elevation plants after a late spring start, and plunge faster into autumn for an early finish.
23rd				Daisy fleabane & Calla lily
25th				Small bedstraw & Black elderberry
26th				Pipewort
27th				White water lily
29th				Common bladderwort and Nodding bur marigold
31st				Evening primrose
August 1st	64	14:46		Green woodland orchid
3rd				Meadowsweet, Pickerel weed and Elliptical St. Johnswort
4th				Sundews and Pondweed (*P. epihydrus*)
5th				Skullcap & Water lobelia
8th	63	14:28		
12th				Swamp loosestrife and Steeplebush
15th	62	14:10		Bog goldenrod
17th				Closed gentian
18th				Eel grass & Bugleweed
22nd	61	13:53		Large-leaved goldenrod, Jointweed & Claspingleaf pondweed
23rd				Water milfoil
25th				Rough bedstraw & Northern willow herb
26th				Joe pye weed; First Red maple leaves turn red on dying branches or stressed trees.
29th				Marsh bellflower
30th				Water smartweed, Mint & Swamp beggars-ticks.
Sept. 1st	59	13:14		Most goldenrods & asters

TABLE 2 (cont.)

Sept. 4th 8th 11th 14th 15th 22nd	 57 55 52	 12:55 12:35 12:09	 Median date of first frost (frost occurs in all months in frost pockets); most thunderstorms end. Autumnal equinox, September 20th to 23rd; fall begins; daylight decreasing at greatest rate.	Clearstem Panicled aster Autumn ladies' tresses Color change develops rapidly in most hardwoods.
October 1st 7th 8th 9th 10th 15th 20th 22nd	 48 45 42 39	 11:44 11:23 11:01 10:38	 Days cooling off at greatest rate, October 20th to 23rd. Last Sunday in October: Eastern Standard Time returns. Turn clocks back one hour.	Rapid leaf fall begins. Witch hazel Heart-leaved aster Most hardwood leaves down, except Trembling aspen. Larches in full needle color and Huckleberries in full leaf color.
Nov. 1st 8th 15th 22nd	 35 32 30 27	 10:13 9:53 9:38 9:22	 Mean daily temperature begins to fall below freezing. Most rain changes to snow. Lakes freeze. Small ponds as early as late October in some years.	
Dec. 1st 8th 15th 22nd	 23 21 19 18	 9:06 8:59 8:56 8:54	 Earliest sunset: 4:18 PM. Shortest days: December 20th to 23rd: 8h 54m. Winter solstice.	

TABLE 3
PHENOLOGY: MAPPING THE SPECIES

Mapping of individual plant species becomes much easier when the time of year that each species is particularly conspicuous and outstanding on the landscape is determined. A hilltop with a good view or a canoe in the middle of a lake provide excellent sites for plotting species distribution on topographic maps. (Early refers to the first week of a month, mid refers to the second and third weeks of a month, and late refers to the fourth week of a month).

Species	Color	When to map in a "Normal" or "Average" year on the Northern Adirondack Upland (elevation 1500 to 2000 feet)
White ash	Purple foliage Real purple, not wine-red as in Red maple	Mid-September
Red maple	Red flowers	Late April to mid May
	Wine-red leaves (not scarlet)	Late August to early October
Trembling aspen	Yellow green leaves expanding	First week in May before other hardwoods leaf out
	Yellow leaves mature	Second week in October after other hardwood leaves have dropped, but before Larch's peak yellow
Balsam poplar	Bronze expanding leaves	Mid-May during leaf out. The balsam odor may be overpowering on a mild, drizzly, windless day
	Bronze-green foliage	Less obvious all summer than in May and before leaf drop
	Bronze leaves before leaf-fall	Late September-early October
Bigtooth aspen	White expanding leaves and twigs	First of June
Eastern larch	Yellow needles	Third week in October after almost all hardwoods have dropped leaves
Huckleberry	Scarlet leaves	Mid-October after hardwood leaf-drop, especially under open Red pine stands along exposed lakeshores. Shrubs about 3 feet tall.
Witchhobblebush	White flowers	2nd week in May. Shrubs 3 to 10 feet tall.

Champlain Valley to Whiteface summit in New York;
Burlington to summit of Mount Mansfield in Vermont;
Coastal Maine to Mount Washington summit in New Hampshire;
Hudson Valley to summit of Slide Mountain, in the Catskills, New York.
High elevation weather stations in the eastern U.S. are rare. Therefore we must use the data from each of them.

Growing degree days—abbreviated from now on as G.D.D.s—are calculated by finding the positive differences in temperature between the daily mean and 40°F summed up over the whole year. Hence, on a day with a mean temperature of 52°, there are 12, i.e. 52 less 40, G.D.D.s. On a day which averages exactly 40°F, there are no G.D.D.s. On a day with a mean temperature below 40°, say 37°, the number of G.D.D.s is still zero. There are no negative G.D.D.s. In agriculture, the base temperature for calculating G.D.D.s is 50 rather than 40; this is because agricultural crops don't do much below 50°F. In contrast, native plants, adapted to shorter growing seasons and cooler air, can be active but sluggish down to 40°. G.D.D.s range from 4200 along Lake Champlain to 1400 atop Whiteface Mountain.

Growing season and frost-free season are often confused and interchanged. In actuality, the two are quite different. Frost-free season is the average number of days between the last vernal frost and the first autumnal frost. Growing season for any given locality is longer than frost-free season since plants often do things before the mean date of last spring frost and after the mean date of first fall frost. An example is the mean flowering date for spring beauty (*Claytonia caroliniana*), May 3 (See Table 2) while the mean date of last frost in spring is about May 22, nearly three weeks later. The mean date of maximum leaf drop for most hardwoods is October 5, while the mean date of first frost in the fall is about September 15, nearly three weeks earlier. Frost-free season on the Upland, except in frost pockets, is barely four months, 120 days, while growing season is more like 5½ months, 170 days. Note that the major leaf-out push is about May 15, near the time of the average last spring frost, and color change begins in earnest about September 15 at the average time of the first fall frost.

The mean date of both last spring and first fall frosts for any site can be calculated by dividing the number of frost-free days by two and counting the quotient of days on either side of the period July 21–23. For example, at Paul Smiths, the mean frost-free period is 110 days; 55 of these days occur before July 21 and 55 of them occur after July 23. Hence, the last spring frost occurs on May 26, and the first fall frost on September 16. This calculation formula will not prove useful for trying to determine the mean frost-free season in frost pockets where frost can occur in any month of the year.

Mean annual temperature, growing degree days, and frost free season are equally good measures of climate for determining vegetational distribution. Mean annual temperature is the simplest to use. The correlation between M.A.T. and G.D.D.s is almost 100%. Thus, one could use G.D.D.s as well, but the four-digit numbers are more clumsy and more time-consuming to calculate. The relationship is that for each 1°F fall in M.A.T., corresponding to about 333 feet of ascent, the number of G.D.D.s to base 40°F falls between 193 and 200 (see Table 4 for the relations of elevation, M.A.T., and G.D.D.s).

The relation between M.A.T. and frost-free season is not so clear because frost-free season does not depend only on elevation and latitude. Local effects as well influence frost-free season. Such local effects are caused by proximity to the ocean or large lakes, frost pockets in valley bottoms, and "heat islands" of large cities. Generally, for each 1° fall in M.A.T., 333 feet ascent, frost-free season falls between 5.84 and 6.0 days.

Table 2 presents the phenology of a number of plants on the Adirondack Upland at elevation 1500 to 2000 feet. But what happens to the phenology of plants if they are growing at significantly higher or lower elevations than 1500 to 2000 feet? There are notable differences.

Simply put, growing season decreases with increasing elevation, not only at the vernal season end but also at the autumnal season end. Thus, plants at higher elevations must start activities late in spring and finish activities early in fall, being telescoped by winter from both ends. Because of this, all activities: leaf out, flowering, photosynthesis, seed or spore set, leaf color turn, and leaf fall must be accomplished in great haste at the higher elevations. Being evergreen helps and many plants at high elevations are evergreen. The dates that high-elevation plants catch up with and get ahead of the Upland plants are July 21–23, the warmest days of the year. However, after July 23, the high-elevation plants plunge rapidly into autumn.

How much later in spring and how much earlier in autumn do the high-elevation plants perform activities than their cousins on the Upland? The higher the elevation, the later the start, and the earlier the end of the growing season. Because the differences in rate varies from year to year, only long-term averages are offered in the following table:

TABLE 4
ELEVATION, GROWING SEASON, AND PLANT PHENOLOGY

Elevation in Feet	Mean Annual Temperature °F	Growing Degree Days Annual, 40°F	Season (approx).		Phenology of Deciduous Species			
			Frost-Free Days	Growing Days	Leaf Out Begins	Full Leaf	Color Change Begins	Leaves All Down
6288—Mt. Wash. N.H.	27	800						
5344—Mt. Marcy	29	1100	55	120	6/1	6/8	9/5	9/24
4867—Whiteface	30	1400	60	126	5/30	6/6	9/6	9/26
4500	31.5	1600	68	132	5/29	6/5	9/7	9/28
4000	33	1800	75	138	5/26	6/2	9/8	9/30
3500	34.5	2100	81	144	5/24	5/31	9/10	10/2
3000	36	2400	89	150	5/21	5/28	9/11	10/4
2500	37.5	2700	96	156	5/19	5/26	9/12	10/6
2000	39	3000	103	162	5/16	5/23	9/14	10/8
1500–1800—Upland	40	3200	110	170	5/14	5/21	9/15	10/10
1000	42	3400	120	178	5/11	5/18	9/17	10/12
500	43.5	3900	130	186	5/9	5/16	9/18	10/14
95—Lake Champlain	45	4200	140	192	5/6	5/13	9/20	10/17

Note that the values in Table 4 (page 33) are averages. Values are accurate from the Lake Champlain level to the Upland and for the summits of Whiteface and Mount Washington where weather stations exist. The values for the intermediate elevations, 2500 to 4500 feet, and for Mount Marcy are estimates.

By examining the table, one can estimate the rates of change of various climatic and phenological factors with increasing elevation.

(1) Mean annual temperature falls at the rate of about 3°F for each 1000 feet of ascent.
(2) Growing season falls at the rate of about 16 days for each 1000 feet of ascent.
(3) Frost-free season falls at the rate of about 14 to 20 days for each 1000 feet of ascent.
(4) Growing degree days fall about 600 for each 1000 feet of ascent.
(5) Spring comes 5 to 10 days later for each 1000 feet of ascent. To determine the date of the arrival of spring and fall, one must choose a species and one event, such as opening of fir buds, and trace that event's spring progress up the mountainside. Then a second species and its event, such as flowering of goldthread, must be traced up the mountainside in spring. A third species and its event must be traced up the mountainside, and so on. After many species and many events are studied, one can take an average of the delay of events with increasing elevation, i.e., the arrival of spring. The reverse can be done in the autumn as events "move" down the mountainsides.
(6) Fall comes 5 to 10 days earlier for each 1000 feet of ascent.

Many plant species grow over a large range of elevations and latitudes, and thus are acclimated to a wide range of mean annual temperatures, growing degree days, and frost-free seasons. However, individual plants within a single species are acclimated to the elevation and latitude of their particular sites (see the elevation range information for many of the species described individually in Chapter 5).

GROWING SEASON, LATITUDINAL CHANGE, AND PHENOLOGY

In addition to elevation, latitude is a major consideration when determining m.a.t. If one selects weather stations at the same, or very similar, elevations over a large range in latitudes and examines M.A.T. data, one finds that the M.A.T. drops 1.5°F for each 1° latitude. Since 1° of latitude equals about 69 miles, then 1.5° latitude equals about 105 miles. M.A.T.s range from Key West, Florida, 24.5° north latitude, with 78°F, to Goose Bay, Newfoundland, 53.5° north latitude, with 32°F. These two stations are both near sea level, elevational effects being nonexistent. The 29° difference in latitude, about 1900 miles distant in a north-south line, corresponds to a difference of 44°F in M.A.T., or 1.5°F per degree latitude. Intermediate weather stations along and near the Atlantic Coast show a simple, linear relationship so that the m.a.t. is easy to interpolate.

Knowing the rates of M.A.T. with both elevation, −3°F/1000 feet, and latitude, −1.5°F/degrees latitude, one can estimate quite accurately the m.a.t. for any site in eastern North America with data from one nearby weather station.

If one is to compare weather stations in the Adirondack Region in order to determine elevation M.A.T. relations, one must see whether the weather stations have a latitudinal difference large enough to affect M.A.T. also. Hence, when one studies southern vs. northern Adirondacks, e.g. Lake George vs. Dannemora, latitude becomes an important factor.

Table 5 presents the differences in growing season factors at localities with the same, or nearly the same, elevation, varying only in latitude. By holding elevation constant, one can study latitudinal effects alone on growing season and phenology. The localities are arranged with increasing northward latitude.

Rates of change over latitude are as follows:

Mean annual temperature falls 1.5°F for each degree latitude northward (69 miles).

Frost-free season falls 20 to 22½ days for each degree latitude northward (69 miles).

Growing degree days fall 400 for each degree latitude northward (69 miles).

Spring arrives 4 days later for each degree latitude northward (69 miles).

Fall arrives 4 days earlier for each degree latitude northward (69 miles).

TABLE 5
LATITUDE AND GROWING SEASON

Locality	Latitude North	Air Miles S to N only	Elevation Feet of Weather Station	Mean Annual Temp, °F	GDDs Annual 40°	Frost-Free Days	Growing Season Months
Key West, FL	24° 30'	0	9	78	10650	365	12
New York City	40° 30'	1104	132	54	6019	180-223	8
Albany, NY	42° 40'	1254	277	49	5040	160-175	7
Burlington, VT	44° 30'	1380	331	45	4270	148	6
Montreal, QUE	45° 30'	1449	187	43	3900	158	5
Quebec City	47° 00'	1553	296	39	3130	138	4.5
Goose Bay, Labrador	53° 30'	1901	144	32	1780	?	3

Combining latitudinal effects with elevational effects on phenology, we can estimate that spring comes 27 days later to Paul Smiths and Saranac Lake than to New York City. Because of elevation, with the former some 1500 feet higher than New York City, the arrival of spring is delayed some 7½ to 15 days. Because of latitude, some four degrees further north, spring is delayed an additional 16 days. The total is somewhere, combined, about 23 to 31 days. The method for testing this is to observe when those species, native to both Paul Smiths and New York City flower. For example, Red maple flowers during the first week of April in New York City and the first week of May at Paul Smiths. [Ignore transplants, as these introduced plants run on different clocks (see introductions below).]

Knowing the relations between mean annual temperature, growing degree days, elevation, and latitude, and having available the U.S. National Weather Service data from the nearest weather station, one can easily predict the climate at any locality.

INTRODUCTIONS

Each autumn, the cottonwoods, silver maples, box elders, Norway maples, and crack willows planted as ornamentals in the Adirondack Upland villages such as Saranac Lake turn color and drop their leaves later than the native Adirondack hardwoods by as much as two to three weeks. Green leaves are common on these five species well into late October, while the leaves in the woods outside the villages are all down. This is due to the fact that the first three species have been transplanted from areas in the U.S. well south of the Adirondacks and the latter two species from Europe. All five species have been brought in from areas with longer growing seasons and later leaf fall. They have not adjusted to the short Adirondack growing season even after living here for fifty to nearly one hundred years; they still maintain the biological clocks of perhaps southern New York, Connecticut, New Jersey, Pennsylvania, or Europe.

Even more surprising is that three of these species, box elder, Norway maple, and silver maple, have become naturalized, reproducing on their own the saplings adhering to the same clock as that of their parents. The silver maples planted in the Village of Saranac Lake are still green on October 10 while native silvers along the Saranac River a few miles downstream toward Bloomingdale have already dropped their leaves. At the other end of the growing season, sugar maples transplanted to Paul Smiths from perhaps southern New York leaf out a week or more ahead of native sugars and are more subject to late spring frosts. If one knows growing season lengths and phenology for trees native to the eastern states in parts of the country south of the Adirondacks, one can estimate the place of origin for transplants by observing how much earlier in the spring or how much later in the fall they leaf out, bloom, fruit and leaf-drop in the Adirondacks. This may be explained by the fact that the date of coloration and leaf fall for any individual tree is probably genetically determined over thousands of years as a result of its ancestors growing in a particular locality. When a tree is transplanted by people to another locality of greatly differing elevation and/or latitude, the tree will maintain its original coloration and leaf drop schedule, and so will its offspring. One may hypothesize that it must take many generations for a transplanted population to begin to adjust to the new growing season.

PRECIPITATION

Mean annual precipitation is defined as the average quantity of rain plus melted snow, sleet, and hail that fall on a locality over a whole year. It is measured in inches or in centimeters, and is an average. Drought and flood years cause considerable departure from this average. For New York State and the Northeastern United States as a whole, the mean annual precipitation is about 40 inches. Ranging from 26 and 29 inches in the northern Champlain Lowlands and the Lake Ontario shore east of the Niagara River respectively to 60 inches at Winnisook Lake in the Catskills at elevation 2660 feet, precipitation in New York State is probably still greater atop the highest Catskill peaks (4000 to 4180 feet), but no weather stations are located at these elevations to document the deluge. Mount Washington, New Hampshire, receives 70 inches a year and the Catskill summits may receive even more.

In contrast, the summits of the Great Smoky Mountains in North Carolina and Tennessee collect an average of 90 inches. The rain forests of the Olympic Peninsula in Washington State are inundated by 100 to 200 inches. Below 20 inches, forests give way to grasslands with scattered trees, called woodlands or savanas; and while 15 inches will support only grassland prairies, less than 10 inches supports only desert vegetation.

Mean annual precipitation for the greater Adirondack region recorded at a number of weather stations is listed below in increasing order. The wettest stations

TABLE 6
MEAN ANNUAL PRECIPITATION
FOR THE GREATER ADIRONDACK REGION

Station	Mean Annual Precipitation in Inches	Elevation in Feet
Chazy	26	200
Albany	32	277
Burlington	32	331
Watertown	33 & 40 †	497
Oswego	35	292
Marble Mt. ‡	36	1980
Lowville	36	900
Gabriels	36 & 37 †	1750
Canton	37	414
Moira	37	400
Tupper Lake	37	1680
Lake Placid	39	1864
Indian Lake	40	1705
Montreal	41	187
Wanakena	41	1510
Lake George	42	350
New York City	42	132
Whiteface Summit	49	4867
Highmarket	51	1786
Number Four	52	1571
Gomer Hill, Tug Hill	54	2175

are at the highest elevations in the High Peaks, e.g. Whiteface Summit at 49 inches. They also occur at moderate elevations in the southwestern Adirondacks and on the Tug Hill Plateau, e.g. Highmarket, Number Four, and Gomer Hill at 51 to 54 inches. The driest station is Chazy in the northern Champlain Lowland at 26 inches. The elevations listed are for the weather stations which may or may not be located within the hamlets, villages, or cities. The symbol † indicates that weather records were kept during two different periods at perhaps different but nearby localities, hence the two values. The symbol ‡ indicates that the Atmospheric Sciences Research Center field station is located on Marble Mountain, a northeast spur of Whiteface Mountain.

In all cases throughout the whole Adirondack region, the mean annual precipitation is adequate for all species of native plants, and it certainly can support a continuous, closed-canopy forest. Forests are lacking in the region naturally only where bedrock is at or near the surface, or where there is open water such as a marsh or lake. The lack of forest is caused by edaphic (soils) factors, not precipitation. Some exceedingly sandy outwash soils are drought-prone so that such plants as witchhobble-bush, sharp-leaved aster, polypody fern, climbing buckwheat and striped maple will begin to wilt after a week to ten days without precipitation in summer. Most of these have large or membranous, thin leaves which lose water pressure before other plants do so. (Droughts do not occur every year, but only once in several years on an irregular basis).

Mean annual snowfall, the average for the year over many years, depends on mean annual precipitation and mean annual temperature. Snowfall is higher in areas with high mean annual precipitation; it is also higher in areas with lower mean annual temperatures because of high elevation or high latitude or both. In the lower mean annual temperature localities a larger percentage of the mean annual precipitation falls as snow, and therefore a smaller percentage falls as rain. The snow season is thus extended beyond winter into late autumn and early spring. Table 7 arranges Adirondack regional weather stations from the lowest to the greatest mean annual snowfall.

The symbol † indicates that weather records were kept during two different periods at perhaps different but nearby localities. Hence the two values.

Snowfall on the summits of the high peaks such as Whiteface and Mount Washington accumulates primarily in protected depressions, ravines, and cirques; here, the deep snowpack may persist until June. On the exposed ridges, the snow accumulates to much less depth because much of it is blown off by the wind; it melts off here first in the spring.

An average ten inches of snow melts down to the equivalent of one inch of rain. Hence, for example, Tupper Lake's 88 inches of snow yearly average melts down to about 9 inches of liquid water. Since the total mean annual precipitation is 37 inches, then 9 inches, 24% of that, falls as snow, and the remaining 28 inches, 76%, as rain.

TABLE 7
SNOWFALL

Station	Mean Annual Snowfall-Inches	% of Mean Annual Precipitation Falling as Snow	Elevation, in Feet, of Weather Station
New York City	30	16%	132
Chazy	39	15%	200
Albany	50	16%	277
Burlington	65	20%	331
Lake George	72	17%	350
Canton	73	20%	414
Oswego	88	25%	292
Tupper Lake	88	24%	1680
Indian Lake	90	23%	1705
Watertown	98 & 100†	30% & 25%†	497
Gabriels	98 & 123†	27% & 33%†	1750
Lowville	101	28%	900
Moira	106	29%	400
Montreal	112	27%	187
Marble Mt., Whiteface	114	32%	1980
Lake Placid	127	33%	1864
Wanakena	136	33%	1510
Number Four	141	27%	1571
Highmarket	160	31%	1786
Adams Center	171	?	400 ca.
Mt. Washington, N.H.	180	26%	6262
Whiteface	220	45%	4867
Gomer Hill/Tug Hill	225	42%	2175

    Snowfall, like mean annual precipitation, has little influence upon the distribution of plants in the lowlands and on the Adirondack Upland, but has considerable effect on plants at the higher elevations above 2500 or 3000 feet where it is greatest. The evergreen spires of spruces and fir have tapered, conical crowns which aid in shedding excess snow, and thus reduce the chances of limb breakage. Those broad-crowned evergreens associated with northern hardwood forests, white and red pines and hemlock, could receive great snow damage if they grew at an elevation 3000 to 4000 feet. Notably, they grow at lower elevations where snowfall is less prolific.

    Ice rime, the condensation of liquid water droplets on frozen objects such as branches and needles, is most prevalent at the highest elevations, generally above 3000 feet and, when extremely heavy, can break limbs. Most of the time, however, the latter can withstand some ice rime.

    Acid deposition, popularly called "acid rain," is a relatively new phenomenon. First noticed in the 1960s and early 1970s, it is currently under intense study. Many scientists and technicians from a number of universities have converged on Whiteface

Mountain in an attempt to find the cause(s) of red spruce decline above 3000 feet. The mountain's soils, forests, air, water, and precipitation are being sampled and examined with unprecedented scrutiny. There is no need here to summarize the vast volume of literature published over the last ten or fifteen years on the effects of acid precipitation on lakes, ponds, fish, forests, and concrete. [See Follos (1986), Ford (1990), Gaffney (1984), Ketchledge (December 1987 and February-March 1988), Randorf et al. (1987), and Whitehead et al. (1986).]

WIND

Wind speed averages 9 to 10 miles per hour over the year for most New York State cities (more at Buffalo where the wind sweeps off Lake Erie at 12.5 miles per hour) and has no effect in most places on vegetation. However, mean wind speed influences vegetation on exposed mountain summits, many upper slopes above about 3000 feet, tops of high cliffs, and along the lee, mostly east, shores of lakes and open fields. Wind speed averages 17 to 22 m.p.h. atop Whiteface Mountain, probably more in the winter. On the upper slopes of many high peaks and in exposed places at lower elevations, flagged trees form a conspicuous part of the landscape. Nearly all the branches point in a single direction, opposite that of the prevailing wind. Between approximately 4000 feet elevation and timberline on most of the high peaks, wind-generated fir waves can be seen for miles around as alternating bands of dead and gray trees with live and green ones. These cyclical waves, passing any given point about every 60 to 90 years, have been fully studied by Sprugel (1984).

Wind has not only a stunting, limb-breaking, and, in extreme cases, felling effect on shrubs and trees but equally, or even more importantly, a dehydrating effect, especially in winter. When the ground is frozen, water lost through leaves, twigs, buds, and branches due to wind cannot be replaced. For woody plants, winter is the dry season despite the fact that precipitation is nearly equally distributed throughout the year. Evergreens have evolved to minimize water loss with features such as a thick waxy epidermis, sunken and thus protected stomates and, a few, with hairy surfaces to reduce wind speed and thus the evaporation rate. Thick conifer needles have minimum surface area for their volume over which to lose water. In contrast, broad-leaved deciduous trees such as birches, maples, beech, poplars, cherries and ashes have leaf blades with vast surface areas and rapid water loss. These hardwoods lose their leaves in the autumn primarily to minimize water loss during the winter. Plants of the Heath Family or Ericaceae, evergreen shrubs in bogs such as bog rosemary, Labrador tea, bog laurel, leatherleaf, and cranberries have leaves with water-retaining features similar to those of evergreen conifers despite the fact that they grow in water. In winter, although the bog is frozen, the leaves are still on and exposed to the wind except during periods of deep snow. Small leaf size, great thickness, high wax, rolled-under margins, and sunken stomates (Labrador tea, in addition, has wooly hairs on the lower surface of the leaves) all help reduce water loss.

Eastern white pine is now much more common in the Adirondacks than it was in previous centuries. Abandoned fields, burns, logging sites, and other human disturbances have eliminated the forest in many areas, giving the shade-intolerant white pine a chance to expand. This is true for other pioneers such as poplars, red cherry, paper and gray birches, larch, and black spruce. Where were most of the white pines prior to the 19th century. If one stands on a summit such as Saint Regis or Ampersand Mountain today and tries to ignore areas of major human disturbance, one notes that white pines are most concentrated around lakes and ponds, with greatest abundance on the east shores. White pines require full sunlight to become established; thus they follow disturbances which create open areas. Blowdowns were and still are most common on the east shores of lakes and ponds due to predominantly westerly winds. Lake country is white pine country because of wind.

A map of the natural, not plantation, distribution of red pine in the Adirondacks (p. 99) at first may present a puzzle. Soil studies usually helpful in understanding tree distribution fails in this case. Red pine occupies a diversity of soil sites, provided that all are well-drained. For example, it grows on Picnic Point on Paul Smith's Campus (shallow outwash sands over anorthosite bedrock); Bluebird Cabins, White's Pine Camp, and Osgood Pond sites (deep, sandy outwash); Kellogg Point on Campus (deep, gravelly outwash); Owl's Head near Keene and Catamount Mountain east of Franklin Falls (shallow till over bedrock); east slopes of Whiteface Mountain (deep glacial loamy tills); and Willsboro Point (deep to shallow silt loams over limestone).

A more careful examination of the distribution map reveals that red pine lines the east shores of lakes and ponds, populates the islands, and dominates peninsulas. Further inland, away from the water, red pine is absent because, shade intolerant, it cannot compete with red spruce, balsam fir, eastern hemlock, yellow birch, beech, sugar maple, and red maple. Some red pine sites are not alongside bodies of water at all, but on steep slopes and/or hilltops with bedrock near or at the surface instead. Some of these slopes and summits have burned, while others have not.

All red pine are on the most exposed, windswept places. Cook, Smith, and Stone (1952) noted this in their studies of the distribution of red pine in New York State, but did so without emphasis. Apparently, red pine can withstand the dehydrating winter winds better than any other tree species.

Red pine stands are by nature open stands, allowing much sunlight to penetrate between the distant crowns, and permitting shade-intolerant shrubs and herbs to thrive. These smaller plants do not require red pine as a companion, for they will grow in other sunny areas without this conifer. Huckleberry, shadbush, pipsissewa, blueberries, cow-wheat, bunchberry, trailing arbutus, and wintergreen are common. Shrubs typical of open, sunny wetlands: mountain holly, wild raisin, sheep laurel, and Labrador tea grow high and dry, surprisingly to some people, under red pine. Even some shade-intolerant pioneer trees: paper birch, bigtooth aspen, trembling aspen, and red cherry become established under red pine. Because of the partial sunlight on the forest floor, red pine reproduces in these stands and should continue, barring human interference, to do so for centuries or millennia to come. The red pine has probably been on these sites for thousands of years because no shade-tolerant climax tree species can survive here. The oldest red pines found locally are on Picnic Point. The growth rates of these wind-exposed trees is very slow. Near constant severe winds cause limb breakage and contortion. The tree with the most contorted

TABLE 3 (cont.)

Species	Color	When to map in a "Normal" or "Average" year on the Northern Adirondack Upland (elevation 1500 to 2000 feet)
Red oak (White oak)	Purple leaves (occasionally leaves turn brown)	Late September
	Tan to brown dried leaves	November and December—many dead leaves persist on twigs
	(Purple-brown dried leaves. Red oak lacks the purple cast.)	(November and December—many dead leaves persist on twigs in the Champlain-Lake George Lowlands.)
Sweet gale	Blue-green leaves	Summer—a shrub of 2 to 3 feet high along edges of ponds and slow streams
Wild raisin and Nannyberry	Stench of wet, black leaves ready to drop or already fallen creates butyric acid which has the odor of rancid butter.	From late August to mid-October; much of the Adirondacks reeks from the putrid stench.

GROWING SEASON, ELEVATION CHANGE, AND PHENOLOGY

Growing season, mean annual temperature, growing degree days, and frost-free season decrease with increased elevation.

Mean annual temperature (abbreviated as M.A.T.) is the average temperature of a weather station or locality for a whole year. In the greater Adirondack region, the M.A.T. ranges from 45°F at Lake Champlain (elevation 95 feet) to 30°F atop Whiteface Mountain at 4867 feet. An estimate for the summit of Mt. Marcy at 5344 feet (where there is no weather station) is about 29°F. One can estimate the M.A.T. for any elevation in the Adirondacks by using Tables 4 and 5.

The average cooling rate with ascent is about −3°F per 1000 feet, or −1°F for each 333 feet of ascent. This is a variable rate, dependent on several atmospheric factors including wind, cloudiness, and humidity. On a cloudy day with wind, rain, or snow the air is well mixed and the rate is barely −2°F/1000 feet. On a dry, sunny, windless day the rate is about −4°F/1000 feet. The maximum possible rate, called the adiabatic lapse rate, is 5°F/1000 feet. But this rate rarely occurs in nature.

The average rate, about −3°F/1000 feet, is calculated by comparing the temperature data from several weather stations with a range of elevations of at least 4000 feet, but all with the same, or nearly the same, latitude. Thus, differences in temperature due to latitude are eliminated. Examples of such groups of weather stations strung out over great elevational, but not latitudinal ranges are:

branches was established about 1810, and a neighbor, even earlier, about 1750. Some red pines on Kellogg Point, at the west end of the Paul Smith's College campus, were established about 1825, still well before settlement in the Paul Smiths area.

LIGHT INTENSITY AND DAYLENGTH

Plants have different strategies for adapting to various environmental factors. These include light intensity and daylength.

1. Shade tolerant: plants which can grow in the shade, although growth rates may be best in partial sun, include Sugar maple, Beech, Hemlock, Hop hornbeam, Striped maple, Witchhobblebush, Wood sorrel, Shining clubmoss, Canada mayflower, Balsam fir, and Red spruce. Chokecherry, a shrub, leafs out in May about one week before the hardwoods, sugar maple and black cherry, under which it often grows and thus takes advantage of the early spring sun.
2. Shade intolerant: plants which must receive full sun at least for a good portion of the day include: Aspens, Balsam poplar, Red cherry, Paper birch, Larch, Red and White pines, Bracken, Milkweed, Canada goldenrod, Rough-leaved goldenrod, Raspberry, and Blackberry.
3. Vernal: plants which hurriedly leaf, flower, and set seed to take advantage of the early spring sun before the hardwoods are in full leaf include Trout lily, Spring beauty, Dutchman's breeches and Squirrel corn. By June the above-ground plant parts have withered. (Except for the latter two, the plants are ecologically similar but unrelated.)
4. Vernal or Spring, with Aestival or Summer Flowering: Wild leek (*Allium*) leafs out with the Vernals before the hardwoods do, and the leaves wither by June. But, and herein lie differences from the Vernals, the flowers emerge in mid-summer in the shade and fruits persist into early autumn. There is a period in early summer when neither leaves nor flowers is visible, and one could think the two organs to be of two different species.
5. Evergreens: Evergreens take advantage of early spring and late autumn sun and above-freezing temperatures, while the deciduous hardwoods are leafless. Evergreens include not only the conifers, but also Christmas, Marginal shield, Braun's Holly, Polypody, and Crested shield ferns, all Clubmosses, Mosses, Liverworts, and such flowering plants as Wintergreen, Trailing arbutus, False violet, Goldthread, Creeping snowberry, Partridgeberry, Shinleaf (*Pyrola*), Twinflower, and Pipsissewa.
6. Green bark: Some woody plants have green stems which allow them to take advantage of early spring and late autumn sun and mild temperatures and, thus, to extend their photosynthesis season. In effect, these twigs, branchlets, and branches are evergreen accessories to leaves. Examples are Trembling and Bigtooth aspens, Mountain holly and Blueberry (*Vaccinium angustifolium*).

Long day plants flower during early summer; others, short day plants, flower in early spring or early fall; while still others, day neutral plants such as dandelions, flower throughout the growing season no matter what the daylength. (Consult any general botanical work for a complete discussion).

Daylength is constant for any one date, e.g. April 22nd has 13 hours and 51 minutes of daylight (See Table 2) each year, varying by a minute or two. Yet, there are differences in when plants do things from year to year, especially in the spring, and much less so in the fall. A cool April will slow down the dormancy break, while a warm one will accelerate leaf-out and, in those species flowering vernally, blooming. For example, a warm spell in early April can cause willows and alder flower buds to open partly. But if a cold spell follows before flowering can occur, the partly-opened buds will stall until the next warm spell brings out the flowers. (Once partially opened, buds cannot retract). Differences in phenology due to temperature in spring can be as much as one to two, or even three, weeks from year to year.

In the exceptionally mild autumn of 1971, the weather did not turn cold until early November; yet all deciduous trees had turned color and dropped the bulk of their leaves by October 10th, right on schedule. Coloration and leaf-fall were not delayed until the colder early November period. In the fall of 1976, the first snow arrived during the second week of October and stayed permanently on the ground through the winter; the trees had not turned color nor dropped their leaves ahead of normal time in September in preparation for an early winter. In the fall of 1977, heavy rains and constant cloudiness delayed the first frost until mid-October, a month later than normal; yet the trees turned color and dropped their leaves on schedule. One exception to the phenological consistency of autumn occurred in 1983. For some reason, color change and leaf fall were a week or so later than normal with temperatures and precipitation not unusual that summer.

Daylength cannot explain leaf coloration and leaf-fall timing because daylength does not take into account the effect of elevation on autumnal arrival. Every year the beech, sugar maple, and yellow birch on the upper slopes of Saint Regis Mountain turn color and drop leaves before those trees on the Paul Smith's College campus, 1000 feet below, by about a week. Likewise, these three species on campus turn color and drop leaves about a week earlier than do those along Lake Champlain, 1500 feet lower. Daylength varies only as a function of latitude and time of year, not elevation. And these three localities, Saint Regis Mountain, Paul Smith's Campus, and Lake Champlain, are all at the same latitude, 44° 23' north. There is no daylength difference among them.

How can these plants predict the individual weather patterns for each autumn and adjust their leaf fall accordingly? They cannot. They must turn color and drop leaves according to long-term climatic average temperatures for their locality, modified only very slightly by the temperatures and precipitation of each particular summer and autumn.

WINTER SEVERITY

Trees cannot predict the severity of a winter. Neither can the National Weather Service, at least not yet with any degree of accuracy. The weather varies greatly from year to year because it is partly random. Plants do things on long-term climatic averages and are often caught unprepared as we are. The snows of May 19, 1976, September 16, 1986, and October 4, 1987 caught the Adirondack hardwoods with their leaves on.

Trees and shrubs must make preparations many months in advance. The winter buds which open in May are preformed the previous summer. Pines require two-and-a-half years in advance to produce their seeds, while other genera of conifers and most hardwoods require one-and-a-half years from the time of bud formation.

June 1982 produced an unusually large crop of red maple fruits. In fact so many buds went to flower and fruit production instead of twig and leaf production that the leaves came out a month late, and many trees had sparse foliage throughout the whole growing season. This heavy fruiting was not the result of preparation for an unusually hard winter, 1982–1983, to follow. The flower buds on these red maples had been formed during the not unusual summer of 1981, one year before. The winter of 1982–1983 was not unusual. The cause of the heavy fruiting remains inexplicable.

LIGHTNING

Lightning most frequently strikes white pines because these trees emerge far above the average level of the forest canopy and are thus good targets. Mature trees of most species other than white pines in the Adirondacks attain heights of 70 to 80 feet, while white pines often reach 120 feet, with 165 feet being a local record. It is not uncommon to see huge old white pines with a spiralling lightning scar running up and down the trunk. Of course, other species of trees can be targets, too, especially if isolated as in the middle of a field.

ASPECT AND VEGETATION

Aspect, the direction which a slope faces, has little effect on vegetation in the Adirondacks, although in other portions of the United States, particularly in the western states, and to a lesser extent from New Jersey southward, it becomes a major factor in determining the distribution of species. North and northeast slopes are coolest and moistest, while southwest slopes, which receive the afternoon sun, are warmest and driest; east, southeast, and west slopes are intermediate, while northwest slopes are coolest but driest due to northwest winds.

In the Adirondacks, the snowpack may take a little longer to melt in the spring on the north and northeast slopes than on other aspects, but the plants growing on these seem to catch up quickly with their cousins on the southerly slopes. Level places with dense spruce, fir, and/or hemlock usually lose their snowpack later than northeast slopes with hardwoods, so that the species of trees present on a site, and their shading, influence snowmelt more than aspect.

One aspect of aspect which does have an effect on Adirondack vegetation is that south and southwest slopes burn more readily since the vegetation on them is warmer and drier than that on any other slopes. Jenkins and Loon Lake Mountains, for example, are two of many Adirondack peaks which have been severely burned on the south and southwest slopes yet remain unburned on the north. Hence the presence of pioneer species, even red oak on occasion, tends to be more common on the south and southwest slopes, but only indirectly due to aspect; their presence is directly due to forest fire.

In New York State, vegetation is determined much more by soil and site history than by aspect.

Sundew (*Drosera rotundifolia*)

Unlike Venus Flytrap's, Sundew leaf hairs bend slowly over and down onto their prey. The sticky droplets, produced by glands at the ends of leaf hairs, trap and digest insects. © Jim Kraus 1992.

(The climatic data for this chapter was obtained from the National Oceanic and Atmospheric Administration (formerly the National Weather Service), Lull (1968), and several New York State publications including Dethier (1966), Dethier and Vittum (1967), and Frederick et al. (1959).)

Chapter 4
PLANTS AND SOILS

Most plant species, including trees, can adapt to and grow over a wide diversity of soil types.

Kudish

Bog laurel (*Kalmia polifolia*). The stamens are sprung and fit into pockets in the petals. When an insect alights on the flower, the stamens rise and powder it with pollen. © Jim Kraus 1992.

Plants cannot find all sites with the proper combination of factors capable of supporting them. Therefore, particular plant species do not always appear on such sites. Most plant species, including trees, can adapt to and grow over a wide diversity of soil types. Only where there are such conditions as extremely wet, extremely dry, extremely sandy or extremely shallow do some species fail to germinate, reach maturity, and/or reproduce. Seeds and spores travel with a great degree of randomness. Because a particular species is absent on an ideal site does not necessarily mean that there is something wrong with the site. Site alone does not determine plant distribution. History, chance, competition with other plants, and site do.

To begin to understand a site, one must know about the geology, soils, climate, shade tolerance, plant biology, and site history. To reconstruct vegetational development one must also examine dispersal mechanisms for seeds and spores used by plants since deglaciation as well as at present.

PLANTS AND SOIL MINERALS

The relationship between lime and tree distribution, so long considered a strong and dominating one by botanists, is weak at best in New York State. This weakness is true not only in the Adirondacks, but in other regions as well. For example, in the lower Hudson Valley between Bear Mountain and Albany, there is similar vegetation: red oak, black oak, chestnut oak, white oak, shagbark hickory, bitternut hickory, eastern red cedar, white pine, hemlock, pitch pine, etc. throughout on well-drained sites. These trees occur on soils derived from all kinds of bedrock including sandstone, shale, granitic gneiss, and limestone. In Newfoundland, balsam fir-paper birch forests grow on soils derived from all kinds of rock: shale, sandstone, schist, granite, quartzite, gabbro and, yes, limestone.

Because most mineral soils on the Adirondack Upland are acid, one might at first believe that all the species of plants growing on them require, or are limited to, acid soils. This is not the case. Many of these plants will grow equally well on neutral or alkaline soils derived largely from limestone. Newfoundland and Labrador support balsam fir, black spruce, and white spruce, vegetation common to both this Canadian Province and the Adirondacks, on limestone soils. This illustrates further the wide range of soil parameters over which many species can thrive.

New evidence has been found that soils have been slowly and naturally acidifying since deglaciation. Buffering carbonates have been leaching out of the upper horizons of glacial deposits so that there has been a lowering of the pH over thousands of years. Eventually, some sort of equilibrium might be reached (See the study by Ford (1990) on New Hampshire for details). Acidification caused by human activities could accelerate the rate of this natural weathering process.

Lime-containing rocks, carbonates as opposed to silicates, provide lots of clay and a higher pH for prolific nutrient-supplying power. Lime-containing rocks are (1) limestone, (2) dolostone or dolomitic limestone, a high-magnesium limestone, (3) marble, a recrystallized and metamorphosed limestone, and (4) dolomitic marble, a metamorphosed dolomite. The predominant mineral in all these rocks is calcite, and its chemical composition is calcium carbonate.

Most marble areas occur in the northwestern Adirondacks, specifically in a belt from Canton to Gouverneur to Natural Bridge, at lower elevations than on the Upland. On the Upland, local marble areas with "rich site" species and high humus pH include Mount Pisgah and Mount Brewster in Saranac Lake. There are also marble areas around Rich Lake in Newcomb and on Moose Pond southwest of Santanoni Peak. There, too, the vegetation is no different from that of non-marble areas. Then also there are non-marble areas such as Mount Baker, Catamount, Scarface Mountain, and the Cobbles west of Franklin Falls where the vegetation consists of "rich site" species.

There is no correlation between the concentration of mafic minerals and soil fertility. Mafic minerals are iron and/or magnesium-bearing, and are most abundant in the Adirondacks in rocks such as gabbro, norite, amphibolite, basalt, camptonite and their metamorphosed equivalents. They are least abundant in anorthosite, granitic gneisses, and charnockitic gneisses. Examples of mafic minerals are hornblende, augite, hypersthene, diopside, biotite mica, garnet, magnetite, and ilmenite. Felsic, non-mafic, minerals are the feldspars, quartz, and calcite. Mafic minerals supply more abundant nutrients to plants than non-mafic minerals, as a general rule.

Because of glaciation, minerals from different rocks are mixed together, dragged along, and finally dumped by the ice sheet. Still, the vast bulk of the glacial till on any one site has come in from a distance of only several miles at most. This suggests that any high concentration of mafic minerals will appear as a "shadow" originating from a mafic bedrock ledge source and extending southward or southwestward, most often the direction in which the glacier moved, for only several miles. The concentration of mafic minerals will decrease with increasing distance from the source ledge. A comparison of my field notes, which record the sites where the former "rich site" species are present, with Isachsen and Fisher's (1970) N.Y. State Geologic Map, which records areas of rocks with high concentrations of mafic minerals, reveals that the "rich site" species grow on till derived from non-mafic as well as mafic rocks. Examples of "rich site" species on nonmafic till are Jenkins and Dewey Mountains (both anorthosite), The Cobbles and Azure Mountain (both granitic gneiss), and Mount Baker and Scarface Mountain (both charnockitic gneiss). Examples of "rich site" species on mafic till are Mount Whitney, Mount Pisgah, Heaven Hill, and Mica Hill (the last west of Keene).

PLANTS AND SOIL TEXTURE

Till and Outwash:

Glaciers contain lots of sand, silt, clay, gravel, cobbles, and boulders. As the ice melts, these various-sized crushed rock fragments drop by gravity onto the bedrock surface forming a mantle. This is known as till or ground moraine. No flowing water is involved in its deposition.

When a glacier melts, the meltwater stream carries sand, silt, clay, cobbles, and gravel. When it stops flowing, it deposits these as glacial outwash, glaciofluvial material or, simply, outwash.

The exceedingly sandy outwash soils do nearly prohibit the subtle pioneers (see page 61) from occurring. These soils, developed in glacial outwash, run 94% or more sand, about 3 to 5% silt, and about .5 to 2% clay. Only very rarely are white ash, American elm, and hop hornbeam found on coarse sandy outwash soils—the rare exceptions are on Black Ridge near Piercefield, Goldiana Pond at the southeast base of Saint Regis Mountain, and along the Hideaway Road at Keese's Mills. Basswood and red oak have yet to be spotted on outwash in the Adirondacks.

The sandiest outwash soils also prevent sugar maple from either dominating or being vigorous, although it can be present on such sites in limited numbers. Dominant instead are yellow birch, red maple, red spruce, balsam fir, and hemlock, with occasional beech and black cherry. Examples of outwash sites where sugar maple is present are around the two Paul Smith's College gymnasia, along the Fish Pond truck trail west of Lake Clear Fish Hatchery, west of Route 30 near Mountain Pond, and between Copperas and Whey Ponds near Fish Creek Ponds State Campground.

Coarse, sandy outwash holds little water if well-drained, and thus the trees growing on it are drought-prone. Mineral nutrients are in smaller quantities on outwash in contrast to those on siltier tills.

Sand, Silt, and Clay:

Soil texture can be defined as the proportions, in percentage of total volume, of the different sizes of mineral particles, sand, silt, and clay, in a sample of soil. The United States Department of Agriculture defines particle size classes as follows: Sand particles are the biggest and range from 2.0 mm in diameter, 1/12 inch, down to 0.05 mm, 1/508 inch, the latter near the limit for the unaided human eye to see. Silt particles are intermediate in size, and range from 0.05mm down to 0.002 mm, 1/12,700 inch; they are about the size of most plant and animal cells and require a light microscope to be seen. Clay particles are the smallest, less than 0.002 mm in diameter. The largest clay particles are about as big as bacteria, chloroplasts, and mitochondria and are just visible with a light microscope; the smallest of clay particles are about the size of viruses and require an electron microscope to be seen. Gravel, coarse rock fragments greater than 2mm in diameter, in soil is measured by volume in addition to texture, and is not included with texture.

Silt Statistics:

The relation between silt and clay concentration and tree distribution in the Adirondacks is a weak one. There is such an overlap in the range of silt and clay concentrations, i.e. the standard deviations of these soils upon which each species of tree grows are so high that few conclusions can be drawn. Only the most coarse, sandy outwash soils, those with the least silt and clay, seem to affect distribution, excluding the more demanding trees such as sugar maple, hop hornbeam, white ash, American elm, and basswood. Soil depth to bedrock and depth to water table are far more important in determining tree distribution than silt and clay.

Samples were collected from some of the pits scattered throughout the north-central Adirondacks, B horizons in some cases, C horizons in others, and both B and C horizons in still others. The samples were analyzed for texture in the laboratory by LaMotte settling kits and Bouyoucos hydrometers. The results are presented in Table 8.

Note that the horizontal rows of the table represent silt and clay, each from the two soil horizons, B and C. The vertical columns on the table represent seven different site classes. Two of the seven are out washsite classes: those with sugar maple absent and those with sugar maple present. The five till site classes include those sites upon which each of the following species was present: sugar maple, hop hornbeam, white ash, American elm, and basswood. Of course, many till sites had more than one of these species present. For each combination of silt or clay, B or C horizons, and site class, three values are given: the mean, the standard deviation, and the number of observations, i.e. the number of soil pits.

Note that in the B horizons, outwash soils have less silt than till soils. The coarsest, sandiest outwash with sugar maple absent averages 15.9% silt, while with sugar maple present, the outwash is a little siltier at 19.3%. The till soils range from 25.3% silt on sites where sugar maple is present to 32.2% on sites where basswood is present. The silt concentration under the other species is intermediate. The difference between means of 25.3 and 32.2 is not statistically significant with such high standard deviations. Since the silt difference between sugar maple and basswood sites is not significant, the silt differences between pairs of any of the other three species, hop hornbeam, ash, and elm, will not be significant either. In other words, the till soil B horizons are the same for silt (see page 53).

Outwash has less silt in the C horizon than the tills. The coarsest, sandiest outwash with sugar maple absent averages only 3.5% silt, while outwash with sugar maple present averages 5.7%. The tills range from 12.8% on sugar maple sites to 27.0% on basswood sites, a statistically significant difference. In other words, the tills are different from one another in silt concentration. The trees do not root down into the C horizon as they do in the B horizons above; C horizons have virtually no effect on tree growth and distribution.

Clay Statistics:

For clay in the B horizons, the coarsest, sandiest outwash soils with sugar maple absent, average 2.1% clay, while outwash soils, with sugar maple present, average 2.6%, about the same as clay concentration in the tills. The till soils range from 2.7% clay on sugar maple sites to 3.0% on ash sites—not a statistically significant difference. Basswood sites are intermediate, further suggesting no difference among the tills.

Clay concentration in the C horizon is not anywhere nearly as important as it is in the B horizons above. Trees do not root this far down. Outwash lacking sugar maple is coarsest with a mean of only 1.4% clay, while outwash with sugar maple present has a mean of 1.7%, much like that of the tills. The tills range from 1.8% in basswood sites to 2.1% in hop hornbeam sites. Needless to say, with the means so close, there is no statistically significant difference.

Clay particles have the highest cation exchange capacity, i.e. nutrient retention, of all mineral particles; soils with abundant clay in New York are generally rich and

TABLE 8
SILT AND CLAY STATISTICS

Particle size and horizon	Statistics	Outwash with sugar maple		Till with the following species present:				
		absent	present	sugar maple	hop horn-beam	white ash	Amer. elm	bass wood
Silt in B horizons	mean	15.9	19.3	25.3	29.4	28.2	29.7	32.2
	std. dev.	12.7	15.0	15.6	11.5	11.1	11.7	12.4
	no. obs.	11	20	25	44	51	32	36
Silt in C horizon	mean	3.5	5.7	12.8	16.9	21.8	21.8	27.0
	std. dev.	7.2	6.1	6.2	12.5	14.4	16.4	14.4
	no. obs.	26	52	46	19	16	10	10
Clay in B horizons	mean	2.1	2.6	2.7	3.0	2.9	2.7	2.9
	std. dev.	0.6	1.2	1.2	1.7	1.7	1.1	1.3
	no obs.	10	19	22	40	47	27	33
Clay in C horizon	mean	1.4	1.7	2.0	2.1	1.9	2.0	1.8
	std. dev.	0.7	0.6	0.8	1.1	1.3	1.3	1.3
	no. obs.	13	35	36	16	14	9	8

fertile. However, clay concentration on Adirondack Upland soils runs so low that its effect on tree species distribution is negligible. Also, there is no difference in clay concentration from outwash to till, and from till with only sugar maple and beech to till with basswood-ash-elm. This lack of difference in clay concentration holds true for both the B and C horizons. The mineralogy of Adirondack anorthosite and gneiss rocks permits weathering to clay-sized particles in very limited quantities.

PLANTS AND SOIL DEPTH

Just as soil depth affects the distribution of the subtle pioneers (see page 61) so too can shallow soils on the Upland prevent sugar maple-beech stands from dominating where the subtle pioneers are absent. When the soil is shallower than about ten to twelve inches, trees will be stunted and/or shrubs such as mountain maple, choke cherry, striped maple, red elderberry, hazelnut, meadowsweet, witchhobble, and wild raisin will replace the sugar maple-beech forest. In other places, a mixture of conifers and hardwoods, like that on the coarsest sandy outwash, replaces the sugar maple-beech. It includes red spruce, balsam fir, hemlock, red maple, and yellow birch. A panoramic vista of the Adirondack Upland reveals the dark, evergreen patches totally surrounded by hardwoods on the shoulders or summits of hills under 2500 feet elevation, and probably on shallow soils surrounding bedrock ledges.

PLANTS AND SOIL WATER

Poorly-drained tills, those with water tables less than 36 inches from the surface, and especially those with tables less than 12 inches, restrict the growth of certain tree species but not others. This restriction is probably due to lack of oxygen for root respiration in water-saturated soils. Sugar maple, beech, hop hornbeam, black cherry, basswood, red pine, huckleberry, and sometimes white ash must grow where the water table is deep and the site, therefore, well-drained.

Other plants grow, except in cases of severe drought, where the soil is always wet. These species, requiring a high water table, include speckled alder, cattail, black ash, winterberry holly, and, most of the time, leatherleaf and touch-me-not.

Still other plants, hemlock, balsam fir, American elm, yellow birch, red maple, northern white cedar, Labrador tea, sheep laurel, wild raisin, and often red spruce, have no dependence on a particular type of water table and can grow over a wide range of water table depths.

PLANTS AND SOIL NUTRIENTS

In 1983, sixty samples of forest soils from seventeen sites in the greater Paul Smiths area were collected and sent to the Soil Testing Service operated through the New York State Agricultural Extension and primarily used to assist farmers at Cornell University, Ithaca. Samples were collected from three site classes: outwash, till without basswood, and till with basswood. At least two horizons were sampled from each site, with as many as six from some. Samples provided nutrient data not only from site to site, but also from horizon to horizon with increasing depth on one given site.

Minerals tested for by Cornell were calcium, magnesium, potassium, nitrate, aluminum, ammonium, phosphorus, and iron. Cornell also tested for pH and exchange acidity. Aluminum, although not a nutrient, may play a role in the toxicity of soils due to acid precipitation. (A brief summary of humus nutrient concentration follows in Table 9.) Only calcium and magnesium concentrations along with pH were noticeably higher on the basswood sites; potassium was slightly higher under basswood. The other minerals: phosphorus, nitrate, ammonium, and iron, were either higher on non-basswood sites or about the same throughout all sites. Standard deviations were very high in most cases, indicating great variation from site to site even within each site class.

Most nutrient concentrations decrease drastically with increasing depth so that the humus holds the greatest fertility of all the horizons. Exceptions are iron and aluminum which accumulate most in the B horizons. pH rises on non-basswood sites with increasing depth because the concentration of organic acids decreases downward into the mineral horizons; the mean pH of till and outwash C horizons without basswood is 5.0.

Nitrogen and Bogs

Because of their high acidity, bogs are notoriously low in such essential plant nutrients as nitrogen, calcium, magnesium, and potassium. Therefore, such insectivorous plants as sundews, pitcher plant, and bladderworts, largely those which grow in bogs, consume insects as a supplemental source of nitrogen much as people take vitamins as a supplemental source of nutrition.

Plants growing in bogs, despite the constant wetness, have water-tight leaves approaching those of desert plants (see discussion on this adaptation on page 40).

TABLE 9
MINERAL NUTRIENT ANALYSIS

Values from calcium through aluminum are means in pounds per acre from humus. Exchange acidity means are measured in milliequivalents per 100 grams.

Nutrients in humus	SITE CLASSES		
	Four outwash sites without sugar maple	Five till sites with sugar maple but without basswood	Six till sites with basswood.
Calcium	1200	1220	6450
Magnesium	88	104	337
Potassium	138	188	215
Phosphorus	17	26	16
Nitrate	10	35	16
Ammonium	36	83	46
Iron	9	9	8
Aluminum	31	64	29
pH	3.9	3.8	5.7
Exchange acidity	59	57	23

PLANTS, HUMUS, AND SOIL pH
pH:

pH can be defined as a measure of the concentration of hydrogen ions. The more concentrated the ions, the greater the acidity; the less concentrated the ions, the less the acidity. There are two salient properties of the pH scale. The first is that it is negative—the lower the pH value, the greater the acidity. The second is that the scale is logarithmic. Therefore, a difference in one unit, e.g. from 5 to 4, represents a tenfold difference in hydrogen ion concentration—a major difference in acidity. A difference in two units, e.g. from 7 to 5, represents a hundred-fold difference in acidity, and so on.

The pH scale runs from zero to fourteen. Zero is absolute acidity, all hydrogen ions, while fourteen is absolute alkalinity, no hydrogen ions. These values do not occur in nature. A pH value of 7.0 is neutral. Most soils range from 3 to 8, a 10^5 or one-hundred-thousand-fold difference.

Humus pH, Trees, and Ground Cover

Trees make humus. The small herbaceous, ground cover plants and shrubs contribute some, but the vast bulk comes from trees. Not only do leaves fall, but so do bud scales, dead twigs and branches, sloughed-off outer bark, cones, flowers, fruits, seeds, and whole trees. Fungi, bacteria, and soil animals rot this litter into humus, the most important horizon of the soil for plant nutrition and water absorption, although the enriched B mineral horizons below contribute some. Most of the fine absorbing roots of a tree are in the humus.

The pH of the humus and its nutritional nature depend on what species of tree or trees happen to be overhead on the site dumping litter on it, and also on what species of decay organisms are active.

Humus pH values obtained for sixty species of forest trees, shrubs, and herbs are shown in Tables 10 and 11 where each species is represented by a horizontal line. At the left end of the line is a dot representing the lowest pH that the species was growing upon, while at the right end of the line is a dot representing the highest pH that the species was found on. Somewhere between these minimum and maximum pH dots lies a third dot representing the median pH for each species. The species are arranged with increasing median pH.

The number of sites that each species was present upon is indicated on the Table by the letter n, an equal sign, and a number, e.g. n=33. This number gives a rough idea of how abundant the species is in the north-central Adirondack Upland. For example, beech is common with n=81, woodfern with n=79, and red spruce n=63. Less common are round-leaf violet with n=6, sweet cicely with n=6, and bristly clubmoss with n=5. The number also indicates how accurate the median is. The more sites that a species was present upon, the more accurate the median, e.g., n greater than 30. Those species present on five to ten sites have a very rough median.

(Note that American elm sites include those where the elms have died as well as those where the elms are still living.)

Some plants thrive in high acidity, low pH sites (around 3.5 to 4.5) such as in bogs and under spruces, fir, hemlock, larch, pines, birches, red maple, and beech. Examples are *Sphagnum* (peat) mosses, goldthread, Clinton's lily, wood sorrel, shining clubmoss, sharp-leaved aster, painted trillium, and roundleaf violet.

Other plants thrive in lower acidity, higher pH sites (around 5.0 to 6.5) under sugar maple, white ash, American elm, and basswood. Examples are blue cohosh, maidenhair fern, Canada violet, blue-stemmed and wide-leaved goldenrods.

Still other plants can thrive over a surprisingly wide range of pH, often a full 2 to 3 units, equivalent to 100 to 1000 times more acidic or basic, or more. For example, balsam fir occurs on soils with humus from pH 3.5 to 6.8, a difference of 2000 times. Others are woodfern, Solomon's plumes, Canada mayflower, and purple trillium.

Since most trees and shrubs grow over a wide range of pH values, this extensive overlap can create all kinds of combinations of woody plants growing together on a site. The conifers and certain hardwoods such as red and sugar maples, yellow birch, beech, and black cherry produce acid humus, pH median 4.5 to 4.75. Elm and white ash produce humus with pH from 5.0 to 5.5 median, while basswood produces humus with pH median 5.8. One could say then that the "rich sites" do exist where elm, ash, and basswood occur; but the "rich site" is created by the trees themselves, not by high silt concentrations in the soil.

TABLE 10
TREES AND SHRUBS OVER HUMUS pH
Dots show minimum, median and maximum
n= number of samples species is present over, shown at median.

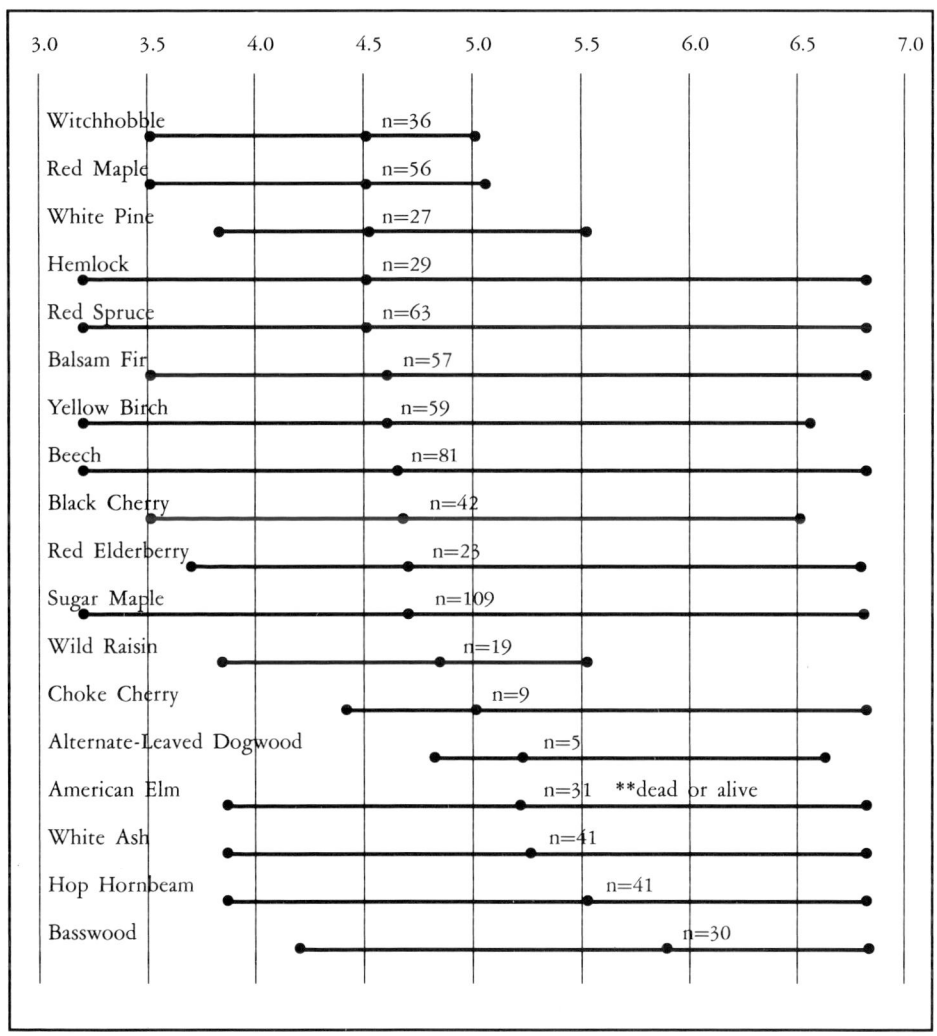

The pH range for most herbs is very broad as is that of woody plants, the extensive overlap creating all kinds of combinations of herbs growing together on a site. Note that the species with apparently small pH ranges results from very small samples. Spring beauty, climbing buckwheat, and bristly clubmoss, for examples, have an n of only five to ten. An increase in the number of sites measured for humus pH for these species would invariably extend the range in both directions. Ranges, as medians, are most reliable for species with at least 30 values.

There are always surprises in the woods which confound our expectations of humus pH and species present.

TABLE 11
HERBS OVER HUMUS pH
Dots show minimum, median and maximum
n= number of samples species is present over, shown at median.

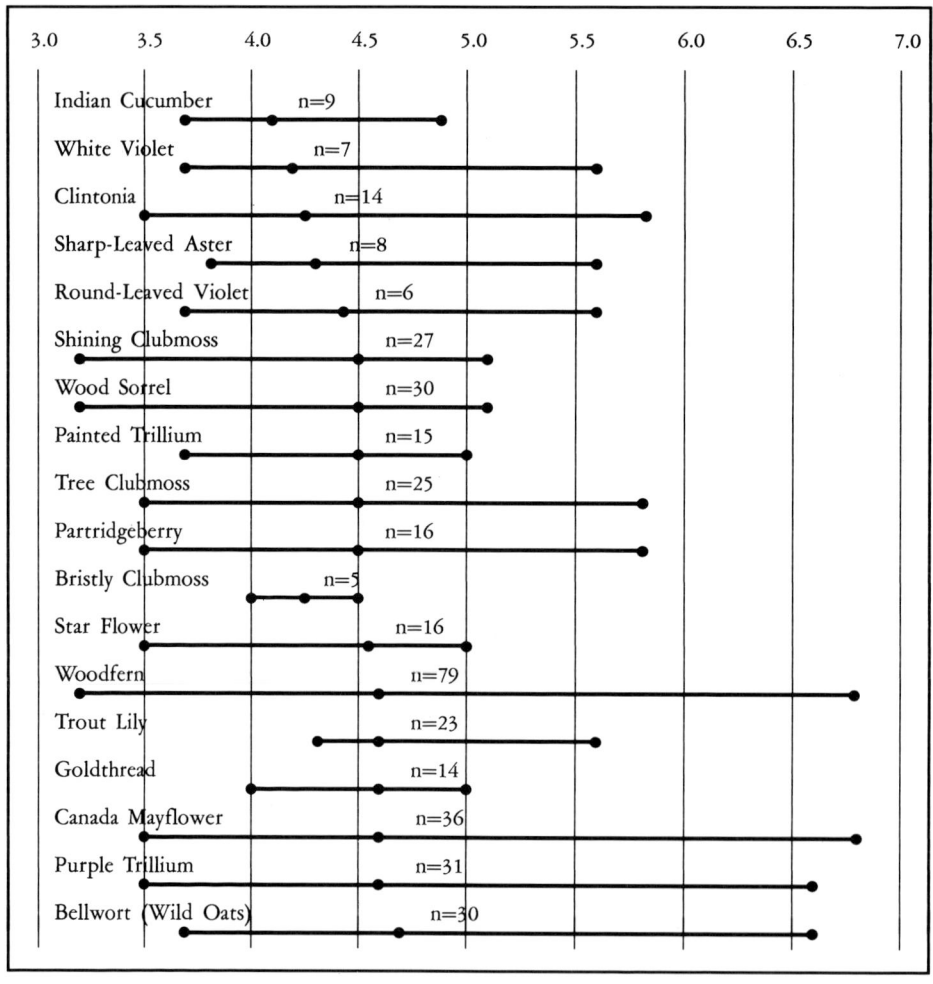

TABLE 11
HERBS OVER HUMUS pH

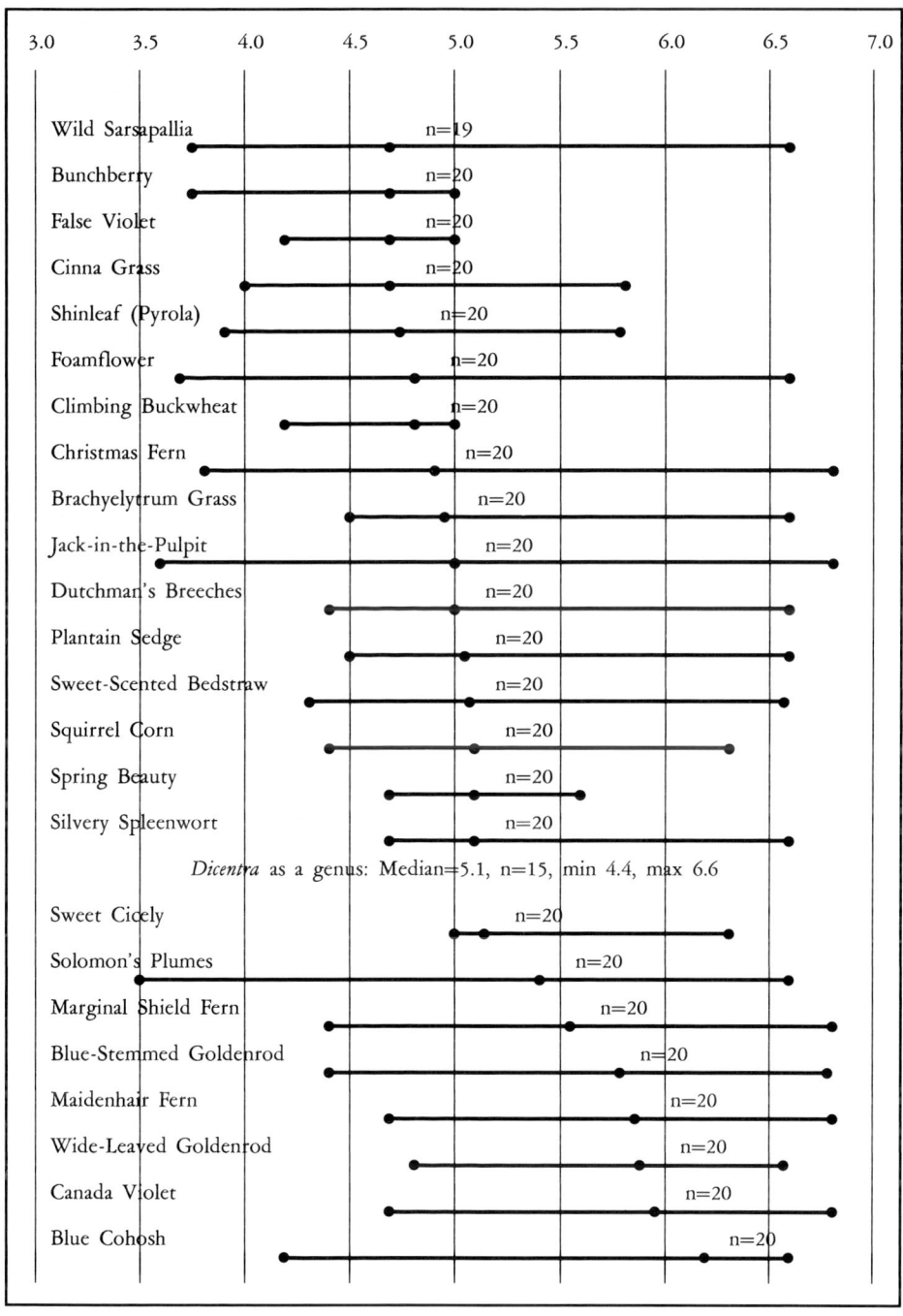

HERB DEPENDENCE ON PARTICULAR TREE SPECIES

Some ground cover herbaceous species appear to be dependent upon, or always associated with, a particular tree species while others do not. The more one observes autotrophic, self-feeding, green herbs in the field, the more one finds that they do not. Many of the higher pH species that one would expect to always be under ash-elm-basswood sometimes are absent. They do always occur under sugar maple, however, and apparently sugar maple humus alone is adequate in providing all the nutrients for their growth. Examples are:

    Christmas fern (*Polystichum acrostichoides*)
    Wild oats (*Uvularia sessilifolia*)
    Spotted touch-me-not (*Impatiens capensis*)
    Canada violet (*Viola canadensis*)
    Plantain-leaved sedge (*Carex plantaginea*)
    Squirrel corn (*Dicentra canadensis*)
    Dutchman's breeches (*D. cucullaria*)
    Silvery spleenwort (*Athyrium thelypteroides*)
    Blue cohosh (*Caulophyllum thalictroides*)

Other species have been found under sugar maple and dead elm, ash and basswood being absent. Since the elms were dead and no longer contributing leaf litter, these plants probably can survive under sugar maple alone:

    Goldie's fern (*Dryopteris goldiana*)
    Maidenhair fern (*Adiantum pedatum*)
    Blue-stemmed goldenrod (*Solidago caesia*)
    Sweet cicely (*Osmorhiza claytonii*)

Further, there is a number of usually higher-pH plants which one would expect to grow under sugar maple, ash, elm, and basswood, but sometimes do not grow under any of them. Whichever tree species these plants grow under: conifers, mountain and red maples, birches, black cherry, and aspens they apparently provide adequate nutrients. In many of these exceptional cases, the herbs are growing along streams in coniferous forests on the Upland. Along the stream, organic acids are washed away and the pH is ordinarily a full unit higher, often about 5, than in the adjacent woods with a pH of about 4. Examples are:

    Marginal shield fern (*Dryopteris marginalis*)
    Nettle (*Laportea canadensis*)
    Enchanter's nightshade (*Circaea alpina*)
    Kidneyleaf buttercup (*Ranunculus abortivus*)
    Tall meadow rue (*Thalictrum pubescens*)
    Bristly currant (*Ribes lacustre*)
    Jack-in-the-pulpit (*Arisaema triphyllum*)
    Foamflower (*Tiarella cordifolia*)
    Wide-leaved goldenrod (*Solidago latifolia*)
    Sweet-scented bedstraw (*Galium triflorum*)
    Milium grass (*Milium effusum*)
    Brachyelytrum grass (*Brachyelytrum erectum*)

More careful observations of which tree species tower over which herbaceous species might tell us which tree species are required to provide adequate nutrients for each herb. It seems that as long as the herb's nutritional needs are met, it makes no difference which tree supplies them.

FROM "THE SILT TRIP" TO "THE SUBTLE PIONEERS"

When writing *Paul Smith's Flora II* (1981), I believed that the distribution of many tree species was directly based on the percentage of silt in the soil, and the distribution of many ground cover plants indirectly based on silt. In the bulk of Chapter 3 of *Flora II*, pages 29 through 92, I presented this thesis, with the very core of the chapter, pages 38 through 42, being a section entitled "The Silt Trip." In essence, I stated that each species of tree had its own minimum threshold silt requirement in the soil and would grow only when the minimum silt concentration was exceeded on the site. The conifers, poplars, cherries, maples (except sugar), birches, and beech could grow on as little as 2 or 3% silt. Sugar maple needed 6%. Sites which included any or all of the following species, hop hornbeam, white ash, American elm, basswood, and northern red oak were called "Rich Sites." Hop hornbeam required 7%, ash and elm 10 to 12%, and basswood the most, 25%. "The Silt Trip" stressed that silt percentage was the major overwhelming controlling factor explaining the distribution of tree species on the Adirondack Upland on well-drained, mineral soils developed in glacial till. It is not.

The idea that "The Silt Trip" is incorrect first struck me as recently as October, 1987, when examining the forest on a hill known as The Cobbles west of Franklin Falls. The whole hill had a similar history (burned over in 1903), and the bedrock was a uniform granitic gneiss. Yet one slope was dominated by sugar maple, white ash, American elm, hop hornbeam, and basswood and the other by red spruce, balsam fir, paper birch, red maple, and mountain maple. Soil tests from both slopes revealed no difference in till mineralogy nor texture—they were the same. Then why was the vegetation so different?

Further study of summits and shoulders of hills such as Creighton and Buck near Paul Smiths created additional suspicion that "The Silt Trip" hypothesis was incorrect. There was no difference in silt concentration between the summits where ash, elm, and hop hornbeam are found, and the mid-slopes which are clothed with nearly solid stands of beech and sugar maple. In fact, the "rich site" species are growing in shallow soils surrounding bedrock ledges—far from being deep, rich soils. The sugar maple-beech stands are growing on the soils developed in the deep tills.

A much more careful statistical examination of silt concentration data over "rich sites" and nutrient poor sites revealed that the till soils were the same (See Table 8, page 53 for details).

My hypothesis now is that the "rich site" species, with the exception of hop hornbeam, the most shade-tolerant, are distributed as a result of disturbance, both natural and human. These tree species, when present, are almost always on sites which have been opened directly to sunlight at specific times in the past by such human activities as burns, loggings, clearcuts, and pastures, or on other sites which are chronically and naturally open due to their being places with shallow soils exposed to wind. Basswood, elm, ash, and oak do not necessarily mark the high-silt tills.

The thesis presented in "The Silt Trip," was based on a number of factors. Starting in 1973, Paul Smith's College students in classes in Dendrology, Forest Soils, and Field Ecology dug over 600 soil pits in the greater Paul Smith's area and in the surrounding Adirondacks. Textural analyses for percentages of sand, silt, and clay were run on a good number of them, both on "rich sites" with ash, elm, basswood and oak, and under "poor" sites without them.

Deep-to-bedrock soils are those with a ledge more than 40 inches beneath the surface. C horizon, consisting of purely mineral glacial till and located below the maximum depth of rooting and all biological activity, was sampled at a depth of about three feet.

Shallow-to-bedrock soils are those with a ledge less than 40 inches beneath the surface. B horizons, subsoil layers permeated by roots and other biological activity, were sampled at a depth of between one and two feet.

On the shallow soils there is seldom a C horizon since the B horizons sit directly on the bedrock. On the deep soils, the B horizons sit on top of the C.

Since many "rich sites" are on shallow soils, B horizons were usually sampled under hop hornbeam, white ash, American elm, basswood, and northern red oak. Since the "poor" sites are on deep soils, C horizons were usually sampled under sugar maple and beech. Then the silt concentration of the B horizons from shallow soils was compared with silt concentrations from the C horizons of deep soils. B horizons are invariably higher in silt concentration than C horizons on any Adirondack soils, deep or shallow, till or outwash (See Table 8, page 53 for numerical values).

Recently, when I re-examined the data and compared B horizons of both "rich site" and "poor" site, I discovered that the silt concentration for both is the same (Again, see Table 8, page 53 for a statistical analysis). This discovery led to the realization that the thesis presented in *Paul Smiths Flora II* was incorrect because it was based on the belief that the silt concentration for both "rich" and "poor" sites is different. This led to the assertion that in contrast to the obvious pioneers: paper and gray birches, white pine, aspens, balsam poplar, red cherry, service-berry, and the willows, the following: white ash, American elm, basswood and northern red oak could better be called "the subtle pioneers."

These almost always occur in areas that have been disturbed—areas where the dense, continuous sugar maple-beech forest either has been removed or cannot grow. The distribution of the subtle pioneers is a function of disturbance rather than of silt concentration. Because the subtle pioneers are less shade-tolerant than sugar maple and beech, reproduction under a closed forest canopy of any tree species is limited. Most virogous ash and elm reproduction occurs in open areas, such as along roadsides, fencerows, edges of woods and in abandoned clearings. Basswood seldom reproduces by seed now in the Adirondack Upland. Root sprouting occurs commonly but rarely does seeding. The subtle pioneers are incapable of competing with sugar maple and beech. Existing maturing stands of the subtle pioneers, except on chronically wind-disturbed sites with shallow soils, are being replaced by the more tolerant sugar maple and beech.

The subtle pioneers frequently follow burned-over slopes, such as those on Mount Baker (1908), Haystack Mountain at Ray Brook (1908), The Cobbles (1903), Catamount (1940 and earlier), Jenkins Mountain east slope (1912), and Mountain Pond Ridge (1913). The superior ability of basswood and red oak to sprout from their own stumps after a logging or fire is well-known.

Subtle pioneers also follow areas that were clear-cut or nearly so, for timber or for pasture, such as Mount Pisgah, Brewster Mountain east slope, and Dewey Mountain; this is why they are so abundant around Saranac Lake Village where the forests have been so severely disturbed. Reverend Sarah Baker, author of the Saranac Valley histories, reports that much of the timber which floated down the Saranac River in the mid-nineteenth century to mills in Clinton County came from the Saranac Lake area. Likewise, Heaven Hill, Cobble Hill, and Mount Whitney in the Lake Placid Village area also abound in the subtle pioneers as a result of severe disturbance of forests. And the lower east slopes of Whiteface Mountain above Wilmington, once nearly clearcut for charcoal, are now colonized by subtle pioneers.

Summits and shoulders of hills differ in two major ways from the middle slopes below. First, bedrock outcrops are frequent on top causing open, sunny areas with shallow soils surrounding them. Second, there is more exposure to wind, a fact which results in frequent tree falls, broken limbs, and injured crowns. Shallow soils themselves cause shallow-rooted trees and further wind-instability. These factors combined make continuous dominance by sugar maple and beech difficult on the summits because they allow the subtle pioneers, which can reproduce along the sunny edges of the ledges and in the many sunny spaces among the missing and broken tree crowns, to seed. This continuous, natural disturbance is exemplified on Creighton Hill and Buck Hill in the Paul Smiths area parts of which, probably, have never been logged.

An historic corollary from the "Silt Trip" was that, because of minimum silt requirements, the "rich site" species could never have, since the end of the Ice Age, grown on the less silty tills. Hence, the "rich site" species would have been restricted to only the higher-silt sites over the last perhaps 8000 years, and could not possibly have ever been widespread over all the well-drained tills on the Upland. Thus, the modern, existing, widely-scattered "rich sites" could not be small, remnant fragments of a former, vaster ash-elm-basswood-oak forest.

The new historic corollary from "The Subtle Pioneers" hypothesis is very different. Obviously, the subtle pioneers can presently grow on just about any well-drained till soil on the Upland and not only on those tills with higher silt concentration; they also did in the past. It follows then that the subtle pioneers could have grown almost anywhere on the well-drained tills if there were no competition from either sugar maple or beech. Could the subtle pioneers have been far more widespread in the past several thousand years on till soils, and could we indeed now be looking at small mere remnant fragments of a former, larger, vaster subtle pioneer forest? Probably not, because competition with the more shade-tolerant sugar maple and beech on the deep, well-drained till soils would have prevented such a widespread distribution. If sugar maple and beech were rarer in the Adirondacks at one or several times in the past than they are now (pollen and macrofossil evidence disputes this rarity), then it may be that subtle pioneers appeared everywhere.

The distribution of ash, elm, and basswood cannot be explained by aspect, since these tree species occur on level sites and on slopes which face all directions. A listing of such sites appears here:

Northwest slopes:	Scarface Mountain, Mount Arab, Mount Pisgah, and Mount Baker.
East slopes:	Jenkins Mountain, Brewster Mountain, Mount Pisgah, and Mount Baker.
West slopes:	Debar Mountain, Mount Baker, and Owl's Head.
Southwest slopes:	Mount Baker, Mount Whitney, Mount Pisgah, and Catamount Mountain
Northeast slopes:	Dewey Mountain
South slopes:	Kate Mountain, Mount Pisgah, Mount Baker, Marsh Pond Mountain, and The Cobbles.

(The effect of aspect on the distribution of the subtle pioneers is discussed more fully in Chapter 3, page 44)

Hop hornbeam does not seem to fit the distribution pattern of the subtle pioneers, although it had been included once with the rich site species. It is more like sugar maple and beech in its shade tolerance, but is far less common than the two big dominants. As an understory tree, hop hornbeam will successfully thrive, compete, and reproduce under dense, closed stands of sugar maple and beech as either widely-scattered individuals, or, more likely, as small groves where the subtle pioneers are noticeably absent. These stands are either old growth as those along the Ampersand Mountain trail, or minimally disturbed only by selective logging as those in the Paul Smiths area along the Jenkins Mountain Road west of Barnum Pond Outlet, along the Hideaway Road, Beech Hill, Sugarbush, "Hill 1860," and south of the Brighton Sanitary Landfill at Gabriels. Hence, hop hornbeam is more or less randomly distributed on tills, depending on where the seeds migrate, and is not restricted to major disturbance areas.

REFERENCES

For a fine account of soils of New York State including northern New York, see Cline and Marshall (1977). On a more detailed level, soil surveys are completed most often by counties. These are published jointly by the Soil Conservation Service of the U.S. Department of Agriculture with the New York State College of Agriculture and are available from the agricultural extension offices in each county.

Four references on the geology of the Adirondacks and northern New York are Fisher et al. (1979), Van Diver (1976 and 1980), and Wyckoff (1967). These supply good background material on the region yet are written in a non-technical manner by professional geologists. On a more technical level, consult publications of the New York State Museum and Science Service in Albany, such as Isachsen and Fisher's (1970) Geologic Map of New York and U.S. Geological Survey topographic quadrangles.

Chapter 5

THE ADIRONDACK UPLAND FLORA

The discoveries never end.
Kudish

Blueberry (*Vaccinium angustifolium*) The sepals are fused to the ovary so that when one eats a blueberry one eats flower parts in addition to fruit. © Jim Kraus 1992.

SEQUENCE OF FAMILIES:

Families of plants are arranged as in Mitchell's *A Checklist of New York State Plants* (1986), based on the 1981 classification system of Arthur Cronquist. Genera are arranged alphabetically within each family. Within each genus, species are arranged alphabetically as well. This alphabetization of genera and species is not a natural arrangement, i.e., one based on evolution, but will facilitate the reader's search for a plant in the text.

TABLE 12
Families, Genera, and Species: Native and Naturalized

Family		Genera	Species		
			Native	Naturalized	Total
Ferns, Clubmosses & Horsetails					
1	Lycopodiaceae	1	7	0	7
2	Isoetaceae	1	1	0	1
3	Equisetaceae	1	6	0	6
4	Ophioglossaceae	1	2	0	2
5	Osmundaceae	1	3	0	3
6	Polypodiaceae	1	1	0	1
7	Adiantaceae	2	2	0	2
8	Dennstaedtiaceae	2	2	0	2
9	Aspleniaceae	10	19	0	19
Gymnosperms: Conifers					
10	Taxaceae	1	1	0	1
11	Pinaceae	5	10	1	11
12	Cupressaceae	2	3	0	3
Angiosperms: Dicots					
13	Nymphaeaceae	2	2	0	2
14	Cabombaceae	1	1	0	1
15	Ceratophyllaceae	1	1	0	1
16	Ranunculaceae	9	12	1	13
17	Berberidaceae	1	1	0	1
18	Papaveraceae	1	0	1	1
19	Fumariaceae	2	3	0	3
20	Hamamelidaceae	1	1	0	1
21	Ulmaceae	1	1	0	1
22	Urticaceae	3	3	0	3
23	Juglandaceae	1	1	0	1
24	Myricaceae	2	2	0	2
25	Fagaceae	2	2	0	2
26	Betulaceae	5	9	0	9
27	Chenopodiaceae	2	0	2	2
28	Amaranthaceae	1	0	1	1

TABLE 12 (cont.)
Families, Genera, and Species: Native and Naturalized

Family	Genera	Species		
		Native	Naturalized	Total
Angiosperms: Dicots (cont.)				
29 Portulacaceae	2	1	1	2
30 Molluginaceae	1	0	1	1
31 Caryophyllaceae	6	0	8	8
32 Polygonaceae	3	4	5	9
33 Clusiaceae	2	4	1	5
34 Tiliaceae	1	1	0	1
35 Malvaceae	1	0	1	1
36 Sarraceniaceae	1	1	0	1
37 Droseraceae	1	2	0	2
38 Cistaceae	1	1	0	1
39 Violaceae	1	6	0	6
40 Cucurbitaceae	1	1	0	1
41 Salicaceae	2	9	2	11
42 Brassicaceae	8	3	7	10
43 Ericaceae	10	15	1	16
44 Pyrolaceae	2	2	0	2
45 Monotropaceae	1	2	0	2
46 Primulaceae	2	5	0	5
47 Grossulariaceae	1	4	0	4
48 Saxifragaceae	3	3	0	3
49 Rosaceae	14	28	3	31
50 Fabaceae	6	0	11	11
51 Haloragaceae	1	1	0	1
52 Lythraceae	2	1	1	2
53 Thymelaeaceae	1	1	0	1
54 Onagraceae	3	7	0	7
55 Cornaceae	1	6	0	6
56 Santalaceae	1	1	0	1
57 Celastraceae	1	1	0	1
58 Aquifoliaceae	2	2	0	2
59 Euphorbiaceae	2	1	1	2
60 Vitaceae	2	2	0	2
61 Polygalaceae	1	1	0	1
62 Aceraceae	1	5	2	7
63 Anacardiaceae	2	2	0	2
64 Oxalidaceae	1	1	1	2
65 Geraniaceae	1	1	0	1
66 Balsaminaceae	1	2	0	2
67 Araliaceae	2	4	0	4
68 Apiaceae	7	6	1	7
69 Gentianceae	1	1	0	1
70 Apocynaceae	1	1	0	1

TABLE 12 (cont.)
Families, Genera, and Species: Native and Naturalized

Family	Genera	Species		
		Native	Naturalized	Total
Angiosperms: Dicots (cont.)				
71 Asclepiadaceae	1	2	0	2
72 Solanaceae	1	0	1	1
73 Convovulaceae	1	1	0	1
74 Menyanthaceae	1	1	0	1
75 Hydrophyllaceae	1	1	0	1
76 Boraginaceae	2	0	2	2
77 Verbenaceae	1	1	0	1
78 Lamiaceae	7	5	5	10
79 Callitrichaceae	1	1	0	1
80 Plataginaceae	1	1	2	3
81 Oleaceae	1	2	0	2
82 Scrophulariaceae	9	8	4	12
83 Orobanchaceae	1	1	0	1
84 Lentibulariaceae	1	4	0	4
85 Campanulaceae	2	5	0	5
86 Rubiaceae	4	9	1	10
87 Caprifoliaceae	5	15	0	15
88 Valerianaceae	1	0	1	1
89 Asteraceae	34	41	23	64
Angiosperms: Monocots				
90 Alismataceae	1	2	0	2
91 Hydrocharitaceae	2	2	0	2
92 Potamogetonaceae	1	9	0	9
93 Najadaceae	1	1	0	1
94 Araceae	2	2	0	2
95 Lemnaceae	1	1	0	1
96 Xyridaceae	1	1	0	1
97 Eriocaulaceae	1	1	0	1
98 Juncaceae	1	6	0	6
99 Cyperaceae	9	41	0	41
100 Poaceae	26	24	14	38
101 Sparganiaceae	1	2	0	2
102 Typhaceae	1	1	0	1
103 Pontederiaceae	1	1	0	1
104 Liliaceae	11	14	0	14
105 Iridaceae	2	2	0	2
106 Smilacaceae	1	1	0	1
107 Orchidaceae	9	12	1	13

TABLE 13
Summary of Tally of Families, Genera, and Species

Taxon	Family	Genera	Species		
			Native	Naturalized	Total
Ferns and Fern Allies	9	20	43	0	43
Gymnosperms	3	8	14	1	15
Angiosperms: Dicots	77	214	276	91	367
Monocots	18	72	123	15	138
All Taxa	107	314	456	107	563

INFORMATION ON EACH SPECIES:

There are nine categories of information possible for each species, although few species will have all nine categories of information.

1. Scientific name of each plant. (An asterisk is used to denote naturalized species).
2. Synonymy, if any, in the scientific name.
3. Common name/s of each plant.
4. Site. Indicates pH range, water table depth, shade tolerance, soil texture, nutrients, soil depth to bedrock, disturbance, etc. It also indicates the kind of places that the plants grow in rather than specific locations or stations, although some selected stations may be listed.

    "The College" or "The Campus" always refer to Paul Smith's College Campus, the geographic center of the Flora area.

    The abbreviation "V.I.C." refers to the Adirondack Park Visitors' Interpretive Center one mile north of Paul Smith's College.

    A table of place-names has been provided in Appendix III to assist the reader in locating stations in and around the Flora area via latitude and longitude.

    Frequent reference is made to the Catskills where soils are derived from shale, sandstone, and conglomerate rocks. Such soils do not occur on the Adirondack Upland where anorthosite and gneiss rocks dominate. It is important to know that most species of plants, especially the deeper-rooted trees and shrubs, can grow outside the Adirondacks on shale, sandstone, and conglomerate soils.

    Frequent reference is also made to the forests further north such as those in Canada, specifically in Newfoundland, Ontario and Québec where many of the Adirondack species grow also on soils derived from limestone. Most species of plants tolerate a wide range of soil types.

5. Elevation. The elevational range in feet above mean sea level is given for those species for which it is known. Stations are given for the minimum and maximum elevations. In many instances, the elevational range of a species extends well below the 1000-foot minimum and/or well above the 4000-foot maximum boundaries of this Flora. An example is Balsam fir which extends over the entire range of elevations in the Flora area. In addition, it can be found over 5300 feet on Mount Marcy and as low as 200 feet in the Chazy area of the Champlain Valley and the Brasher Falls area of the Saint Lawrence Valley. Significantly, many plant species can adapt to the extremely wide range of growing season lengths which, in turn, are dependent on the very wide range of elevations (see Chapter 3).
6. Phenology. Flowering, fruiting, budding, leafing, leaf fall times, and spore or cone production time for non-flowering plants are all examples of phenology. Growing season length may be included here. Because it is not possible for one botanist to visit all the species at all times during the growing season, the phenological observations in some cases are sketchy. Readers will surely extend

the flowering period for many species and fine-adjust the median flowering date.
7. Frequency. Includes abundance and rarity codes. An insight into the abundance of each species in the Flora area can be obtained by knowing in how many different stations the plant occurs. For the common species, no records have been kept for the hundreds, perhaps thousands of stations which each reader is almost sure to locate each time when in a "correct" site: roadsides, pine woods, bogs; etc. Records of the number and location of stations for the less common plants within the 30–mile radius at Paul Smith's College and between 1000 and 4000 foot elevation are presented.

Of course, there are many more stations yet to be noted. And, indeed, a number of reliable-source people have reported additional stations for many of the species as well as additional "new" ones. The discoveries never end.

Rarity codes are listed for those plants which are endangered, threatened, rare, and/or protected in New York State. (See Chapter 6 for details.)
8. Miscellany. This includes information that might be of interest, information not listed in the other categories. Examples are reproduction, age, edibility, toxicity, uses, unusual morphological features; etc.
9. References. References to the Bibliography in this chapter are abbreviated to author/s with the year of publication in parentheses, e.g. Ketchledge (1965). If a reference involves more than a single species, such as several species, a genus, or even a whole family, the reference will be cited at the beginning of the section on the genus or the family.

INFORMATION ON MAPS:

The reasons for species distribution are not always obviously linked to climate, soils, and site disturbance history. Unexpected stations and surprises are everywhere. To illustrate this, maps showing the distribution of a number of selected species, not widespread in northern New York and with unusual geographic distributions, accompany the text in Chapter 5 The maps are derived from observations made through the summer of 1990. Since many stations for each species have not yet been observed, the maps are often preliminary, offering only a suggestion of the total distribution. Maps show a few place-names for locating stations more easily. The thirty-mile radius circle from Paul Smiths is shown on the map in Chapter 1.

There are four different categories of maps, each illustrated by one or several exemplifying species. The map of one species often differs dramatically from the maps of others because the distribution pattern for each species is unique and independent. Some species, however, exhibit some similarities and can be grouped into categories such as remnant populations and advanced guard populations.

Each map includes regions beyond the 30–mile radius, below 1000 feet, and above 4000 feet to show how plant populations of the Adirondack Upland relate to plant populations in the surrounding Saint Lawrence, Lake Champlain, and Lake George Lowlands. The maps show only the present distribution of each species.

Plants that people should avoid:

Some widespread species can physically injure people with thorns, spines, and prickles, but they are not poisonous. Examples are Blackberry, Raspberry, Nettles, Hawthorns, Pasture juniper, Hemp nettle, and some Currants and Gooseberries. Other plants are poisonous to ingest but not to touch. Most of these, other than White snakeroot, are widespread. They include Swamp milkweed, Yew, False hellebore, Baneberries, Sheep laurel, Indian tobacco, Bittersweet nightshade, and Bulb-bearing water hemlock. Only one species, poison ivy, is poisonous to the touch and has a limited distribution in the Adirondacks worthy of mapping.

Remnant populations:

Five remnant species: Larch, Balsam Fir, and Red, Black and White Spruce are common on the Upland, but rare or absent in the Lowlands. With perhaps the exception of Red spruce, they were common in the Lowlands between 12000 and 8000 years ago, but, for the most part, have disappeared (see Chapter 2). The maps show first the relatively few stations in the Lowlands, if any, where one finds the trees, and, second, those stations where one first encounters the species as one climbs onto the Adirondack Upland. For example, where and at what elevation does one first see Balsam fir as one travels from Malone to Paul Smiths, from Watertown to Cranberry Lake, and from Lake George to Schroon Lake? It is important to note that a first glimpse of a species is not solely a function of elevation. It also depends on the kind of site, its disturbance history, its seed travel history, and a whole list of other, many unknown, factors.

Larch is the least rare of the five remnant species in the Champlain and St. Lawrence Lowlands, having persisted in swamps where competition from hardwoods is normally minimal. It is much more common, of course, on the Adirondack Upland, especially in bogs, swamps, burned-over sandy outwash plains, and even abandoned fields.

Balsam fir is scarcer in the Lowlands than is Larch, but, like Larch, occupies swamps. It also grows on high-and-dry sites at about 750 feet elevation. Above about 1000 feet it is abundant.

Red spruce does not appear in the Lowlands, possibly because it cannot do well in the swamps which have sheltered Larch and Balsam fir for about 12,000 years from the hardwood invasions. Red spruce begins to appear, on the Adirondack Upland, at about 750 feet, and is abundant above 1000 feet. There is a problem with the migration history of Red spruce, however. To botanists until a few years ago, its absence meant disappearance from a former presence. Now with studies such as those by Whitehead et.al. (1986), Jackson (1989), and Ford (1990), there is evidence that Red spruce first invaded northern New York and northern New England only about 2000 years ago, very recently in post-glacial terms. If this new evidence is accepted, Red spruce may not yet have advanced into the Lowlands and could hardly be classified as a remnant species along with Larch, Fir, and the other two spruces. (Perhaps it should be moved into the "Puzzling Populations" category described below.)

Black spruce is a predominantly bog species, absent in the Lowlands, possibly because there are no bogs there. It is the last one of the Larch-Fir-Spruces remnant group encountered while climbing onto the Upland, since it rarely occurs below 1300 feet. One notable site exception is a pair of bogs between Warrensburg and Pottersville at about 800 feet where the species grows. Above 1300 feet, bogs and Black spruce become common. The spruce also invades burned-over sandy outwash plains, and sometimes even old fields. It is more abundant on the northern Adirondack Upland, e.g. Tupper Lake to Paul Smiths, than on the southern Adirondack Upland, e.g. Old Forge to Blue Mountain Lake because there are more bogs and outwash plains in the north. This conifer is again common in the alpine zone above 4500 feet.

White spruce is absent in the Lowlands. Unlike the other four species in the remnant category which are widespread and abundant on the Upland this spruce is common locally. Its distribution is unpredictable. One cannot explain its presence by the nature of the site, climate, and recent disturbance history. One would have had to be at the present White spruce stations, say, 10000, 8000, 6000, 4000, and 2000 years ago to figure out why it is here today. (See the text on White spruce further on in this chapter for details.)

Advanced guard populations:

Several species, so common in the Lowlands are rare on the Upland, the reverse situation from the Remnant Populations category. (Consult Chapter 2 for migration routes, as these plants are probably first invading the Upland from the Lowlands.) Many have migrated up the branches of the Ausable River to Wilmington and Keene Valley. Others have migrated up the Boquet, Schroon, and Raquette Rivers. Still others have climbed up the slopes of the Upland to Black Brook and Catamount Mountain.

Butternut clearly shows migration up the Ausable, Boquet, and Schroon Valleys from the Champlain-Lake George Lowland.

Witch hazel, representing shrubs, also follows the valleys clearly, and has a distribution similar to that of Butternut.

Eastern red cedar follows abandoned agricultural lands up the lower slopes of the valleys.

Maple-leaved viburnum is not restricted to the Champlain Lowland and tributary river valley bottoms. It also occurs fairly high up, 2100 feet, on the ridges, e.g., Catamount Mountain, Chapel Pond Pass, and Crane Mountain, but it is always within 25 miles of Lake Champlain and Lake George.

Pitch pine does not follow valley bottoms but rather sand plains.

Heart-leaved aster represents an advanced guard herb with the Lowlands influence (both Champlain and Saint Lawrence) first entering the Adirondack Upland.

Other species with similar distributions of advanced guard but not selected for mapping include Rough-leaved dogwood, American hornbeam, Gray panicled dogwood, White snakeroot, New England aster, and White woodland aster.

Puzzling populations:

These species fit neither the Remnant nor the Advanced guard categories. Their

Black cherry (*Prunus serotina*)
Each pair of brilliantly-colored stipules is part of the leaf. But unlike the blade and petiole which persist all summer, they fall off when the leaf blades are fully expanded.
© Michael Kudish 1992.

distributions have nothing in common except they present real challenges to the plant geographer.

Red pine is quite common on the Upland as a native, but occurs only on specific kinds of sites, free of competition from other species of trees. It is most abundant in the lakes regions of the Upland where it often dominates peninsulas, islands, and east shores. In areas with fewer lakes, it occurs most often around ledges, cliffs, and bluffs.

Red oak is abundant in the Lowlands but rare on the Upland above 1000 feet, occurring in only several widely-scattered localities. (Site detail on these localities is presented later in this chapter.)

Mountain alder occurs only in the eastern Adirondacks and in the High Peaks. It ranges in elevation from 900 feet to timberline at 5280 feet. It is absent west of the line connecting the MacIntyre Range, and Whiteface and Catamount Mountains. No one knows why.

Balsam poplar is common in the northern Champlain Valley and northern half of the Upland. It is near the southern limits of its natural range here and is thus uncommon in the southern Champlain Lowland and southern half of the Upland. South of the Mohawk Valley it is very rare.

Silver maple is restricted as a native to only a few flood plains of the larger rivers, with stations often tens of miles apart. How did the silver maples "find" the flood plains and skip over the intervening miles of rapids?

Three-toothed cinquefoil occurs typically above 4000 feet in the alpine zone. But why is it also widely scattered on the Upland down to elevations as low as 1500 feet? It occurs on open bedrock ledges, sandy roadsides, and even old fields.

CLUBMOSSES AND HORSETAILS

1. LYCOPODIACEAE (CLUB-MOSS FAMILY)
All Clubmosses are protected in New York State.

Lycopodium annotinum L. BRISTLY CLUBMOSS or STIFF CLUBMOSS
- Site: Mostly under spruce-fir, but occasionally under northern hardwoods. Well-drained. Shade-tolerant. In the College area, populations are widely-scattered, not present almost everywhere as is *L. lucidulum*. Becoming more abundant with increasing elevation into the Spruce-fir zone to timberline. pH of humus: n=5; min. 4.0; max. 4.5; median 4.5. The clubmosses, *L. clavatum, L. complanatum*, and *L. tristachyum* are common under pioneering pines and in open, disturbed places but yield to their more shade-tolerant cousins, *L. annotinum, L. lucidulum,* and *L. obscurum* under northern hardwoods and mixed woods.
- Elevation: From 1100 feet on the Dickinson Esker to 4940 feet on Haystack.
- Phenology: Cone past maturity on October 19.
- Frequency: Frequent

Lycopodium clavatum L. WOLF'S FOOT CLUBMOSS, RUNNING CLUBMOSS or STAGHORN CLUBMOSS
- Site: A species intolerant of well-drained areas, along edges of roads, fields and golf courses, etc.
- Elevation: From between 1000 and 1200 feet in Keene Valley to 2500 feet on Azure Mountain and 2990 feet on Noonmark Mountain.
- Phenology: Cones over-mature by October 14, 1981, near Campus.
- Frequency: Occasional

Lycopodium digitatum L. or *Lycopodium complanatum* var. *flabelliforme* Fern. GROUND CEDAR or NORTHERN RUNNING PINE
- Site: Old-field, shade-intolerant species, sometimes persisting into young woods. Well-drained.
- Elevation: From ca. 450 feet on Tongue Mountain to at least 1700 feet at Paul Smiths.
- Frequency: Occasional.
- Reference: Cook, R.E. (1983)

Lycopodium inundatum L. SWAMP CLUBMOSS or BOG CLUBMOSS
- Site: In moist sand, vernally flooded. Sometimes on bog mats. Poorly-drained. Shade-intolerant.
- Elevation: From about 1500 feet at Saranac Lake to 1770 feet at Ferd's Bog.

Frequency: Seven stations

Lycopodium lucidulum Michx. SHINING CLUBMOSS

- Site: Our most common woodland shade-tolerant species. Under northern hardwoods, spruce-fir or hemlock stands. Well or imperfectly drained soils. Only *L. annotinum* is as shade-tolerant as this one. pH of humus: $n=27$; min. 3.2; max. 5.1; median 4.5.
- Elevation: From between 1000 and 1200 feet in Keene Valley to 4161 feet on Phelps Mountain.
- Frequency: Common
- Miscellany: The only non-alpine lacking strobili (spore cones). The sporangia instead are borne in the axils of the regular leaves. Vegetative reproduction is also present in the form of green gemmae or bulblets among the leaves. Evergreen, forming colonies perhaps dozens of years old and tens of feet in diameter, all connected by underground rhizomes. Dr. Mildred Faust stated, in a personal communication, that this species is the only clubmoss which remains the same size above ground for decades; for each new growth increment annually above ground, the rhizome (stem) buries itself for the same length in the humus. Other species do not bury themselves much. Expansion of the colony, then, in *L. lucidulum,* above ground must be due to branching.

Lycopodium obscurum L. TREE CLUBMOSS, PRINCE'S PINE or GROUND PINE

- Site: The most common species on the immediate College campus. Less shade tolerant than *L. lucidulum* and *L. annotinum,* but more so than all the others. This species often follows disturbance. Hence partly-open forest canopies support this species which can also occur under undisturbed stands. Well-drained. pH: of humus: $n=25$; min. 3.5; max. 5.8; median 4.5.
- Elevation: From about 450 feet on Tongue Mountain to 4050 feet on Porter Mountain.
- Frequency: Common.

Lycopodium tristachyum Pursh GROUND CEDAR or NORTHERN GROUND PINE

- Site: Locally abundant in disturbed open sandy areas, e.g. old dumps, old railroad grades, sand pits, roadsides, and severe burns. However, along a new V.I.C. trail this species grows in shaded, dense mixed woods, suggesting some shade-tolerance here—a most unusual phenomenon.
- Elevation: Between 1500 and 1700 feet.
- Frequency: 13 stations.

2. ISOETACEAE (QUILLWORT FAMILY)

Isoetes sp. QUILLWORT
- Site: Growing in shallow water with about half the height of the plant submerged and appearing as clumps of grass. No true local grasses are this aquatic.
- Elevation: From 1540 feet to 1650 feet.
- Phenology: June through August.
- Frequency: 4 stations.
- Miscellany: Our species is still to be determined. Until 1986, all specimens collected bore no megaspores, making identification impossible. Peck lists *I. echinospora* Dur. var. *Braunii* (Durieu) Engelm. Muenscher lists *I. echinospora* Dur. in Osgood Pond. Fernald calls this species *I. muricata* Dur., a synonym.

    In 1986, I found a specimen with megaspores, and proceeded to "key out" the species. The result, *I. melanopoda* Gay & Dur., matches the descriptions but is a Midwestern species! We are either way out of range to the east or we have an outlier!

    Megaspores are about 0.3mm in diameter, either trigonous or with tetrahedral ridges, a few additional irregular folds, and papillose. There are 4 bast fiber groups in the leaves.

3. EQUISETACEAE (HORSETAIL FAMILY)

Equisetum arvense L. FIELD HORSETAIL or COMMON HORSETAIL
- Site: The commonest horsetail, occurring mostly in poorly-drained sites but occasionally in deep, well-drained sands such as behind the Science Building on Campus and at the base of the Administration Building. Often in roadside ditches and at edges of swamps. Moderately shade tolerant to intolerant.
- Elevation: From 210 feet at Chazy to over 2100 feet near Heart Lake.
- Phenology: Median dates for fertile fronds shedding spores, May 1 to 5. Earliest date of appearance of fertile fronds, April 4, 1985.
- Frequency: Common.
- Miscellany: Horizontal rhizomes (stems) connecting the vertical run quite deep beneath the surface in deep, dry sands; the author excavated to 16 inches to find them. This is our only Horsetail with dimorphic fronds—a division of labor—the brown, reproductive, spore-bearing fronds emerging sometimes as early as mid-April, and the green, photosynthetic fronds following in May. Used by early American colonists to clean pots and pans (hence an alternate common name, Scouring rush) because horsetails are among the few vascular plants to absorb silica; silica adds to the support and rigidity of the stems and creates the abrasive quality.

Reference: Register and West (1976)

Equisetum fluviatile L. forma *Linnaeanum* (Doll) Broun or *E. limosum* L. SWAMP HORSETAIL, WATER HORSETAIL or PIPES
> Site: Aquatic, often standing in shallow water, forming colonies of both branched and unbranched stems.
> Elevation: From 100 feet at Point Au Roche to 1630 feet at Paul Smiths.
> Phenology: Cones present June 25, 1975, in Jones Pond.
> Frequency: 4 stations.
> Miscellany: Occasional stems bear an apical small strobilus (spore cone).

Equisetum hyemale L. var. *affine* (Engelm.) A.A. Eat. COMMON SCOURING RUSH or ROUGH HORSETAIL
> Site: Along open, sunny, often moist banks and in poorly-drained level areas, forming dense thickets. Stems unbranched.
> Elevation: From 100 feet along Lake Champlain to 1790 feet on Brewster Mountain.
> Phenology: Cones present September 4, 1976, at Upper Jay.
> Frequency: 4 stations.

Equisetum scirpoides Michx. DWARF HORSETAIL or DWARF SCOURING RUSH
> Site: Drainage imperfect, probably flooded vernally along brooks and seeps. Shade-tolerant.
> Elevation: 1620 feet.
> Frequency: One station near Franklin Falls. (Dr. Ketchledge reports two more stations in the High Peaks.)

Equisetum sylvaticum L. WOOD HORSETAIL
> Site: Perhaps the most shade-tolerant Horsetail as the name implies; well-drained to poorly-drained sites. Widespread in the Adirondacks, but only locally common in isolated patches.
> Elevation: 1460 to about 2200 feet.
> Phenology: Cones present May 21. Plants beginning to turn brown, August 1, 1987, near Brighton Town Hall.
> Frequency: 13 stations.

Equisetum variegatum Schleich. ex Weber & Mohr VARIEGATED HORSETAIL
> Site: A small evergreen Horsetail in poorly-drained areas, in full sun or partial shade, unbranched.
> Elevation: 115 to 1625 feet.
> Phenology: Cones present on August 22, 1974 on Campus.
> Frequency: One station.

FERNS

All ferns, except three, are protected in New York State.

4. OPHIOGLOSSACEAE, (ADDER'S-TONGUE FAMILY)

Botrychium dissectum Spreng. GRAPE FERN or CUT-LEAF GRAPE FERN
- Site: Well-drained sites, but growing in from open sun to dense shade. Sometimes on lawns!
- Elevation: From 1600 to 2058 feet.
- Phenology: Fronds dimorphic, the sterile evergreen and the fertile brown. Both observed together between October 7 and October 15.
- Frequency: Four stations.

Botrychium virginianum (L.) Sw. RATTLESNAKE FERN
- Site: Well-drained sites with higher than average humus pH. Shade-tolerant. Typically, under Sugar maple, and often White ash and Basswood.
- Elevation: From 150 feet at Point au Roche to 2157 feet atop Buck Hill.
- Frequency: Nine stations.

5. OSMUNDACEAE ("FLOWERING" FERN FAMILY)

Osmunda cinnamomea L. CINNAMON FERN
- Site: In poorly-drained areas such as swamps. Intolerant to midtolerant of shade.
- Elevation: From 210 feet at the Chazy Fir Swamp to 3124 feet on Moose Mountain.
- Phenology: The sporogenous fronds appear in the spring with the green fronds but wither by early summer. Sporogenous fronds on Campus, June 9, 1986.
- Miscellany: Fronds dimorphic (separate orange-brown sporogenous vs. green vegetative). Cinnamon fern is edible when the fronds uncoil vernally.

Osmunda claytoniana L. INTERRUPTED FERN
- Site: Northern hardwoods, spruce-fir stands, and swamps. Shade-tolerant to midtolerant. On similar sites and often with Cinnamon fern. But this one can grow on well-drained sites as well; Cinnamon fern is almost exclusively restricted to the poorly drained.
- Elevation: From ca. 450 feet at Tongue Mountain on Lake George, to 4370 feet on Mount Marcy.

Frequency: Common.

Osmunda regalis L. ROYAL FERN
   Site: Locally abundant in poorly and very-poorly drained soils, especially along streambanks and in shallow water. Moderately shade-tolerant.
   Elevation: From 210 feet at Chazy Fir Swamp to 2000 feet.
   Frequency: Five stations, but probably much more frequent.

6. POLYPODIACEAE (POLYPODY FAMILY)

Polypodium virginianum L. or *Polypodium vulgare* L. POLYPODY, COMMON POLYPODY or ROCK POLYPODY.
   Site: Apparently, this species occurs only on vertical cliffs and atop large boulders because it is a favorite deer browse. The fern exists only where it is inaccessible to deer. The students and faculty at E.S.F. Newcomb (Huntington Forest) first recognized this. Could the fact that other ferns, e.g., *Cystopteris fragilis,* which occur only on ledges be attributed to the same cause?
   Polypody leaves crisp up first, as do those of Striped maple and Sharp-leaved aster, to indicate a dry period in summer or fall. Shade-tolerant. Well-drained.
   Elevation: From ca. 450 feet on Tongue Mountain to 2450 feet on Pitchoff Mountain.
   Frequency: Common.

7. ADIANTACEAE (MAIDENHAIR FAMILY)

Adiantum pedatum L. MAIDENHAIR FERN
   Site: Higher humus pH soils under Sugar maple and often White ash and Basswood. Well-drained. Shade-tolerant. pH of humus: n=14; min. 4.7; max. 6.8; median 5.75.
   Elevation: From 210 feet at Chazy Fir Swamp to 2168 feet on Jenkins Mountain.
   Frequency: 15 stations.
   Reference: NY State *The Conservationist,* 1963.

Cryptogramma stelleri (Gmel.) Prantl SLENDER CLIFF-BRAKE or FRAGILE ROCK-BRAKE
   Site: On moist cliffs with marble (calcite) bodies and charnockitic gneiss in the Cascade Notch.
   Elevation: 2150 feet to 2500 feet approximately.
   Frequency: Two stations.

8. DENNSTAEDTIACEAE (TREE FERN FAMILY)

***Dennstaedtia punctilobula* (Michx.) Moore HAY-SCENTED FERN**
- Site: Very abundant where present, usually in large groves, but local. Most common along roadsides and in openings in forest where there is at least some sun for this mid-tolerant. Well-drained. It is reported to spread very rapidly to fill canopy openings.
- Elevation: From ca. 450 feet on Tongue Mountain to Pitchoff Mountain 3240 feet.
- Frequency: One of only three ferns *not* protected in N.Y. State. Common.
- Miscellany: Identified by its powerful aroma when brushed through.

***Pteridium aquilinum* (L.) Kuhn BRACKEN, BRAKE FERN or EAGLE FERN**
- Site: Profuse on disturbed sites especially on the well-drained and in full sun. Intolerant, becoming stunted in partial shade. A persistent pioneer following loggings, burns, and clearings, often retarding the succession of tree species by its density.
- Elevation: From 588 feet at Camp Pok-O-MacCready near Willsboro, to 3800 feet at Lyon Mountain.
- Frequency: The second of three ferns not protected in N.Y. State. Common.
- Miscellany: Young shoots said to be edible by some, but others describe them as carcinogenic.
- References: NY State *The Conservationist,* 1963
  Cook, R.E. (1983)
  R.T. Smith & J.A. Taylor (editors), (1986)

9. ASPLENIACEAE (SPLEENWORT FAMILY)

***Athyrium asplenioides* (Michx.) Desv. or *Athyrium filix-femina* (L.) Roth. LADY FERN**
- Site: In well-drained and imperfectly-drained soils. Shade-tolerant. Northern hardwoods, mixed woods, and spruce-fir.
- Elevation: From 100 feet along Lake Champlain to 3700 feet at Indian Falls.
- Frequency: Common.

***Athyrium pycnocarpon* (Spreng.) Tidestr. NARROW-LEAVED SPLEENWORT or GLADE FERN**
- Site: In a small swamp on Brewster Mountain. Humus pH here is quite high, 6.8. Shade-tolerant.
- Elevation: Ca. 1790 feet.

Frequency: One station.

Athyrium thelypteroides (Michx.) Desv. SILVERY SPLEENWORT
- Site: On higher humus pH sites, mostly on poorly and imperfectly-drained springy areas. Occasionally on the well-drained. Shade-tolerant. Usually under Sugar maple. Often found under Basswood, White ash, and Elm. pH of humus: n=15; min. 4.7; max. 6.6; median 5.1.
- Elevation: From 210 feet at Chazy Fir Swamp to 2157 feet on Buck Hill.
- Frequency: 34 stations.

Cystopteris bulbifera (L.) Bernh. BULBLET FERN
- Site: Cliffs and soils where calcium carbonate (lime) is present, and thus a high pH.
- Elevation: From 1750 feet on Mount Pisgah to 2030 feet at Cascade Lakes.
- Frequency: Two stations.
- Miscellany: Bears vegetatively-reproducing bulblets on lower leaf surfaces.

Cystopteris fragilis (L.) Bernh. COMMON FRAGILE FERN
- Site: Could this fern be a popular deer browse in accessible places (as *Polypodium* is) so that the fern grows mainly in inaccessible places such as vertical ledges? It appears not to have the lime requirement of its cousin, *C. bulbifera.*
- Elevation: From 908 feet on Palmer Hill to 2800 feet on Pitchoff Mountain.
- Frequency: Nine stations.

Dryopteris campyloptera (Kunze) Clarkson or *D. spinulosa* (O.F. Muell.) Watt. var. *americana* (Fisch.) Fern. MOUNTAIN WOODFERN
- Site: As a distinct species from the lower-elevation, smaller, evergreen *D. intermedia,* this is the large, waist-high frost-sensitive deciduous fern of the spruce-fir forests, mostly above 2500 feet.
- Elevation: To 4736 feet on Gothics Mountain.
- Frequency: It is certainly common in the High Peaks.

Dryopteris cristata (L.) Gray CRESTED SHIELD FERN or CRESTED WOOD FERN
- Site: Swamps, quite shade-tolerant.
- Elevation: From 210 feet at Chazy Fir Swamp to 1700 feet at Star Mountain Pond.
- Frequency: 12 stations.
- Miscellany: Evergreen and with erect and leathery leaves, possibly hybridizing

with *D. intermedia* to form Boott's fern, *D. X Boottii.*

***Dryopteris goldiana* (Hook. ex Goldie) Gray GOLDIE'S FERN or GIANT WOOD FERN**
- Site: Higher humus pH sites under Sugar maple and usually White ash and Basswood. Well-drained. Shade-tolerant.
- Elevation: From about 1500 feet E of Franklin Falls to 2200 feet on Blue Hill and Kate Mountain.
- Frequency: 10 stations.

***Dryopteris intermedia* (Muhl. ex Willd.) Gray or *D. spinulosa* (O.F. Muell.) Watt. var. *intermedia* (Muhl.) Underw. WOODFERN, SPINULOSE WOODFERN or FANCY FERN**
- Site: The most common fern in the Adirondacks due to its shade and acid-humus tolerance. Since the Adirondacks are chiefly forested and its soils are chiefly acid, this combination of tolerances would be an advantage to ANY ground-cover plant! It is also an evergreen. (See Chapter 3, on plant strategies for advantages of being evergreen.)

    Once combined with *D. campyloptera* as a single species, the Woodfern is the smaller-fronded, lower elevation, evergreen species, abundant in northern hardwoods, mixed woods, and spruce-fir. Well-drained to imperfectly-drained sites. pH of humus: n=79; min. 3.2; max. 6.8; median 4.6.
- Elevation: From 100 feet along Lake Champlain to 3532 feet on Algonquin Trail.
- Frequency: Common.
- Miscellany: *Dryopteris X bootii* (Tuckerm.) Underw. Boott's Fern. Our plants appear to be hybrids between *D. intermedia* (Muhl. ex Willd.) A. Gray and *D. cristata* (L.) A. Gray, as both parent species are present. Two sites: First site elevation 1630 feet; second site elevation 1650 feet at Paul Smiths, both swamps.

***Dryopteris marginalis* (L.) Gray MARGINAL SHIELD FERN**
- Site: Shade-tolerant on mostly well-drained sites with higher humus pH. Almost always under Sugar maple and often also White ash and Basswood. One exceptional site is on the N shoulder of The Cobbles, W of Franklin Falls, where this species grows under Mountain maple, Paper birch, and Red spruce—no Sugar maple. At the Keese's Mills Quarry, this fern grows on both dry and dripping wet ledges. (See Chapter 3 on advantages of being evergreen.) pH of humus: n=21; min. 4.4; max. 6.8; median 5.55.
- Elevation: From 210 feet at Chazy Fir Swamp to 2911 feet on Crane Mountain.

Frequency: 30 stations.

Gymnocarpium dryopteris (L.) Newm. or ***Dryopteris disjuncta*** (Ledeb.) C.V. Morton OAK FERN.
- Site: On well-drained soils under northern hardwoods, mixed woods or conifers.
- Elevation: From 210 feet at Chazy Fir Swamp to 1700 feet around Paul Smiths, well up into the spruce-fir forest. Possibly to timberline.
- Frequency: Occasional.

Matteucia struthiopteris (L.) Todaro or ***Pteretis pensylvanica*** (Willd.) Fern. OSTRICH FERN or FIDDLE HEADS
- Site: In poorly drained sites. Uncommon in the Paul Smith's area. Becoming much more widespread and common in the eastern Adirondacks as one travels from the Ausable and Bouquet watersheds to the Champlain Lowland.
- Elevation: From Wickham Marsh at ca. 100 feet to about 2270 feet in the Pitchoff-Sentinel Notch.
- Frequency: Infrequent.
- Miscellany: Fronds edible in springtime.

Onoclea sensibilis L. SENSITIVE FERN
- Site: Abundant in mostly poorly-drained areas. Moderately shade-tolerant.
- Elevation: From 210 feet at the Chazy Fir Swamp to 2353 feet along new Algonquin Peak Trail.
- Frequency: One of the three ferns in N.Y. State *not* protected. Common.
- Miscellany: Fronds dimorphic (of two forms—the sterile green and the fertile brown). Poisonous to horses. Common name refers to sensitivity to first autumnal frost, not to touch.

Phegopteris connectilis (Michx.) Watt or ***Dryopteris phegopteris*** (L.) Christens NARROW BEECH FERN, NORTHERN BEECH FERN or LONG BEECH FERN
- Site: Not common in northern hardwoods and mixed woods, but becoming more abundant upwards in the spruce-fir forests to timberline. Well-drained to imperfectly-drained sites. Moderately shade-tolerant.
- Elevation: From between 1000 and 1200 feet in Keene Valley to 4620 feet on Haystack Mountain.
- Frequency: Frequent.

Polystichum acrostichoides (Michx.) Schott. CHRISTMAS FERN
- Site: Common on higher humus pH soils almost always under Sugar maple and often also under White ash and Basswood. Shade-tolerant. Well-drained. Evergreen (see Chapter 3 on this strategy). pH of humus: n=27; min. 3.8; max. 6.8; median 4.9.
- Elevation: From 200 feet at Chazy to 2300 feet on Giant Mountain.
- Frequency: 44 stations.

Polystichum braunii (Spenner) Fee BRAUN'S HOLLY FERN
- Site: An uncommon fern but not exceedingly rare. When found, often around ledges and/or boulder talus but not always. Kate Mountain is an exception where it grows on deep till under northern hardwoods on a site not especially stony. Shade-tolerant. Well-drained.
- Elevation: From about 1700 feet to about 2500 feet.
- Frequency: 6 stations.

Thelypteris noveboracensis (L.) Nieuwl. or *Dryopteris noveboracensis* (L.) Gray NEW YORK FERN
- Site: Woodland fern, midway in tolerance between Hayscented and Woodferns. Well-drained to springy (imperfectly-drained) sites. Mostly under northern hardwoods.
- Elevation: From 100 feet along Lake Champlain to 2300 feet on the MacIntyre Range.
- Frequency: Common.

Thelypteris palustris Schott or *Dryopteris thelypteris* (L.) Gray or (L.) Sw. MARSH FERN
- Site: In swamps. Shade-tolerant.
- Elevation: From 210 feet at the Chazy Fir Swamp to about 2180 feet at Heart Lake.
- Frequency: Occasional.

Woodsia ilvensis (L.) R. Brown RUSTY WOODSIA, RUSTY CLIFF FERN or FRAGRANT WOODSIA
- Site: On anorthosite ledge of a peninsula site.
- Elevation: 1625 feet.
- Frequency: One station.

GYMNOSPERMS: THE CONIFERS

General references on Conifers (as opposed to references on individual species).

Miller and Krall, 1970. (Pine sawfly)
Miller and Krall, 1965. (Spruce Gall Aphids)
Ketchledge, July 1983. (The Spruces)
Silverborg, 1969. (Spruce Canker)
Winch (1959–1960). (Conifers of New York)
Miller and Allen (1971) on *Pinus*.

10. TAXACEAE (YEW FAMILY)

Taxus canadensis Marsh. YEW or GROUND HEMLOCK
- Site: Shade-tolerant shrub on well-drained or imperfectly-drained soils. Under hemlock, mixed hemlock-hardwood, or hemlock-spruce-fir stands. Locally abundant but stations widely-scattered. Limestone soils of Lake Champlain, gneiss and anorthosite soils of Adirondacks, sandstone soils of Catskills and Chateaugay Chasm.
- Elevation: 100 feet on Four Brothers Islands to 2300 feet in Pitchoff-Sentinel Notch.
- Frequency: 27 stations.
- Miscellany: The whole plant is poisonous to people, but deer browse it heavily. Dioecious (only female plants produce cones, the male only pollen).

11. PINACEAE (PINE FAMILY)

Abies balsamea (L.) Mill. BALSAM FIR
- Site: Shade tolerant. Well-drained to poorly-drained. Limestone soils (as at Chazy, Brasher Center and Stockholm) to anorthosite and gneiss in the Adirondacks to the sandstones, shales, and conglomerates of the Catskills. pH of humus with 57 samples: min. 3.5; max. 6.8; median 4.6. At the lowest elevations in the Champlain and Saint Lawrence Lowlands, Balsam fir occurs in swamps as relicts where northern hardwoods have not yet invaded completely. At the middle elevations (1500 to 1800 feet) around Paul Smiths, Balsam fir is most abundant where northern hardwoods competition is least: sandiest outwash, swamps, old fields, shallow soils around bedrock outcrops. Above 4000 feet, it dominates to timberline. Above timberline, it occurs in protected

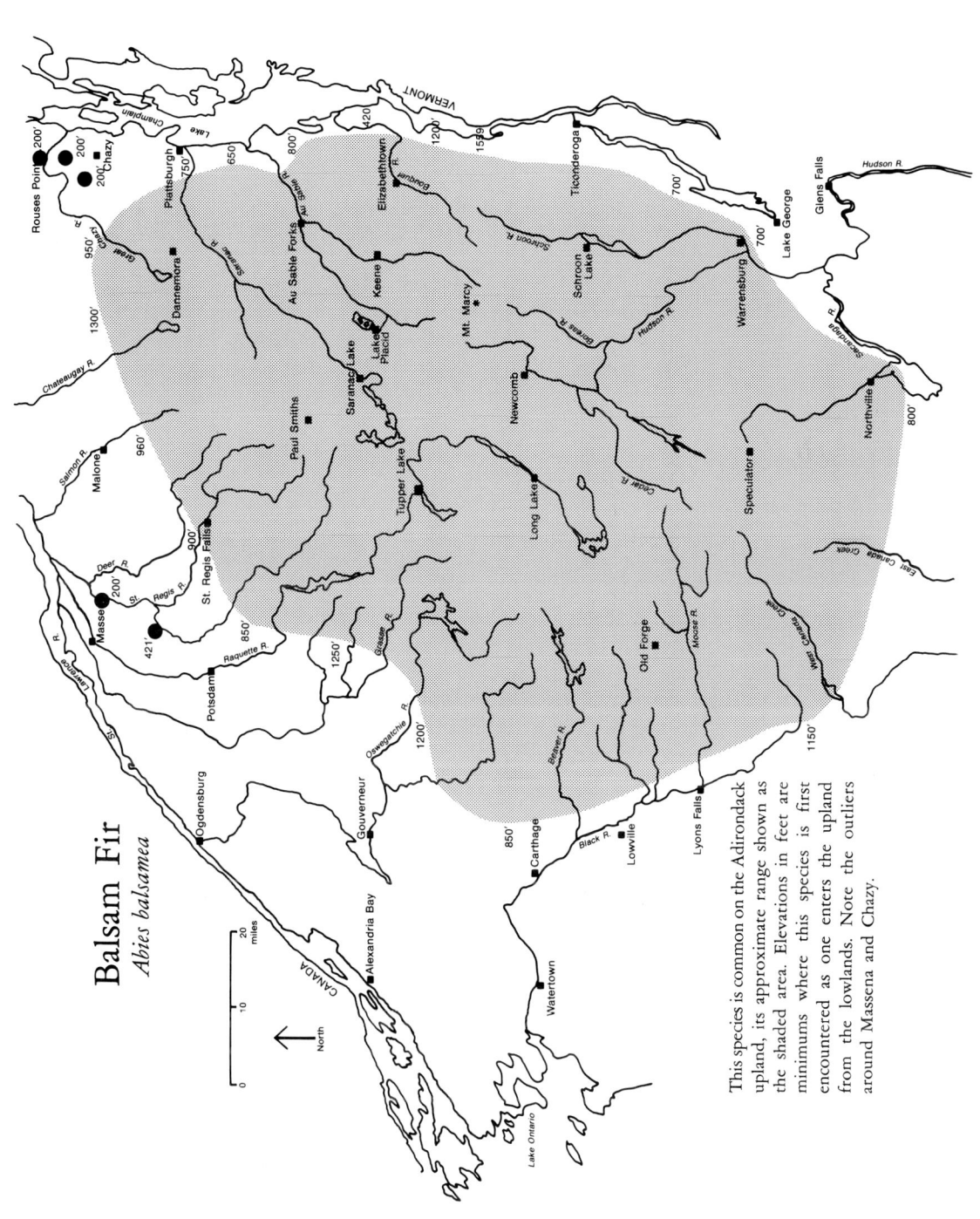

Balsam Fir
Abies balsamea

This species is common on the Adirondack upland, its approximate range shown as the shaded area. Elevations in feet are minimums where this species is first encountered as one enters the upland from the lowlands. Note the outliers around Massena and Chazy.

	places out of the severest winds.
	MAP: The map shows a brief sketch (not complete or anywhere near so) of the native distribution of Balsam fir in the Adirondacks. Note the elevations at which the species enters as one travels onto the Adirondack Upland from the outside.
Elevation:	From 200 feet at Brasher Center and 210 feet at the Chazy Fir Swamp to 5330 feet on Mount Marcy. This range suggests a difference in mean annual temperature of 15 degrees F, or 85 days frost-free season, or 3100 growing-degree days (see page 33).
Frequency:	Common.
Miscellany:	*A. balsamea* var. *phanerolepis*, bearing cones with exserted (sticking out) bracts, is in the subalpine zone; it is midway between (with cone features) *A. balsamea* and *A. fraseri* of the southern Appalachians.
References:	Bakuzis and Hansen (1965)
	Zon (1914)
	Sprugel (1984)

Larix laricina (DuRoi) K. Koch. EASTERN LARCH or TAMARACK

Site:	One of the four conifers abundant in the Adirondacks but uncommon in the Champlain and Saint Lawrence Valleys (Red spruce, Black spruce, and Balsam fir are the other three). Larch is most common and most well-known as an associate of Black spruce in bogs, but neither species makes best growth under such conditions (Ketchledge, 1967 and 1970). Both Larch and Black spruce grow best on well-drained outwash sands which were severely disturbed (as by fire, for example) or on well-drained tills once cleared for fields and now abandoned. Because these species are so shade-intolerant, they are eventually outcompeted by other species on the better sites. Soils: limestone at Helena and Raymondville to Adirondack gneisses and anorthosite.
Elevation:	From 200 feet on the Ausable River Delta below Ausable Chasm and SW of Helena to 4620 feet on Haystack.
Phenology:	Immature upright bright red seed cones along Heron Marsh June 13, 1988.
Frequency:	Common.
Miscellany:	The only native Adirondack DECIDUOUS conifer.
References:	Buck (1985)
	Miller and Krall (1970)

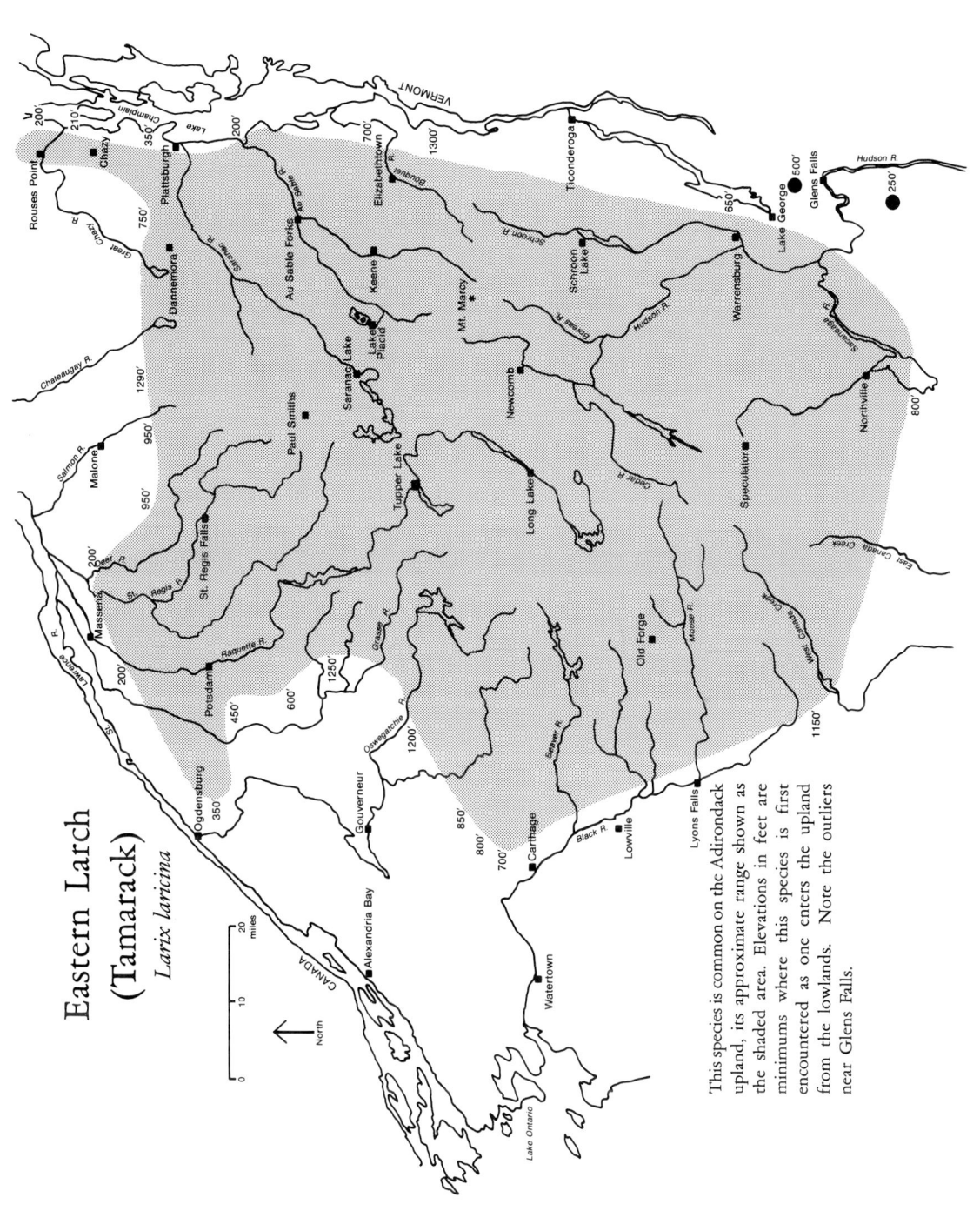

Eastern Larch (Tamarack)
Larix laricina

This species is common on the Adirondack upland, its approximate range shown as the shaded area. Elevations in feet are minimums where this species is first encountered as one enters the upland from the lowlands. Note the outliers near Glens Falls.

Picea glauca (Moench) Voss WHITE SPRUCE or CAT SPRUCE

Site: Cook, Smith and Stone (1973) have described the natural distribution of White Spruce in New York and show many more Adirondack locations; their map is of a small scale, however, and one cannot pinpoint the precise locations unfortunately. These authors suggest that White spruce occurs mostly where there is some marble or lime in the soil, but I have found White spruce independent of lime or marble; it grows where there is lime and also where there is not. The only features in common among the sites that I have seen are a nearly-flat to gently sloping plateau or valley bottom, often near a river, with fields that have been abandoned. White spruce is at the southern limits of its range in the central Adirondacks.

Elevation: 1000 feet between Jay and Wilmington to 2050 feet at Mount Hoevenberg cross-country ski trails.

Phenology: Brightly-colored upright immature seed cones on campus (planted tree), May 27, 1988.

Frequency: 6 stations.

Miscellany: Although commonly planted locally in the Paul Smith's-Saranac Lake area both as an ornamental and for reforestation, White spruce turns out also to be a native. I had been examining a stand of this along the Saranac River above Bloomingdale for about fourteen years debating whether the population was native or escaped from plantation; finally, in September, 1986, after mapping the population all the way down the River to the Moose Pond Bridge below Bloomingdale, I've come to the conclusion that the tree is native. No one could have planted it in some of the inaccessible sites. The species occurs in the woods with other conifers and hardwoods mixed and also pioneers in old fields. Sites are well-drained for the most part, although sometimes imperfectly.

Dr. Ketchledge had brought my attention to a stand on the West Branch Ausable River above the junction of the River Road and Route 86 East of Lake Placid Village. He believed this population to be native and I agree.

References: Cook, Smith & Stone (1973)
Cook & Stout (1959–1960)

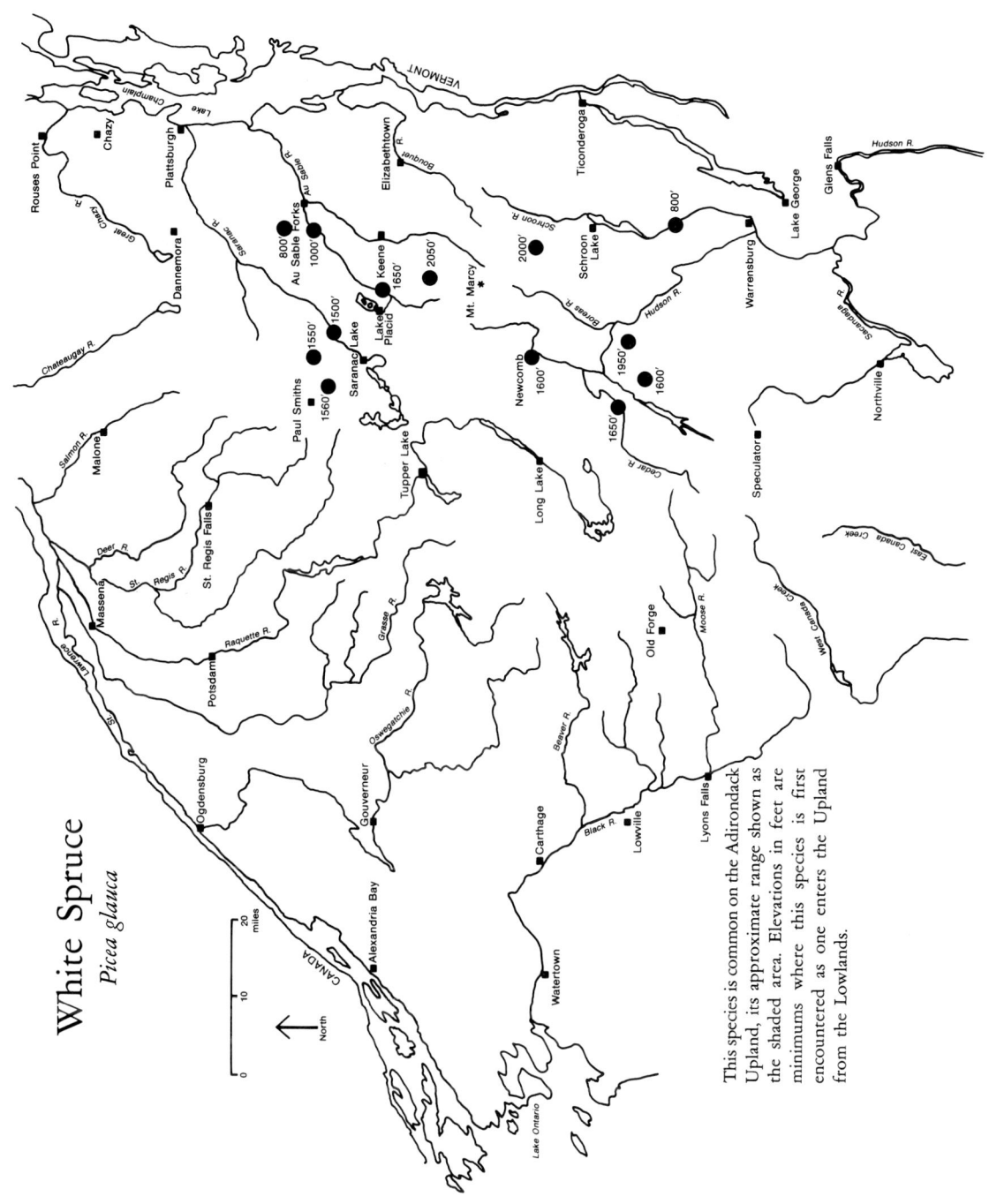

Picea mariana (Mill.) B.S.P. BLACK SPRUCE or BOG SPRUCE

Site:	Most common and well-known in bogs, but Black spruce does not always grow in bogs. It can occur also on open, severely-burned areas such as the sandy outwash plains from Brandon northward toward Quebec Brook and ESE of Derrick along the road to Floodwood and Saranac Inn. It has pioneered on these well-drained to exceedingly well-drained sites. Its only requirement is light and plenty of it (shade-intolerant), not a high water table. In fact, Black spruce may grow faster on a dry sand than in a bog; a spruce cored in the Moose River Plains on well-drained outwash was 13+ inches in diameter ca. 40 feet tall and 60 years old, while Black spruce in bogs can be this age and only several inches in diameter. On the drier, sandy sites, the needles tend to be longer than those in the bogs. To timberline and in the alpine zone where it is common with Balsam fir forming krummholz (a thicket of stunted trees), often rooted in *Sphagnum* moss and thus in an "inverted bog." Soils: limestone north in Canada to gneisses and anorthosite in Adirondacks. "Black Spruce Mountain" W of Prospect Mountain, Lake George: Why the name?
Elevation:	From 800 feet in Jenks Swamp NE of Chestertown to 5320 feet on Mount Marcy.
Frequency:	Common.
Miscellany:	Black spruce has vegetative reproduction by layering in the bogs. This species crosses with Red spruce, producing hybrids along the edges of bogs where both species exist.
Reference:	Morgenstern and Farrar (1964).

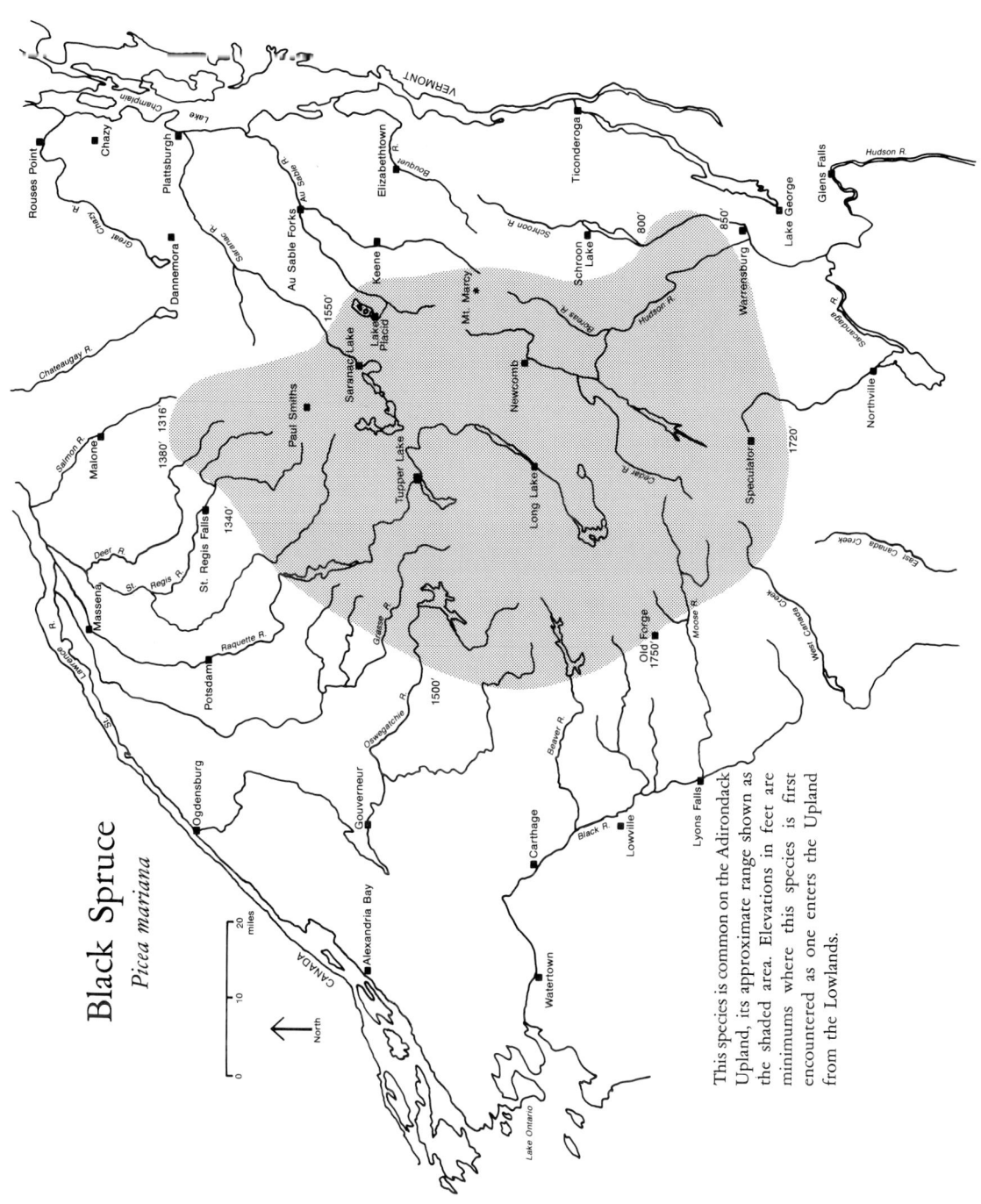

Picea rubens **Sarg.** RED SPRUCE or HE-BALSAM
- Site: The dominant tree from about 2500 to 4000 feet elevation, although the recent decline (see references) is making it less so above 3000 feet. Mostly well-drained. Shade-tolerant. Below 2500 feet, Red spruce is scattered in the northern hardwoods forest, but is most common on sites where Sugar maple and beech cannot grow well: either too sandy or too shallow soils on and near rock outcrops. Red spruce is famous for its being able to grow very slowly, suppressed in deep shade; then rapidly into the canopy when the shading mature trees die. pH of humus: n=63; min. 3.2; max. 6.8; median 4.5.
- Elevation: From 750 feet at three places: Saranac (not Saranac Lake), Parishville and SE of Chestertown along the Northway to 5310 feet on Mount Marcy.
- Frequency: Common.
- Miscellany: Hybridizes with Black spruce.
- References: (many on Spruce decline)
  Ketchledge, (February-March 1988)
  Morgenstern & Farrar (1964)
  Ketchledge (1967)

Pinus banksiana **Lamb.** JACK PINE
- Site: Occasionally planted as an ornamental and mixed in with Scots pine in reforestation stands. Native in the Adirondacks but rare because it is at the southern limit of its natural range. Most stands are in Clinton and NE Essex Counties. A most shade-intolerant pioneer of well-drained to droughty, exceedingly well-drained soils, often very shallow to bedrock and/or very sandy. Soils: sandstone at Altona; gneisses and anorthosite in Adirondacks. Locations: Two stations in the Flora area. The first is on Guideboard Road near Fern Lake Road, northeast of the hamlet of Black Brook at elevation 1040 feet. The second is on old fields on Juniper Hill, about 1½ miles south-southeast of Wilmington, elevation 1240 feet.

  Jack pine is not uncommon below 1000 feet, and thus outside this Flora area, in the East Branch Ausable River Valley from Ausable Forks to Upper Jay, mostly on well-drained abandoned fields. There is also some west of Schuyler Falls. The largest native stand in New York State is one mile south of the hamlet of Altona, on very shallow till overlying the Potsdam sandstone, elevation 800 to 900 feet; this stand last burned ca. 1958.
- Elevation: From 600 to 700 feet in the Ausable Forks-Jay area to 1240 feet on Juniper Hill near Wilmington.
- Frequency: Despite the local frequency of this pine, Birmingham (1988) and

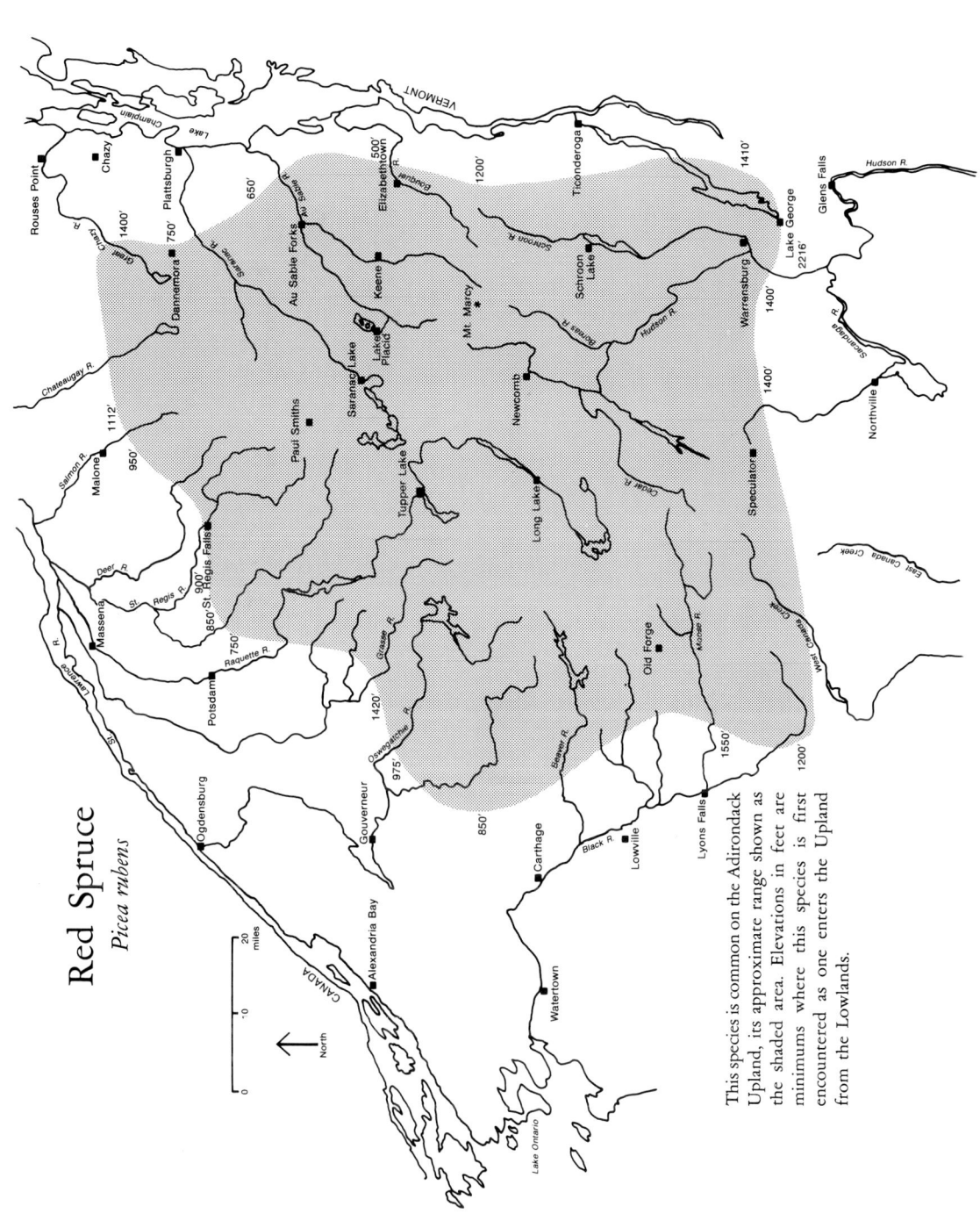

	Mitchell (1986) both list it as rare as a native. Two stations.
Miscellany:	This is a species that has evolved with brush fires taking advantage of the burns to eliminate competition of shrubs and other competing tree species. Cones can persist for decades on the trees, many opened only when subject to the heat (ca. 130° F) of a brush fire. In areas that have not burned, Jack pine can be a pioneer on abandoned fields and pastures.
References:	Littlefield (1960)
	Tappeiner (1973)

Pinus resinosa Soland. ex Ait. RED PINE or NORWAY PINE

| Site: | See Chapter 3. Mostly a plantation-reforestation species in New York State, Red pine locally is also native, especially on well-drained sites. Shallow or deep soils, till or outwash soils, Red pine does well provided there is sunlight and the stand is kept open. Growth is poor to none on poorly-drained sites. Shade-intolerant. |

The ground cover flora under Red pine is unusual (see page 41) in that shade-midtolerants thrive under the semi-open stands.

This is one of the very few species in NY State which seems distributed more by climatic factors than by soils. Note the great elevational range: slightly over 100 feet on Lake Champlain to over 3000 feet. Many of the mountains that Red pine grows on have been burned over, e.g., Crane Mountain, Keene area, Catamount, Pok-O-Moonshine, Coot Hill. Elevation: 100 feet at Point Au Roche and Willsboro Point to 3168 feet on Catamount and 3477 feet on Pitchoff Mountains. Soils: limestone along Lake Champlain to Adirondack gneisses and anorthosite. Sandstones and shales and conglomerates in Catskills. Map, p. 99 for Red pine stations.

Frequency:	Frequent.	
References:	Cook, Smith, Stone (1952)	Richards, Morrow, Stone (1962)
	DeMent and Stone (1968)	Skilling (1973)
	Faber (1980)	Tappeiner (1973)
	Faber (1985)	Yops & Smith (1964)

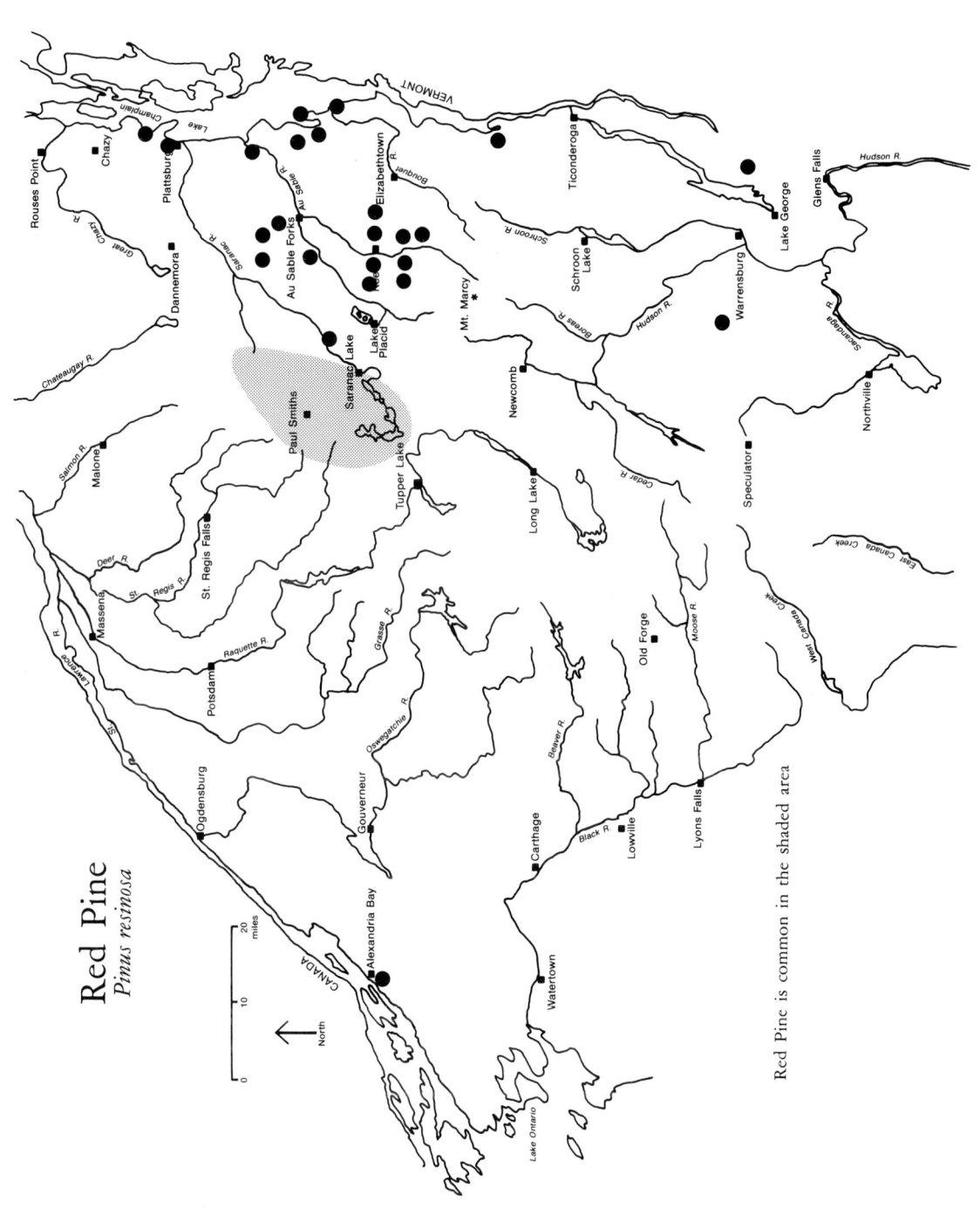

Pinus rigida Mill. PITCH PINE
- Site: In contrast with Jack pine, Pitch is uncommon because it is at the northern end of its natural range. Pitch is locally present in the Plattsburgh-Keeseville area and again between Glens Falls and Albany. One isolated stand occurs near Plessis, in Saint Lawrence County. Pitch is ecologically similar to Jack, occurring also on mainly very shallow-to-bedrock and/or very sandy soils; Pitch replaces Jack south of the Adirondacks, an ecological near-equivalent. Repeated brush fires perpetuate Pitch pine: open cones persist for years on the branches and sprouts arise following fires from the base of the trunk (very few conifers can sprout like this; Redwood is another). Soils: limestone at Cumberland Head and Plattsburgh to sandstone at Plessis to gneisses above Ausable Forks. In limestone areas, pitch pine grows on deep sands and the limestone itself may have little effect on the trees.
- Elevation: We have one station in our Flora at 1040 feet along Guideboard Road at Silver Lake Road, Black Brook. It is found as low as 100 feet at Cumberland Head State Park outside the Flora area.
- Frequency: One station.

Pinus strobus L. EASTERN WHITE PINE
- Site: Commonest on well-drained outwash soils, but also present on well-drained tills. Present on imperfectly-drained soils, but does very poorly with short, yellowing leaves on the poorly-drained. Intolerant to mid-tolerant of shade.

    White pine cannot withstand as well the dehydrating winds of winter that Red pine can and usually "permits" the Red to shelter it somewhat. Despite the scarcity of *Ribes,* Currants and gooseberries, in the Flora area, blister rust is still prevalent.

    This species is most abundant in the Paul Smiths area around the lakes and ponds, bogs and marshes—not dominant on the wetter soils, but abundant further upslope. Most of the old White pines (over 150 years) locally have been established following blowdowns; younger trees have come in following fires, abandoned fields, clearcuts, and selective loggings. Blowdowns are most prevalent along the E, SE, and NE shores of the lakes. Red pines dominate the well-drained shorelines, but are shorter in height and less subject to windthrow than the White pines further inland.

    In the Schroon Valley, White pine dominates following the removal of forests for the pulp and paper industry in the 19th century, much of the removed forests going to Glens Falls to the mills.

    pH of humus: n=27; min. 3.8; max. 5.5; median 4.5. Soils:

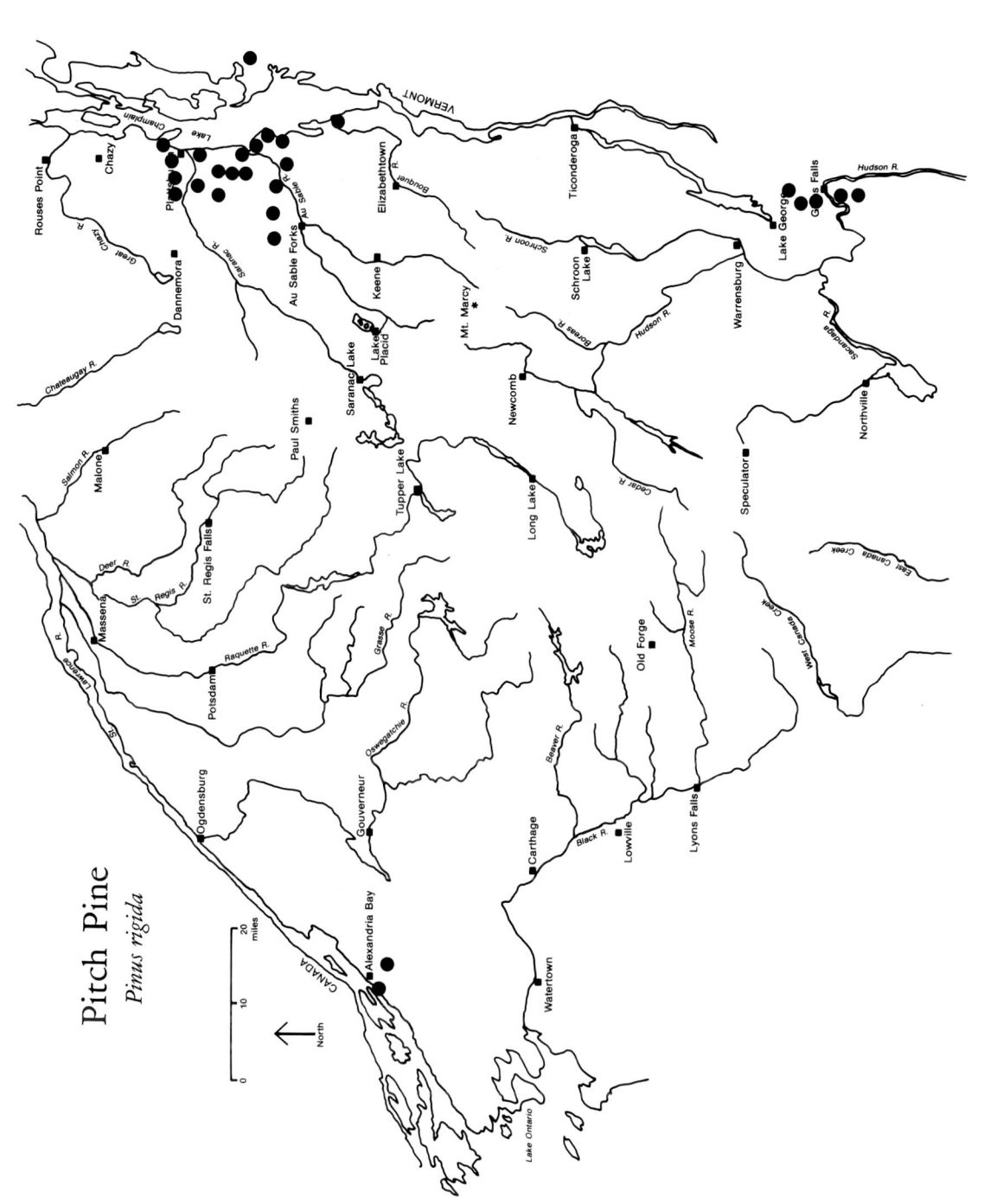

	limestone along Lake Champlain to Adirondack anorthosite and gneisses. Sandstones and shales in Catskills.
Elevation:	From 100 feet along Lake Champlain to 2760 and 3168 feet on Catamount Mountain.
Phenology:	Pollen cloud on Easy Street, June 18 through June 22.
Frequency:	Common.
Miscellany:	The Adirondacks' tallest conifer (and the tallest tree in the eastern United States) very rapidly-growing and reaching diameters of 52 to 55 inches at breast height, and heights of 120 to 165 feet in 300 to 350 years. See also page 244.
References:	Greason (February, 1986) Risley & Hastings (1961)
	Isherwood (1986) Sayles (1959)
	Miller and Krall (1969) Sievers (1955–1956)
	Pangburn (December, 1981)

* *Pinus sylvestris* L. SCOTS PINE or SCOTCH PINE

Site:	The species has become naturalized and thus will reproduce on its own provided that there is adequate light for this shade-intolerant. Well-drained. Atop Jenkins Mountain at elevation 2500 feet, the source of these Scots pines is the vast plantations to the north around Follensby Junior and Mountain Ponds. Winds must have carried the seeds southward or southeast-ward for a mile or more up to the summit of Jenkins.
Miscellany:	Introduced from northern Europe and Asia around the turn-of-the-century and used primarily for reforestation in burned-over, sandy-soiled areas. Secondary use is as an ornamental. Naturalization is frequent along the edges of plantations, especially along roadsides, and in gravel and sand pits.
Reference:	Faber (1980)

Tsuga canadensis (L.) Carr. EASTERN HEMLOCK

Site:	Our most shade-tolerant conifer and characteristic of the Northern hardwoods forest rather than of the spruce-fir. Rare above 2500 feet. Well-drained to poorly-drained till or outwash. In those areas of the northern hardwood forest where the soils are shallow over bedrock ledges, Hemlock often dominates where Sugar maple and beech cannot; Hemlock is often accompanied here by Yellow birch, Red spruce, Fir, and Mountain maple.

Hemlock seems to reproduce better not in its own shade but rather under the shade of the hardwoods. Because of the dense shade under Hemlock and acid infertile humus, ground cover species are few in number and ground cover sparse to absent. A diversity of spring wildflowers cannot grow under a stand of

	Hemlock. Seeds like those of Yellow birch, Red spruce, and Fir are small and germinate best not on hardwood leaf litter, but on raised portions of the forest floor such as on mossy moist stumps, logs, and boulders. pH of humus: n=29; min. 3.2; max. 6.8; median 4.5. Soils: limestone along Lake Champlain to gneisses and anorthosite of the Adirondacks. Sandstone and shale in Catskills.
Elevation:	From 100 feet along Lake Champlain to 2511 and 2752 feet on Crane Mountain.
Frequency:	Common.
Miscellany:	Hemlock is slow-growing and long-lived, reaching diameters of 4 feet and ages of 400 years.
References:	Faust (1978) Simmons (1971)

12. CUPRESSACEAE (CEDAR or CYPRESS FAMILY)

Juniperus communis L. PASTURE JUNIPER or COMMON JUNIPER

Site:	Shrub forming dense stands in old fields and pastures; also on exposed, open, shallow soils adjacent to bedrock ledges below about 2500 feet elevation. Shade-intolerant. Well-drained to exceedingly well-drained. Locally common but often miles apart in stations. (Painful to walk through.) Soils: limestone soils of Champlain Valley to anorthosite and gneisses of the Adirondacks.
Elevation:	From 100 feet along Lake Champlain to ca. 5300 feet on Mount Marcy.
Frequency:	17 stations.
Reference:	Livingston (1972)

Juniperus virginiana L. EASTERN RED CEDAR or SAVIN
- Site: Most abundant in the Champlain Valley where it is at the northern end of its range but migrating also up the East Branch Ausable Valley to about Keene. Shade-intolerant and well-drained. Occasionally with its dwarf cousin, *J. communis*. A common pioneer of abandoned fields. Limestone soils along Lake Champlain to gneiss of Adirondacks, and the shales and sandstones of the Schoharie Valley.
- Elevation: 100 feet along Lake Champlain to 1700 feet on East Hill, Keene.
- Frequency: Only two stations above 1000 feet and those both in Keene.
- References: Hirt & Silverborg (1963)
  Livingston (1972)

Thuja occidentalis L. NORTHERN WHITE CEDAR or ARBOR VITAE
- Site: Most abundant in swamps and along streams and lake shores, although a high water table is NOT essential for growth. Attains elevations at least as high as 4270 feet but, then, where water is flowing over the bedrock; here, organic acids can be washed away (pH higher). In places where the water movement is purely vertical, organic acids accumulate, and the pH is lower. Note the deer-browse lines (no foliage within about 5 feet of the ice) at pond and lake edges; this deer favorite is high in Calcium. Shade-tolerant. Well to very poorly-drained sites. Soils: limestone along Lake Champlain to anorthosite and gneiss of the Adirondacks.
- Elevation: From 100 feet along Lake Champlain to 4270 on Wright Peak. (Ketchledge reports it almost to the summit of Algonquin over 5000 feet.)
- Frequency: Common.
- Miscellany: This species grows very slowly—about as slowly as Hemlock. Increment cores show an average of 12 inches increase in diameter per century, so that trees 30 inches in diameter at breast height (4½ feet above ground) and 60 feet high can be 250 years old. One stunted tree on Picnic Point on Campus, its roots above lake level on a ledge (and thus subject to drought stress) was 5 inches in diameter and 105 years old.
- References: Drahos (1958–1959)
  Musselman et al. (1975)

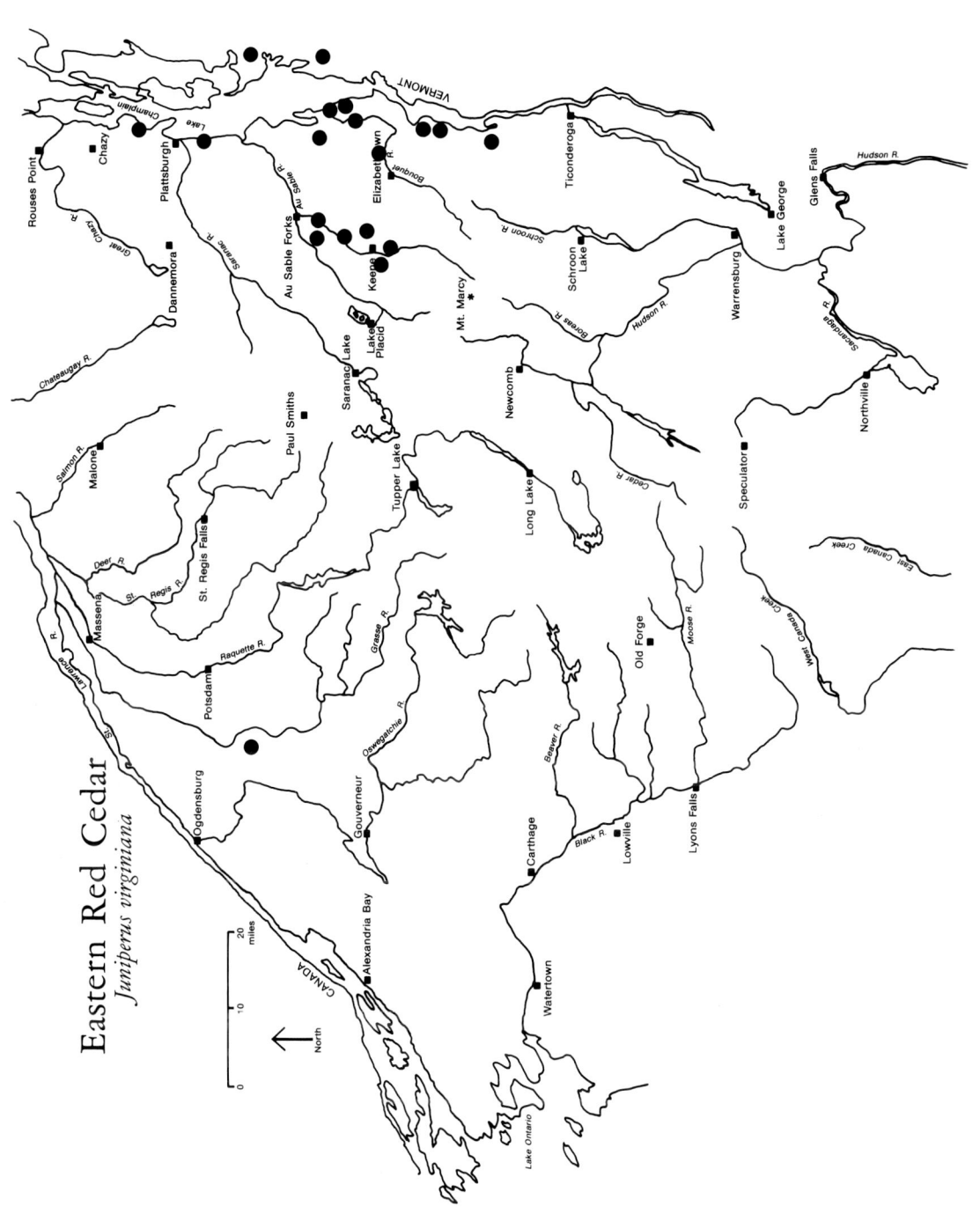

DICOTYLEDONEAE: THE DICOTS

13. NYMPHAEACEAE (WATER LILY FAMILY)
Reference on Water Lily family: Mitchell & Beal (1979)

Nuphar luteum (L.) Sibth. & Sm. YELLOW POND LILY, SPATTERDOCK or BULL-HEAD LILY
    We have two subspecies:

ssp. *macrophyllum* (Small) Beal or *Nuphar advena* (Ait.) R.Br. ex Ait.
- Site: This is the broad-leaved subspecies, an aquatic rooted in shallow water.
- Elevation: From 1500 feet along the Saranac River to 2780 feet in Bartlett Pond on the east slopes of Mount McKenzie.
- Phenology: In flower between June 5 and August 19.
- Frequency: Common.
- Miscellany: Rhizome can be used as a potato substitute.

ssp. *pumilum* (Timm.) Beal or *Nuphar microphyllum* (Pers.) Fernald
- Site: This is the small pond lily, in the Osgood River about 2 miles below White's Pine Camp, where the stream turns from north to west.
- Elevation: 1650 feet.
- Phenology: In flower July 19.
- Frequency: One station.
- Miscellany: Flowers float on surface.

Nymphaea odorata Dryand. ex Ait. WHITE WATER LILY or FRAGRANT WATER-LILY
- Site: Less common than *Nuphar advena* but in similar situations. Rooted in shallow water.
- Elevation: 1500 to 2008 feet.
- Phenology: In flower July 18 to August 21.
- Frequency: Frequent.

14. CAMBOMBACEAE (WATER-SHIELD FAMILY)

Brasenia schreberi Gmel. or *B. purpurea* or *B. peltata* in Peck (1899) WATER SHIELD
- Site: Rooted in shallow water.
- Elevation: From 1500 to 1700 feet.
- Frequency: Occasional.
- Reference: Mitchell & Beal (1979)

15. CERATOPHYLLACEAE (HORNWORT FAMILY)
Reference: Mitchell & Beal (1979)

Ceratophyllum demersum L. COONTAIL or HORNWORT
- Site: Observed by Dr. Alfred Ernest Schuyler in Lower Saint Regis Lake.
- Elevation: 1619 feet.
- Phenology: On neither June 18 nor August 17, 1986 were either flowers or fruits present.

16. RANUNCULACEAE (BUTTERCUP OR CROWFOOT FAMILY)
Reference on the family as a whole:
Mitchell & Dean (1982)

Actaea pachypoda Ell. or *Actaea alba* of American authors, not (L.) Mill. WHITE BANEBERRY or DOLL'S EYES
- Site: Shade tolerant, well-drained, usually under northern hardwoods.
- Elevation: From about 1500 feet in the Saranac Lake area to 2690 feet on Cascade Mountain. Probably occurs much lower than 1500 feet as well.
- Phenology: In flower June 6.
  In fruit August 8 to 25.
- Frequency: 13 stations.
- Miscellany: Fruits poisonous. On Baxter Mountain in Keene Valley, and on Baldface Mountain some fruits were pink, not white.
- Reference: Schottman (July-Aug., 1990)

Actaea spicata L. ssp. *rubra* (Ait.) Hulten or *Actaea rubra* (Ait.) Willd. RED BANEBERRY
- Site: Shade tolerant, well-drained, under northern hardwoods, probably requiring humus with a pH higher than that supporting *A. pachypoda*. Scarce in the Paul Smiths area, but more common toward the eastern Adirondacks.
- Elevation: From 100 feet on Four Brothers Islands in Lake Champlain to 2500 feet above John's Brook Loj.
- Phenology: In fruit between July 9 and August 17.
- Frequency: 13 stations.
- Miscellany: Fruits poisonous.
- Reference: Schottman (July-Aug., 1990)

Anemone quinquefolia L. WOOD ANEMONE, WINDFLOWER or FIVE-LEAVED ANEMONE
- Site: Well-drained sites, mostly under northern hardwoods.
- Elevation: 1550 feet.

Phenology: A vernal, the leaves wither with hardwood leaf maturation. In flower May 26, 1985 in the Raquette Falls area (between Long Lake and Axton).
Frequency: One station.

Aquilegia canadensis L. COLUMBINE
Site: Most often growing on bedrock ledges, wet or dry, ranging from sandstones, anorthosite, metasedimentaries, to limestone.
Elevation: 200 feet at Long Sault Dam on the Saint Lawrence River to 2700 feet on Pitchoff Mountain.
Phenology: In flower May 12, 1985, at Altona, elevation ca. 700 feet.
Frequency: Six stations.

Caltha palustris L. MARSH-MARIGOLD or COWSLIP
Site: Poorly-drained areas, usually in the sun, such as swamps, marshes, and edges of streams.
Elevation: From 200 feet at Alice Lake near Chazy and in Plattsburgh to 1630 feet at Brighton Town Hall.
Phenology: In flower, earliest date, April 24.
In flower, median date, May 17.
In flower, latest date, mid-June.
In fruit June 26.
Frequency: Nine stations, but probably many more.
Miscellany: Sap an irritant. Some people eat the young leaves (before flowering) as salad greens.

Clematis virginiana L. VIRGIN'S BOWER
Site: Poorly-drained, sunny areas such as edges of meandering streams and in swamps. Not common on the Upland.
Elevation: From 210 feet at the Chazy Fir Swamp to 2073 feet on Crane Mountain.
Phenology: In flower July 19.
In fruit September and October.
Frequency: 11 stations.

Coptis trifolia (L.) Salisb. or *C. groenlandica* (Oeder) Fern. GOLDTHREAD
Site: A most common, shade-tolerant species on well to poorly-drained sites, often with *Sphagnum* in black spruce swamps, but also under mixed conifer-hardwood stands and northern hardwoods. To timberline under balsam krummholz. Indicates acid humus. pH of humus: n=14; min. 4.0; max. 5.0; median 4.6. The new trifoliate leaves uncoil as a fern does each spring, replacing the old evergreen leaves of the previous year. Name from the golden yellow roots. See section on evergreen herb

	strategy, page 58.
Elevation:	From 210 feet at Chazy Fir Swamp to Haystack Mountain 4940 feet.
Phenology:	In flower, earliest date, May 12.
	In flower, median date, May 20.
	In flower at elevations above 4000 feet June 14 to 30.
	In fruit August.
Frequency:	Common.
Reference:	Schottman (1985, July).

Hepatica nobilis Mill. var. *acuta* (Pursh) Steyerm. or *Hepatica acutiloba* DC. with sharp-lobed leaves. HEPATICA or LIVERLEAF

Site:	Well-drained, shade-tolerant, under rich-site northern hardwoods, often under basswood, sugar maple, white ash, American elm, and hop hornbeam. On tills derived from metasedimentary rock, often with calcite, and from limestone. Higher humus pH assumed.
Elevation:	From 100 feet at Point Au Roche and 200 feet at Barnhardt Island to 2190 feet on Brewster Mountain.
Phenology:	One of the earliest to flower in spring.
	Heaven Hill: April 13, Lake Placid.
	Barnhardt Island, Massena: April 16, beyond the Flora area.
Frequency:	Five stations.
Reference:	Smith, Forman, and Boyd (1989).

Ranunculus abortivus L. KIDNEYLEAF BUTTERCUP

Site:	Infrequent woodland species, mostly under sugar maple and rich-site hardwoods. Shade-tolerant. Well to imperfectly drained. Present at Camp Canaras on Upper Saranac Lake, without Sugar maple, on outwash soils.
Elevation:	From 210 feet at Chazy Fir Swamp to 2760 feet on Cascade Mountain.
Phenology:	In flower May 14 through 17.
Frequency:	Eight stations.

* *Ranunculus acris* L. TALL BUTTERCUP or COMMON BUTTERCUP

Site:	Native of Eurasia, but an abundant naturalized species sometimes penetrating far into native woods along logging roads and foot trails. Well to poorly-drained, but intolerant of shade.
Phenology:	In flower from June 11 through July.
Frequency:	Common.
Miscellany:	Sap an irritant.

Ranunculus recurvatus Poir. ex Lam. HOOKED BUTTERCUP

Site: Northern hardwoods often under sugar maple, in more imperfectly drained areas and along rills, not common. The most shade-tolerant *Ranunculus* along with *R. abortivus*.
Elevation: From about 1700 feet at Paul Smiths to 2360 feet on old Algonquin Trail.
Phenology: In flower June 14.
In fruit June 15 through July 20.
Frequency: 8 stations.

Ranunculus trichophyllus Chaix ex Vill. or *R. aquatilis* L. var. *capillaceus* (Thuill.) DC WHITE WATER CROWFOOT
Site: An aquatic with flowers floating on the surface, but leaves dissected and submerged.
Elevation: North Branch Saranac River above Rainbow Lake, 1670 feet.
Phenology: Flowers September 9.
Frequency: One station.

Thalictrum pubescens Pursh or *Thalictrum polygamum* Muhl. TALL MEADOW RUE
Site: In open, sunny, poorly-drained areas such as swamps, marshes, and edges of meandering streams. A real northern species, even more abundant in swamps in Canada. Abundant under speckled alder and with *Calamagrostis* grass.
Elevation: 100 feet at Point Au Roche to 3089 feet on Lyon Mountain.
Phenology: In flower, Paul Smith's College Sugarbush, June 26.
Fruit already dropped by October 17.

17. BERBERIDACEAE (BARBERRY FAMILY)
Reference: Mitchell (1983)

Caulophyllum thalictroides (L.) Michx. BLUE COHOSH
Site: One of the indicators of highest humus pH, most often growing under Sugar maple-White ash-Basswood. Shade-tolerant. Well-drained to imperfectly-drained. pH of humus: n=7; min. 4.2; max. 6.6; median 6.2.
Elevation: From 200 feet at Barnhardt Island near Massena to 2100 feet on Brewster Mountain.
Phenology: In flower May.
In fruit September.
Leaves first emerging April 11 through 16 near Massena.
Frequency: 15 stations.
Miscellany: Fruit wall breaks down as the seeds mature so that the blue ripe seeds are borne naked as a gymnosperm's.

Reference: Schottman (May 1989).

18. PAPAVERACEAE (POPPY FAMILY)
Reference: Mitchell (1983)

* *Chelidonium majus* L. CELANDINE
   Site: Naturalized from Europe. Saranac Lake.

19. FUMARIACEAE (FUMITORY FAMILY
Reference: Mitchell (1983)

Corydalis sempervirens (L.) Pers. PALE CORYDALIS or PINK CORYDALIS
- Site: Well-drained, on exposed bedrock ledges. Shade-intolerant.
- Elevation: From 588 feet at Willsboro's Camp Pok-O-MacCready to 2380 feet on Mount Baker.
- Phenology: In flower June through early September.
  In fruit late August into September.
- Frequency: 7 stations.

Dicentra canadensis (Goldie) Walp. SQUIRREL CORN
- Site: Vernal of well-drained northern hardwood sites, mostly under Sugar maple. pH of humus: n=5; min. 4.4; max. 6.3; median 5.1.
- Elevation: From ca. 1700 feet to Buck Hill 2157 feet. Probably to much lower elevation.
- Phenology: The strategy here, as with its cousin, Dutchman's breeches, is to leaf out and flower early while there is still light on the forest floor; once the hardwood leaves are fully expanded, Squirrel corn's and Dutchman's breeches' leaves wither—usually in June. Spring beauty and Trout lily have the same strategy.
  In flower, earliest date, April 19.
  In flower, median date, May 5.
  In flower, latest date, May 21.
- Frequency: Six stations.
- Reference: Schemske (1978).
  Schottman (1990).

Dicentra cucullaria (L.) Bernh. DUTCHMAN'S BREECHES
- Site: Very similar in its ecology to *D. canadensis*, and often growing with its cousin. pH of humus: n=5; min. 4.4; max. 6.6; median 5.0.
- Elevation: From 200 feet at the Chazy to Buck Hill summit 2157 feet.
- Phenology: Median date of flower buds, April 23.
  In flower, earliest date, April 26.

In flower, median date, May 5.
  In flower, latest date, May 17.
  In fruit, May 29.
Frequency: Seven stations.
References: Macior (1970).
  Schemske (1978).
  Schottman (May, 1990).

20. HAMAMELIDACEAE (WITCH-HAZEL FAMILY)

Hamamelis virginiana L. WITCH HAZEL
Site: Shrub of well-drained sites, tolerant of shade, primarily a plant of more southerly oak forests. Common along the Lake Champlain Valley and its tributaries such as the lower Ausable, not venturing into the central Adirondacks. Note the Saint Lawrence Valley station near Colton on the map. Soils: from limestone soils along Lake Champlain to Adirondack gneisses and anorthosite to Catskills shales and sandstones.
Elevation: From 100 feet at Plattsburgh along Lake Champlain to 1134 feet in Keene Valley, and 1980 feet on Prospect Mountain.
Phenology: In flower October 2 through 12.
  This is the only woody plant which flowers autumnally; the capsules require a full year to ripen. Then the seeds are shot out explosively with considerable pressure to a distance of several yards.
Frequency: Two stations: one in Keene Valley and the other in Wilmington.
References: Mitchell (1988)
  Ryan (1986)

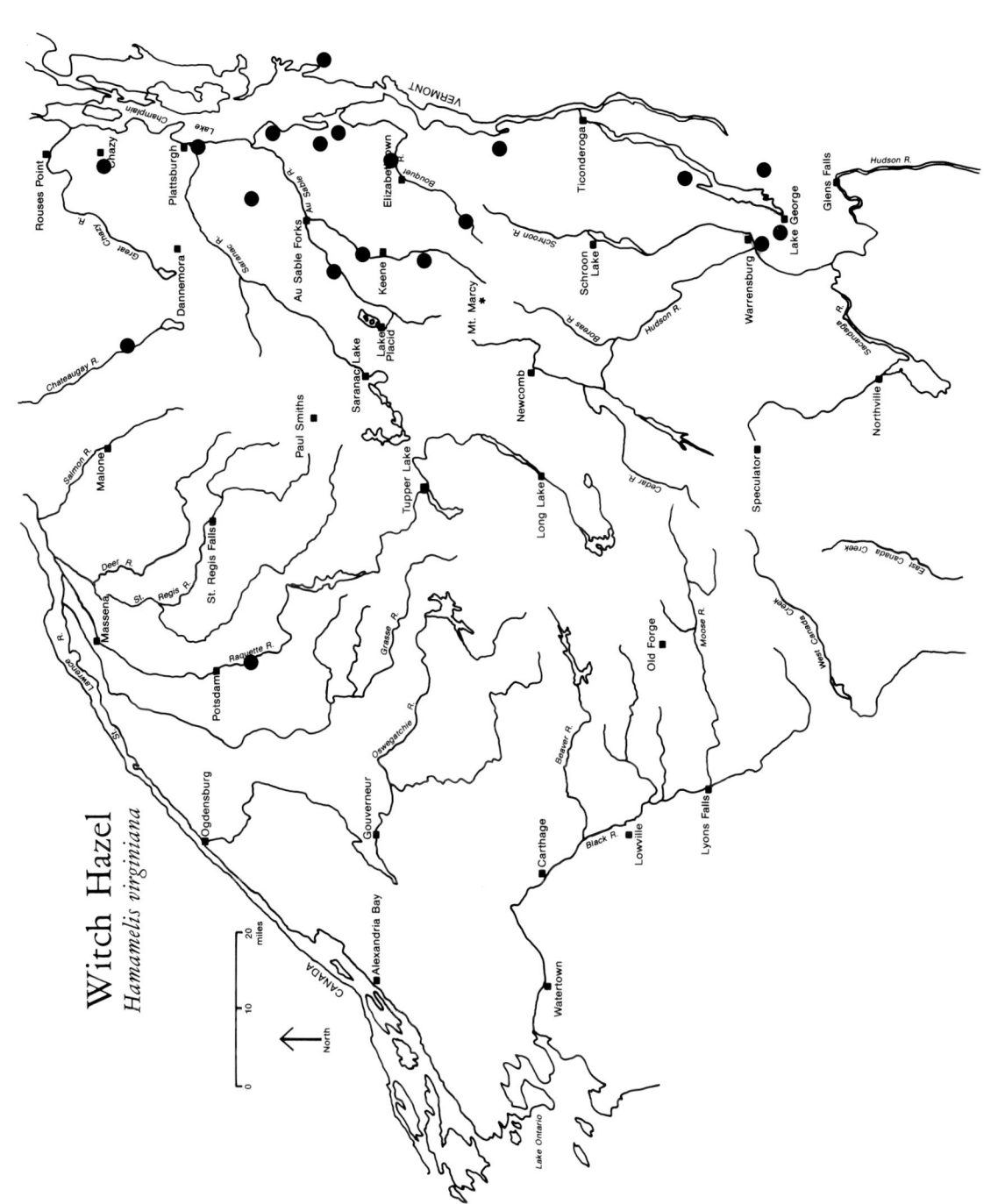

21. ULMACEAE (ELM FAMILY)

Ulmus americana L. AMERICAN ELM or WHITE ELM

Site: One of our more demanding species, probably equal to White ash, less so than Basswood, but more so than Sugar maple. Mid-tolerant of shade. Elm can grow from well-drained soils on hilltops to poorly-drained soils in swamps and floodplains. Elm leaves decay fast and are rich in calcium and other nutrients. Soils: limestone along Lake Champlain to Adirondack gneisses and anorthosite. (In the Catskills on shales and sandstones). Soil texture:

% Silt:
    B horizons: n=31; min. 10.0; max. 57.4; med. 29.0
    C horizons: n=10; min. 3.2; max. 50.0; med. 17.2

% Clay:
    B horizons: n=26; min. 1.4; max. 4.8; med. 2.5
    C horizons: n= 9; min. 0.6; max. 4.3; med. 1.9

pH of humus: n=31; min. 3.8; max. 6.8; med. 5.1. The 31 samples taken include sites where the elms were dead, as elm death is not caused by the soil. These sites once DID support elm.

Elevation: From 100 feet along Lake Champlain to 2600 feet (dead) on Sawyer Mountain near Indian Lake.

Phenology: Elm flowers early and fruits in late spring, as do Red and Silver maples, Poplars and Willows.

Frequency: 41 stations.

Miscellany: In the 1960s, the Dutch Elm disease took a large toll of Elms locally affecting not only native trees but also those that had been planted along Saranac Lake Village streets. In the latter case, abundant saplings offer hope that the Village once again will have its Elms.

References: On Elms in general:
Ketchledge (1964, April-May)
Miller and Allen (1972)
Mitchell (1988)
Northeastern Forest Expt. Station (1970)

On Dutch Elm Disease:
Bishop (1970)   Pack (1935)
Lanier (1975)   Pellettieri (1968)
Moore (1965)   Sullivan (1968) & (1975)

22. URTICACEAE (NETTLE FAMILY)
Reference: Mitchell (1988)

Laportea canadensis (L.) Wedd. NETTLE
- Site: On higher humus pH sites, especially in springs and on sites where ground water percolates downward. Uncommon in the flora area, but becoming increasingly more common at lower elevations and eastward. Most frequently under Sugar maple, but also in the Spruce-fir forest on Cascade Mountain and on Snowy Mountain.
- Elevation: From Camp Pok-O-MacCready at Willsboro, 588 feet, to ca. 3000 feet on Snowy Mountain and 3650 feet on Cascade Mountain.
- Phenology: In flower August 8.
- Frequency: 21 stations.
- Reference: Schottman (1987, August).

Pilea pumila (L.) Gray CLEARWEED or CLEARSTEM
- Site: Two sites on Mount Pisgah. Wet cliffs and waterfalls.
- Elevation: From 100 feet along Lake Champlain to about 1700 feet on Mount Pisgah.
- Phenology: In flower August 18, 1974, and September 22, 1973, on Mount Pisgah.

Urtica dioica L. ssp. *gracilis* (Ait.) Selander NETTLE
- Site: Well-drained sites, in partial shade to sun: thus, moderately shade-tolerant. Weedy, but native, often coming in on disturbed areas such as at old dumps and log decks.
- Elevation: From 100 feet along Lake Champlain to 1780 feet on Raquette Lake.
- Phenology: In immature fruit in Paul Smith's area August 26 through September 8. In fruit October 27 on Four Brothers Island, Lake Champlain.
- Frequency: Two stations.
- Reference: Schottman (1987, August).

23. JUGLANDACEAE (WALNUT FAMILY)

Juglans cinerea L. BUTTERNUT or WHITE WALNUT
- Site: Primarily a species of the Champlain Valley, but it has migrated up the East Branch Ausable River as far as Keene Valley. Well-drained. Shade intolerant to mid-tolerant, often found in abandoned rock quarries. Soils: from limestone on Lake Champlain to Adirondack gneisses and anorthosite. (In the Catskills, on shales and sandstones.)
- Elevation: From 100 feet along Lake Champlain to 1300 feet on Baxter Mountain in Keene Valley.
- Frequency: Two stations
- References: Greason (November-December, 1986).
  Ketchledge (1963, August-September).
  Mitchell (1988).

24. MYRICACEAE (THE WAX MYRTLE FAMILY)
Reference: Mitchell (1988)

Comptonia peregrina (L.) Coult. or *Myrica asplenifolia* L. SWEET FERN
- Site: An aromatic shrub of well to exceedingly well-drained sites in full sun (shade intolerant). Soils: limestone soils of Champlain Valley to Adirondack gneisses and anorthosite.
- Elevation: From 200 feet on Ausable River delta to 2300 feet on Buck Mountain on Lake George.
- Frequency: Four stations, but much more common nearer Lake Champlain.
- Miscellany: This species is sometimes planted along steep roadbanks to slow erosion.

Myrica gale L. SWEET GALE
- Site: Aromatic shrub of shorelines, marshes, and bogs. Intolerant of shade. One of the few woody plants that seems obligated to grow in high water table sites. I've not yet seen it high and dry.
- Elevation: 588 feet at Long Lake, Camp Pok-O-MacCready, Willsboro to 1850 feet along the Chubb River.
- Phenology: In flower May 4 to May 10.
- Frequency: Common.
- Miscellany: It can be recognized from a canoe by its size (2 to 3 feet tall), its color (a distinct blue-green) and its position (usually along the edge of the stream). The whole genus and family are aromatic; Bayberry is a cousin.

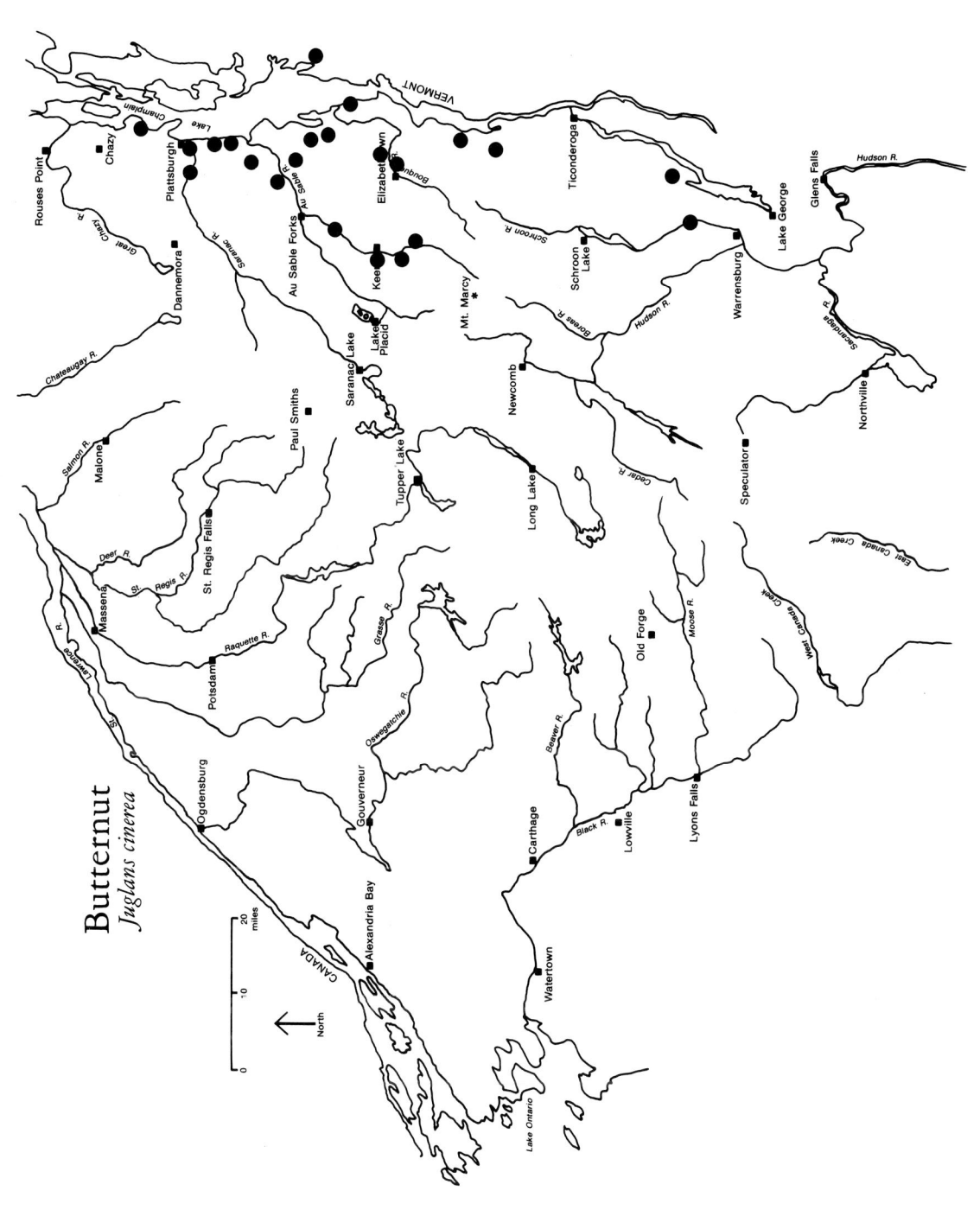

25. FAGACEAE (BEECH FAMILY)

***Fagus grandifolia* Ehrh. AMERICAN BEECH**
- Site: The second dominant of the Northern hardwoods forest, equally as tolerant of shade as Sugar maple. Less demanding than Sugar maple, however, and thus occurring on poorer (e.g., sandier and/or shallower) sites. Beech litter decays very slowly and is lacking in nutrients (almost pure carbohydrate); this in conjunction with dense shade creates scarce ground cover. A diversity of spring wildflowers cannot grow under a stand of Beech. Well-drained sites only. pH of humus: n=81; min. 3.2; max. 6.8; median 4.65. Soils: limestone of Champlain Valley to Adirondack gneisses and anorthosite. (In the Catskills, on shales and sandstones).
- Elevation: 100 feet along Lake Champlain to 3470 feet on Snowy Mountain. Uncommon above 2500 feet.
- Phenology: Male flowers falling after pollination on Campus, 5/30/88.
- Miscellany: Beech is presently suffering from a combination fungus-scale insect disease which appears to attack mainly mature trees; in 1975 the disease had spread as close to Paul Smiths as Keene Valley, and in 1978 to the Old Forge area. Now it is well-established locally (see references below). Important food source for squirrels, chipmunks and other mast-eating mammals and birds. Nuts edible to humans too. Are the semicircular cotyledons, appearing in early May when the seeds germinate, edible?

   Beech frequently reproduces vegetatively by root sprouts, so that a large number of what look like saplings established from seed may not be at all! In fact, much of the "reproduction" may not be sexual reproduction but an asexual clone.

   The herb, Beech drops (*Epifagus virginiana*) is a root or mycorrhizal fungus parasite.
- References: Buzzard (1970) Tirrito (1989)
  Simmons (1971) Webster (1975)
  Stock (1983) on the Beech Bark disease

***Quercus rubra* L. RED OAK or NORTHERN RED OAK**
- Site: The bulk of the distribution of red oak in the Adirondacks is mostly limited to elevations below 1000 feet, following up the main river valleys such as the East Branch Ausable to Keene Valley, the West Branch Ausable to High Falls Gorge, the Schroon to Schroon Lake, and the Sacandaga to Wells. On the Upland above 1000 feet, red oak is rare as a native; several sites have been found and the first three listed have been studied. These locations are tens of miles apart, with no native oak

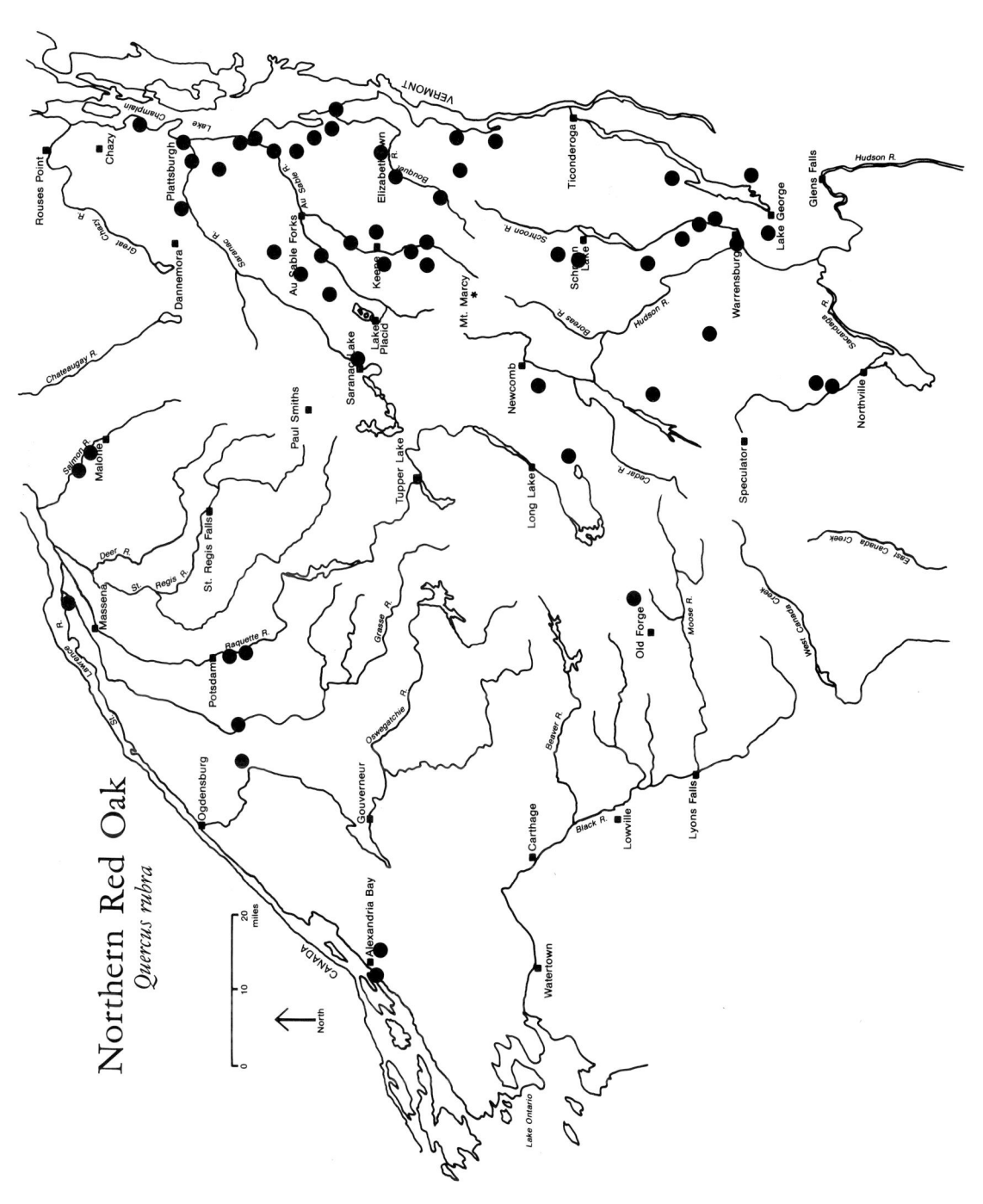

between:
1. Mount Baker just east of Saranac Lake
2. the southwest base of Blue Mount at Blue Mountain Lake
3. Catamount Mountain southwest of Silver Lake
4. Goodnow Mountain, Newcomb
5. Chimney Mountain, southeast of Indian Lake
6. Limekiln Lake area, east of Inlet
7. Crane Mountain, but this station might be a part of the larger Hudson-Schroon Valleys population.

What do all these sites have in common? Several things:

1. Many are on southwest aspects of elevation to 1900 to 2200 feet, with steep slopes of 10 to 30 degrees.

2. All have been disturbed, many burned and others either heavily logged or clearcut. This favors the moderately shade-tolerant oak. Many of the older oaks are multi-stemmed, suggesting stump sprouting following logging and/or fire.

3. Often, the oak is accompanied by Hop hornbeam, White ash, American elm, and/or Basswood. The higher-pH ground cover plants are often present. The tills are well-drained, stony, and bouldery, with frequent outcrops. Deeper pockets of till between the outcrops support the forest.

How red oak migrated to these isolated sites had been a puzzle until an article by Johnson & Adkisson (1986) entitled *Airlifting the Oaks* solved the mystery. Blue jays! These are among the best and fastest oak distributors, caching many acorns but consuming only some of them. Blue jays also transport beech and chestnuts. Johnson & Adkisson calculate that oaks advanced northward about 10,000 years ago through the eastern and central United States at an average rate of 380 yards per year. This equals 5 years per mile. Leaves, like beech, are slow to decay and are of little nutrient value to the humus. Soils: limestone soils along Lake Champlain to Adirondack gneisses and anorthosite. Potsdam sandstone at Cadyville. (In the Catskills, on sandstones and shales).

Elevation: From Lake Champlain at 100 feet to 2540 feet on the Crows in Keene, and one sapling on Catamount Mountain at 2760 feet.

Phenology: In half-leaf in Keene May 29, 1983; this seems later than average.

Frequency: Four stations in the Flora area: Mount Baker, Catamount Mountain, Keene Valley, and Wilmington.

Miscellany: Acorns bitter, inedible to people. Leaves poisonous to eat.

	Planted as an ornamental in Saranac Lake and becoming naturalized in the Village.
References:	Flint (1972)
	Headstrom (1958)
	Johnson & Adkisson (1986)
	Ketchledge (1963, February-March)

26. BETULACEAE (BIRCH FAMILY)

Alnus incana (L.) Moench ssp. *rugosa* (DuRoi) Clausen *or Alnus rugosa* (DuRoi) Spreng. SPECKLED ALDER or TAG ALDER

Site:	Profusely abundant in swamps forming dense thickets to 25 feet high, especially on flats along slow, meandering streams. This is one woody plant that appears to require a high water table (poorly-drained sites) at least for part of the year. (Once during a dry summer while visiting some people who were planning to construct a house with alders growing on the proposed sewage site, I said, "Don't," as drainage is poor in the spring.)
	In times when the rivers are not in flood, the ground surface upon which the alders stand is 12 to 24 inches above the river level. About 1982, beavers dammed Weller Brook near the Paul Smith's College Campus and raised the water level between 1 and 2 feet. By September, 1985, most of the alders were dead. Speckled alder, hence, must require a very fine-line balance of water table height - not too low, not too high! Shade-intolerant. Roots with nitrogen-fixing actinomycete nodules look like tiny cauliflowers or broccoli, but orange. Soils: limestones along Lake Champlain to Adirondack gneisses and anorthosite. (In the Catskills, on sandstones and shales).
Elevation:	100 feet along Lake Champlain at Wickham Marsh to 2780 feet at Bartlett Pond on MacKenzie Mountain.
Phenology:	This is the first plant to shed pollen in the spring.
	The earliest recorded by this author is March 31, 1977. In 1988, some plants were shedding pollen on April 10 along the Chubb River.
	In a "late" spring, no pollen was shed until April 29, 1989.
Frequency:	Common.

Alnus viridis (Chaix) DC ssp. *crispa* (Dryand. ex Ait.) Turrill or *Alnus crispa* (Dryand. ex Ait.) Pursh GREEN ALDER or MOUNTAIN ALDER

Site:	Well-drained to poorly-drained but shade-intolerant, often forming dense thickets.
Elevation:	900 feet at the base of Pok-O-Moonshine Mountain and 1120 feet at the Flume in Wilmington to 5280 feet on Mount Marcy.
Frequency:	15 stations below 4000 feet.
Miscellany:	This shrub, super-abundant in the Canadian Maritime Provinces and common in Maine, is at the extreme south western end of its natural range in the eastern U.S. Soper and Heimburger (1982) offer a range map (p. 96) for this species; it extends northward and westward from the headwaters of the Ottawa River to Lake Superior and Hudson Bay. Fernald (p. 538) mentions an outlier in western North Carolina! It occurs only in the High Peaks area and does not venture westward onto the Upland around Saranac Lake or Paul Smiths. The western boundary can be traced on a map from Catamount Mountain (E of Franklin Falls) to Whiteface Mountain's west flank, and to Mts. Wright and Algonquin. There is no edaphic (soils), no climatic, nor disturbance history to explain this western "front". Is the species advancing westward, retreating to New England, or is the "western front" holding steady?
Reference:	Soper & Heimburger, 1982.

Betula alleghaniensis Britt. or *B. lutea* Michx. f. YELLOW BIRCH or SILVER BIRCH

Site:	The third dominant of the northern hardwoods, although less shade-tolerant (only mid-tolerant) than Sugar maple and Beech. Well to poorly-drained sites (Sugar maple and Beech do not occur in the poorly). Can exist on talus, shallow soils, over bedrock ledges or on boulders. Often dominant with Mountain maple in the transition between northern hardwoods and spruce-fir, between ca. 2800 and 3000 feet. Seeds germinate with difficulty on hardwood litter and thus do best on mossy logs, stumps, and boulder as does Hemlock. Hence, Yellow birch seedlings do not compete with Beech and Sugar maple until the plants are sapling size. Soils: limestone along Lake Champlain to Adirondack gneisses and anorthosite. (In the Catskills, on sandstones, shales, and conglomerates). pH of humus: n=59; min. 3.2; max. 6.6; median 4.6.
Elevation:	100 feet along Lake Champlain to 3413 feet on Whiteface Mountain. Uncommon above 3000 feet.
Frequency:	Common.

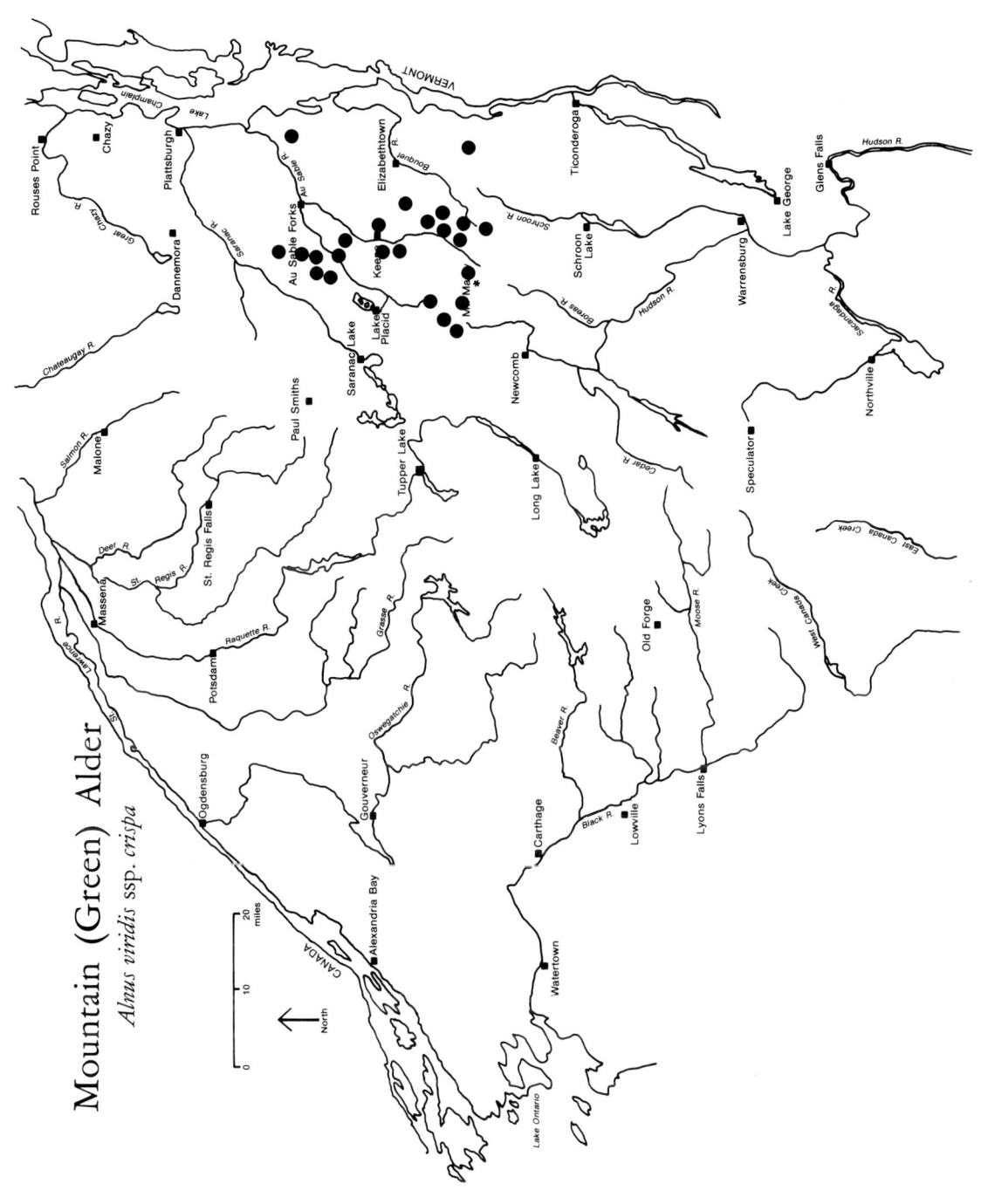

Miscellany:	Reaching the greatest size and age for any Birch in New York: to four feet in diameter and over 250 years.
Reference:	Ketchledge (1962, October-November) for all NY Birches.

Betula pumila L. SWAMP BIRCH

Site:	A large population of this northern birch exists in Spring Pond Bog, in the Boreal Heritage Preserve west of Derrick. Here, the shrub reaches heights of ten feet and grows in *Sphagnum*. The second station in the Flora area was about 1½ miles west of the Brighton-Santa Clara Town line along the Keese Mills Road; the large shrub once growing here appeared to have been naturally seeded. Could the seed source have been at Spring Pond Bog, 8½ miles distant by air? The shrub was removed in the spring of 1990 during the widening of the Keese's Mills Road. There are two reports of it in Saint Lawrence County, west of Madrid and near Lake Bonaparte, beyond our Flora area.
Elevation:	1550 feet.
Frequency:	This species is listed as rare in New York by Mitchell (1986) and Birmingham (1988). We now have one station in the Flora area.
Reference:	Ketchledge (1982, September).

Betula papyrifera Marsh. PAPER BIRCH, WHITE BIRCH or CANOE BIRCH

Site:	Commonest in disturbed areas, especially in the spruce-fir above 2500 feet where it depends on blowdown and canopy openings for reproduction. Intolerant of shade. Well to imperfectly drained. Similar to Yellow birch in its seed germination habits. To timberline with balsam fir. Abundant at lower elevations following burns, abandoned fields, blowdowns, clearcuts, and other disturbances. Soils: limestone along Lake Champlain to Adirondack gneisses and anorthosite. (In the Catskills, on sandstone, conglomerate, and shale).
Elevation:	100 feet along Lake Champlain to 5320 feet on Mount Marcy.
Frequency:	Common.
Miscellany:	Var. *cordifolia* (Regel) Fern. is considered by some to be a different species, but many consider it only the heart-leaved, red-barked, high-elevation variety of Paper birch. I am not sure that I'd recognize it even as a variety; I do not distinguish it from the Paper birch that we have around Campus. Those trees which have pioneered after the 1903 and 1908 fires are approaching maturity now; they are about to be replaced by the shade-tolerant species.

Reference: Miller and Krall (1970).

Betula populifolia Marsh. GRAY BIRCH
- Site: The Flora area's second white-barked birch, slenderer, shorter-lived, and smaller than Paper. Occasional in quarries, gravel pits, along roadsides and railroads, and in other disturbed places, often with Balsam poplar. Well-drained. Shade-intolerant. Very common in the Saint Lawrence Valley in the Nicholville-Potsdam area. This is a more southerly, lower-elevation birch than is Paper, being most abundant along the coastal plain from Nova Scotia to Delaware and "weedy". Soils: limestone along Lake Champlain to Adirondack gneisses and anorthosite. (In the Catskills, on shale and sandstone).
- Elevation: 100 feet along Lake Champlain at Valcour to 2261 feet on Cat Mountain south of Cranberry Lake.
- Frequency: Occasional.
- Reference: Miller and Krall (1970).

Carpinus caroliniana Walt. AMERICAN HORNBEAM, HORNBEAM, MUSCLEWOOD, BLUE BEECH, WATER BEECH or IRONWOOD
- Site: A small, shade-tolerant understory tree of primarily poorly to imperfectly-drained sites, but can grow on the well-drained as well. Most common along river banks, flood plains, and in swamps. Native along the Champlain Valley with a "tongue" as viewed on a map up the East Branch Ausable Valley as far as Keene Valley. Soils: limestone along Champlain Valley to Adirondack gneisses. (In the Catskills, on shales and sandstones).
- Elevation: From 200 feet at Lake Alice (Chazy) to 1020 feet in Keene Valley.
- Phenology: In immature fruit June 19, above Wadhams.
- Miscellany: The very hard, dense wood surprisingly rots very rapidly. Trees that are dying will be totally gone within a decade - no trace of them left.

Corylus cornuta Marsh. BEAKED HAZELNUT
- Site: Shrub locally abundant but never common over large areas. One may push through a ton of it and not see it again until miles later. Well-drained. Frequently coming in under Pine plantations. Shade-tolerant. Soils: limestone in Champlain Valley to Adirondack gneisses and anorthosite. (In the Catskills, on sandstones and shales).
- Elevation: 100 feet along Lake Champlain in Plattsburgh to 2800 feet on Pitchoff Mountain.

Phenology: Pollen shed on April 17, 1977, on Whisker Hill, east of Tupper Lake.
Frequency: Frequent.
Miscellany: Nuts edible.
References: Slate (1961)
Tappeiner and John (1973)

Ostrya virginiana (Mill.) K. Koch HOP HORNBEAM, IRONWOOD or HARDHACK

Site: This species may be more demanding of soils than Sugar maple, but less so than White ash, American elm, and Basswood. Well-drained. Shade-tolerant understory tree often with Striped maple. pH of humus: n=41; min. 3.8; max. 6.8; median 5.5 Soils: limestone in Champlain Valley to Adirondack gneisses and anorthosite. (In the Catskills, on sandstone and shale). Soil texture:

% Silt:
B horizons: n=44; min. 6.7; max. 57.4; med. 30.5
C horizons: n=19; min. 2.1; max. 47.1; med. 11.9

% Clay:
B horizons: n=40; min. 0.9; max. 11.4; med. 2.8
C horizons: n=16; min. 0.6; max. 4.3; med. 2.0

Elevation: 100 feet along Lake Champlain to 2500 ca. on Little Porter Mountain.
Frequency: 57 stations.
Miscellany: The tree is apparently short-lived, as the individuals which began growing following the 1908 burn on Mount Baker had reached maturity in the 1970s; the largest, oldest ones died in the mid-1980s. The life-span must be close to that of people. The largest trees are about 1 foot in diameter and 60 feet tall, but such size as this is rare.

27. CHENOPODIACEAE (GOOSEFOOT FAMILY)

* *Chenopodium album* L. LAMB'S QUARTERS or WHITE GOOSEFOOT
Site: Common European weed around buildings on Campus and in Saranac Lake.
Phenology: In fruit October 27.
In flower, on Campus, August.
Miscellany: Used for salad greens.
Reference: Ives (1975).

* *Salicornia europea* L. GLASSWORT
Site: Scattered at several sites along the railroad right-of-way between

Gabriels and Lake Clear Junction. In cinders. Typically a salt-marsh species.
Phenology: In flower August 16, 1974.
Miscellany: European.

28. AMARANTHACEAE (AMARANTH FAMILY)

* *Amaranthus retroflexus* L. PIGWEED
- Site: European weed. Saranac Lake.
- Phenology: In fruit October 27, 1985 on Four Brother Island "C" in Lake Champlain.
  In immature fruit September 1975 at Hotel Saranac in Saranac Lake.

29. PORTULACACEAE (PURSLANE FAMILY)

Claytonia caroliniana Michx. WIDE-LEAVED SPRING BEAUTY or CAROLINA SPRING BEAUTY
- Site: Common vernal species on well-drained sites, mostly under northern hardwoods. pH of humus: n=5; min. 4.7; max. 5.6; median 5.1.
- Elevation: From 200 feet at Chazy to 2180 feet at Heart Lake.
- Phenology: One of the earliest plants to flower in spring, the leaves withering by early summer.
  Earliest date of flowering April 16.
  Median date of flowering (from 20 observations) May 3.
  Latest date of flowering May 21.
  Fruiting (with chipmunks digging up the rhizomes) May 28 and 29.
- Miscellany: The flowers have an unpleasant stench.

* *Portulaca oleracea* L. PURSLANE
- Miscellany: Common European naturalized weed.
- Reference: Zimmerman (1976).

30. MOLLUGINACEAE (CARPETWEED FAMILY)

* *Mollugo verticillata* L. INDIAN CHICKWEED or CARPETWEED
- Site: Paul Smiths and Saranac Lake, on sand and cinders.
- Miscellany: European prostrate weed.

31. CARYOPHYLLACEAE (PINK OR CARNATION FAMILY)

* *Dianthus deltoides* L. MAIDEN PINK

Site: Old golf course at Paul Smiths.
Phenology: In flower August 18, 1978.
Miscellany: Naturalized from Europe.

* *Saponaria officinalis* **L.** BOUNCING BET or SOAPWORT
Site: Gabriels and Saranac Lake.
Miscellany: European weed. Poisonous. Sap can be used to make soap.

* *Silene latifolia* **Poir.** or *Lychnis alba* **Mill.** EVENING LYCHNIS, EVENING CATCHFLY or WHITE CAMPION
Phenology: In flower, 1st of season, June 9, 1988 at Whitefathers, Lake Kushaqua.
Miscellany: Common European weed.
Reference: Schottman (July, 1989).

* *Silene vulgaris* **(Moench) Garcke** or *Silene cucubalus* **Wibel** BLADDER CAMPION
Phenology: In flower at Paul Smiths June 16.
Miscellany: Occasional European weed.
Reference: Schottman (July, 1989).

* *Spergula arvensis* **L.** SPURREY
Site: Saranac Lake and Gabriels.
Phenology: In flower July 13.
A few flowers still present on November 5, 1975 at Gabriels.

* *Spergularia rubra* **(L.) J. & C. Presl.** SAND SPURREY
Site: Paul Smith's College Campus Quadrangle and near Essex Hall.
Phenology: Section of grass-free quadrangle was pink with flowers on June 15, 1986.
Miscellany: European weed. Not observed in previous years, but present again in 1987.

* *Stellaria graminea* **L.** STITCHWORT or COMMON STITCHWORT
Site: Paul Smiths, McColloms, Saranac Lake, Gabriels.
Phenology: In flower June 22 and 24.
Miscellany: Occasional European weed.

* *Stellaria media* **(L.) Vill. (not Cyrillo)** CHICKWEED
Phenology: Flowering throughout the whole growing season.
Miscellany: Abundant European weed.

32. POLYGONACEAE (BUCKWHEAT FAMILY)
Reference on the whole Buckwheat Family: Mitchell & Dean (1978)

Polygonella articulata (L.) Meisn. JOINTWEED
- Site: Open, dry, sandy, burned or otherwise disturbed places. In railroad ballast, Gabriels to Lake Clear to Saranac Inn and Saranac Lake.
- Elevation: 1500 to 1700 feet.
- Phenology: In flower August 16 through 28.
- Frequency: Three stations.
- Miscellany: Native annual, nearly leafless when flowering.

Polygonum amphibium L. WATER SMARTWEED
Of two varieties:
1. var. *stipulaceum* Colem.
    - Site: This is an aquatic, floating native.
    - Frequency: 5 stations.
2. var. *emersum* Michx.
    - Site: This is a terrestrial, erect, several feet high, and a cosmopolitan (also native to Europe), found on the floodplain of Saranac River opposite Mount Pisgah.
- Elevation: From 1500 to 1700 feet for both varieties.
- Phenology: In flower August 19 through September 9.

* *Polygonum aviculare* L. KNOTGRASS
- Site: Abundant prostrate lawn and sidewalk crack weed. Often replaces grass on poor sections of lawns.
- Miscellany: From Europe.

Polygonum cilinode Michx. CLIMBING BUCKWHEAT
- Site: Our only native woodland *Polygonum*, common in openings and on rock ledges, and tolerating moderate shade. Well-drained. A vine, mostly in northern hardwood forests and becoming profuse after loggings when some sun penetrates the canopy; the growth rate must be quite rapid. Behind maintenance buildings on Campus temporarily after landfills, tolerating bare sand in full sun with nil organic matter. pH of humus: n=9; min. 4.2; max. 5.0; median 4.8. Wilting on July 24, 1985, after 8 days lacking rain, on 25-foot high ledge at extreme east end of The Tongue.
- Elevation: From 1500 to 2327 feet on Merwin Hill.
- Phenology: In flower, 1st of season, at Paul Smith's College Gym June 13, 1988.

  In fruit September 7.
- Frequency: Common.

* *Polygonum cuspidatum* Sieb. & Zucc. JAPANESE KNOTWEED
    Site: Oriental ornamental escape in Saranac Lake such as at Denny Park and in several other locations.
    Phenology: Grows exceedingly rapidly, several inches per day, reaching a thicket 10 feet high by end-of-summer.
                    In fruit October 1973 at Denny Park.

* *Polygonum persicaria* L. LADY'S THUMB
    Phenology: In flower from July through the end of August.
    Miscellany: Common European weed.

Polygonum sagittatum L. TEARTHUMB
    Site: Poorly-drained, open areas such as flood plains and marshes.
    Elevation: 900 feet at Pok-O-Moonshine to 1630 feet at Heron Marsh.
    Phenology: In flower September 7.
    Frequency: Two stations.

* *Rumex acetosella* L. SHEEP SORREL
    Site: Profuse European weed on lawns, old fields, and wasteplaces everywhere.
    Phenology: In flower from June 9 through early summer.
    Miscellany: Leaves good for salads, adding sourness.

* *Rumex crispus* L. CURLY DOCK
    Miscellany: European weed, common.

33. CLUSIACEAE or HYPERICACEAE or GUTTIFERAE (SAINT JOHN'S-WORT or MANGOSTEEN FAMILY)

Hypericum canadense L. CANADIAN SAINT JOHNSWORT
    Site: Edge of Bigelow Road, between D&H RR grade and Oregon Plains Road, near Bloomingdale. With *Hypericum ellipticum* in a moist spot.
    Elevation: 1560 feet.
    Phenology: In flower July 15.

Hypericum ellipticum Hook. SAINT JOHN'S-WORT
    Site: Poorly-drained sites.
    Elevation: 1500 to 1700 feet.
    Phenology: In flower July 15 through August 23.
                    In fruit September 7.
    Frequency: Four stations.

Hypericum mutilum L. SAINT JOHN'S-WORT
    Site: In wet spots.
    Elevation: From 700 feet at Upper Jay to 1710 feet at Onchiota.
    Frequency: Three stations.

* *Hypericum perforatum* L. SAINT JOHN'S-WORT or COMMON SAINT JOHN'S-WORT
    Site: The Flora area's most common species, an abundant European weed that has taken over fields, roadsides, and wasteplaces. All the native *Hypericum* require high water tables, but this one grows in well-drained sites; it shares the full-sun requirements of the native species, however.
    Elevation: To 2880 feet.
    Phenology: In flower July 11 through summer.
    Miscellany: Poisonous to animals.

Triadenum virginicum (L.) Raf. or *Hypericum virginianum* L. MARSH SAINT JOHN'S-WORT
    Site: In swamps, bogs, and marshes, and on stumps and logs emerging above the water level. Shade-intolerant.
    Elevation: From 1500 to 2000 feet in the local Flora area.
    Phenology: In flower early summer.
                In fruit August.
    Frequency: Common.

34. TILIACEAE (LINDEN FAMILY)

Tilia americana L. BASSWOOD or AMERICAN LINDEN
    Site: One of our most demanding trees, creating and indicating the highest humus pH. Well-drained soils. Moderately shade-tolerant. Inceptisols rather than Spodosols. High calcium requirement; leaves fall with abundant calcium and rot rapidly in autumn, forming a rich mull high-pH humus along with Sugar maple, Elm, and White ash. Soils: from limestone along Lake Champlain to Adirondack gneisses and anorthosite. (In the Catskills, on shales and sandstones). pH of humus: n=30; min. 4.1; max. 6.8; median 5.8. Soil Texture:
        % Silt
            B horizons: n=36; min. 9.4; max. 59.2; med. 31.4
            C horizons: n=10; min. 3.2; max. 50.0; med. 24.9
        % Clay
            B horizons: n=32; min. 0.9; max. 6.8; med. 2.8
            C horizons: n= 8; min. 0.6; max. 4.3; med. 1.6
    Elevation: From 100 feet along Lake Champlain to 2580 feet on Chimney Mountain.

Phenology: In flower June.
In fruit September-October.
Frequency: 28 stations.
Miscellany: This species commonly sprouts but rarely seeds.

35. MALVACEAE (MALLOW FAMILY)

*** *Malva moschata* L. (Malvaceae) MUSK MALLOW**
Site: European garden escape, becoming common at Paul Smiths.
Phenology: In fruit August 30.

36. SARRACENIACEAE (PITCHER-PLANT FAMILY)

Sarracenia purpurea L. PITCHER PLANT
Site: In bog mats. Insectivorous, with bacteria in the "pitchers" which digest insects' proteins as a nitrogen supplement in otherwise low-nitrogen (and generally low-nutrient) peat soils.
Elevation: From 1500 to 1770 feet.
Phenology: In flower summer.
Frequency: Protected in N.Y. State, but frequent in our area.
References: Argo (1964) Heslop-Harrison (1978)
Ashley & Gennaro (1971) Ketchledge (Oct.-Nov., 1983)
Buttrick (1980) Selkow (1976)
Doeffinger (1978) Stauffer (1972)
Eisner (1967) Zahl (1961)

37. DROSERACEAE (SUNDEW FAMILY)

Drosera intermedia Hayne SPATULATE-LEAVED SUNDEW
Site: With *Xyris*, on the *Sphagnum* mat along Bog Pond.
Elevation: From 1630 feet at Bog Pond to 1770 feet at Ferd's Bog.
Phenology: In flower August 4.
In fruit late August.
Frequency: Protected in N.Y. State.
Miscellany: Insectivorous.

Drosera rotundifolia L. ROUND-LEAVED SUNDEW
Site: Bog mats with Pitcher plant.
Elevation: 1500 to 1860 feet along shore of Lake Placid.
Phenology: In flower bud July 17.
In flower August 4.
Frequency: Protected in N.Y. State, but frequent in the Flora area.
References: Argo (1964)
Ashley & Gennaro (1971)   Hugo (1989)

Buttrick (1980)　　　　　Ketchledge (Oct.-Nov., 1983)
Doeffinger (1978)　　　　Selkow (1976)
Eisner (1967)　　　　　　Stauffer (1972)
Heslop-Harrison (1978)　　Zahl (1961)

38. CISTACEAE (ROCKROSE FAMILY)

Lechea maritima Leggett ex BSP. BEACH PINWEED
- Site: Black Brook sand plain, in old gravel pit along Guideboard Road near junction with Silver Lake Road.
- Elevation: 1040 feet.

39. VIOLACEAE (VIOLET FAMILY)
References on Violets in general:
Beattie & Lyons (1975)
Russell (1957)
Schottman (1986, June)

Viola canadensis L. CANADA VIOLET
- Site: Our most demanding violet species, requiring higher pHs under at least Sugar maple. Well-drained, often growing with *V. pubescens*, and under Basswood-White ash-Elm-Hop hornbeam. pH of humus: n=8; min. 4.7; max. 6.8; median 5.95. Shade-tolerant.
- Elevation: From 600 feet at Pierrepont to 2157 feet on Buck Hill.
- Phenology: In flower between May 12 and May 29.
  On occasion, flowers in the autumn. August 8 on Baldface Mountain.
- Frequency: Ten stations.
- Reference: Cook, R.E. (1983)

Viola conspersa Reichenb. AMERICAN DOG VIOLET
- Site: Found growing among the ties of the old railroad behind Saranac Lake's Will Rogers Hospital.
- Elevation: 1540 feet.
- Phenology: In flower June 4.
- Miscellany: Listed as *L. labradorica* Schrank in Peck with *V. canina* var. *muhlenbergii* Gray as a synonym.

Viola incognita-macloskeyi or *Viola incognita* Brainerd X *Viola macloskeyi* F.E. Lloyd ssp. *pallens* (Banks ex DC.) M.E. Baker SWEET WHITE VIOLET
- Site: Often in poorly-drained areas such as along brooks. pH of humus: n=7; min. 3.7; max. 5.6; median 4.2.

Elevation:	From 1100 feet on the Dickinson Esker to 2830 feet on McKenzie Mountain.
Phenology:	In flower, earliest date, April 18. In flower, median date, May 8. In flower, last date, June 4.
Frequency:	Common.
Miscellany:	Apparently, *V. incognita* and *V. macloskeyi* have either fully hybridized in the Flora area or are forms of the same species. Flowers with features of both species (e.g., bearded and beardless petals) exist in the same plant or clump. *Viola blanda* Willd. is a synonym for *V. macloskeyi* ssp. *pallens*.

Viola pubescens Ait. DOWNY YELLOW VIOLET

Site:	Mostly on richer sites (higher pH) under Sugar maple and often under Basswood, White ash, Elm, and Hop hornbeam. Well-drained. Shade-tolerant.
Elevation:	From 200 feet at Lake Alice near Chazy to 2157 feet on Buck Hill.
Phenology:	In flower from May 12 through June 2.
Frequency:	Four stations.

Viola rotundifolia Michx. ROUND-LEAVED VIOLET

Site:	Well-drained sites and shade-tolerant under northern hardwoods and/or conifers. pH of humus: n=6; min. 3.7; max. 5.6; median 4.45.
Elevation:	From between 1000 and 1200 feet in Keene Valley to 2200 feet on Pitchoff Mountain.
Phenology:	This species flowers so early (before the leaves have fully expanded) that one often is too late for and thus misses flowering. It is the first violet to flower in the Flora area, synchronous often with *Claytonia* and *Erythronium*. In flower, earliest date, April 19. In flower, median date, May 6. In flower, latest date, May 26.
Frequency:	Frequent.

Viola sororia Willd. or *Viola papilionacea* Pursh. COMMON BLUE VIOLET

Site:	Often growing with *V. incognita-macloskeyi* in wet places in hardwood forests and sometimes cultivated.
Elevation:	1400 feet near Deer River to 1700 feet. Probably much lower also.
Phenology:	In flower, earliest date, April 19. In flower, median date, May 15.

40. CUCURBITACEAE (GOURD or CUCUMBER FAMILY)

Echinocystis lobata (Michx.) Torrey & Gray SPINY WILD CUCUMBER
- Site: Flood plain of Saranac River at Denny Park, and at several other locations around Saranac Lake.
- Phenology: In fruit October 1973.
- Miscellany: Fruits are dry and hollow, inedible when mature. A vine, possibly native, but also possibly introduced from southern New York where it is definitely native.

41. SALICACEAE (THE WILLOW FAMILY)
References on Poplars and Popples in general:
Ketchledge (1962, June-July)

* *Populus alba* L. WHITE POPLAR or SILVER POPLAR
- Site: Occasional European ornamental in Saranac Lake which sometimes becomes naturalized. I recall saplings at the east end of Park Avenue, near the entrance to American Management Association in Saranac Lake.

Populus balsamifera L. BALSAM POPLAR, BALM-OF-GILEAD, TACAMAHAC or POPPLE

- Site: Locally common in abandoned quarries, gravel pits, roadsides, old fields, railroad embankments, and other disturbed places. Well-drained to imperfectly-drained. Soils: limestone soils along Lake Champlain to Adirondack gneisses and anorthosite. Intolerant of shade.
- Elevation: From 100 feet along Lake Champlain to ca. 2850 feet on Whiteface.
- Frequency: 19 stations.
- Miscellany: The Flora area is near the southern limits of the natural range of this plant; the bulk of the population lies well to the north in Canada.

Populus grandidentata Michx. BIGTOOTH ASPEN, LARGETOOTH ASPEN or POPPLE

- Site: Not as abundant as Trembling aspen, but reaching greater age and larger size. Bigtooths adjacent to the old sawmill site on Campus were almost 100 years old in 1988 and at least two feet in diameter. Shade-intolerant pioneer, but also occurring scattered in semi-open Red pine stands as part of the "climax". Well-drained. Soils: limestone soils of the Champlain Valley to Adirondack gneisses and anorthosite. (In the Catskills, on shales and sandstones).
- Elevation: 100 feet along Lake Champlain to 3130 feet on Pitchoff Mountain.
- Phenology: The two aspens do not hybridize because of a phenological barrier: Trembling flowers 2 to 3 weeks before Bigtooth. Buds open the latest of all native trees in the Flora area. It is not until the first week in June in most years that Bigtooth presents its wooly-white expanding leaves, although the flowers occur a little earlier. Fruits in late June.

Populus tremuloides Michx. TREMBLING ASPEN, QUAKING ASPEN or POPPLE

- Site: Abundant pioneer on open disturbed sites on all soils except the very poorly-drained. Shade-intolerant. Reproduces by vegetative clones as well as by seed; in this way it can rapidly "conquer" a new opening in the forest and take it over without having to wait for maturity to set seed. Aspens, before human disturbance opened up so much of the Adirondacks in the 19th Century, were opportunists, perpetuating only by "hopping" from one disturbance opening to the next. Soils: from limestone in Champlain Valley to Adirondack gneisses and anorthosite. (In the

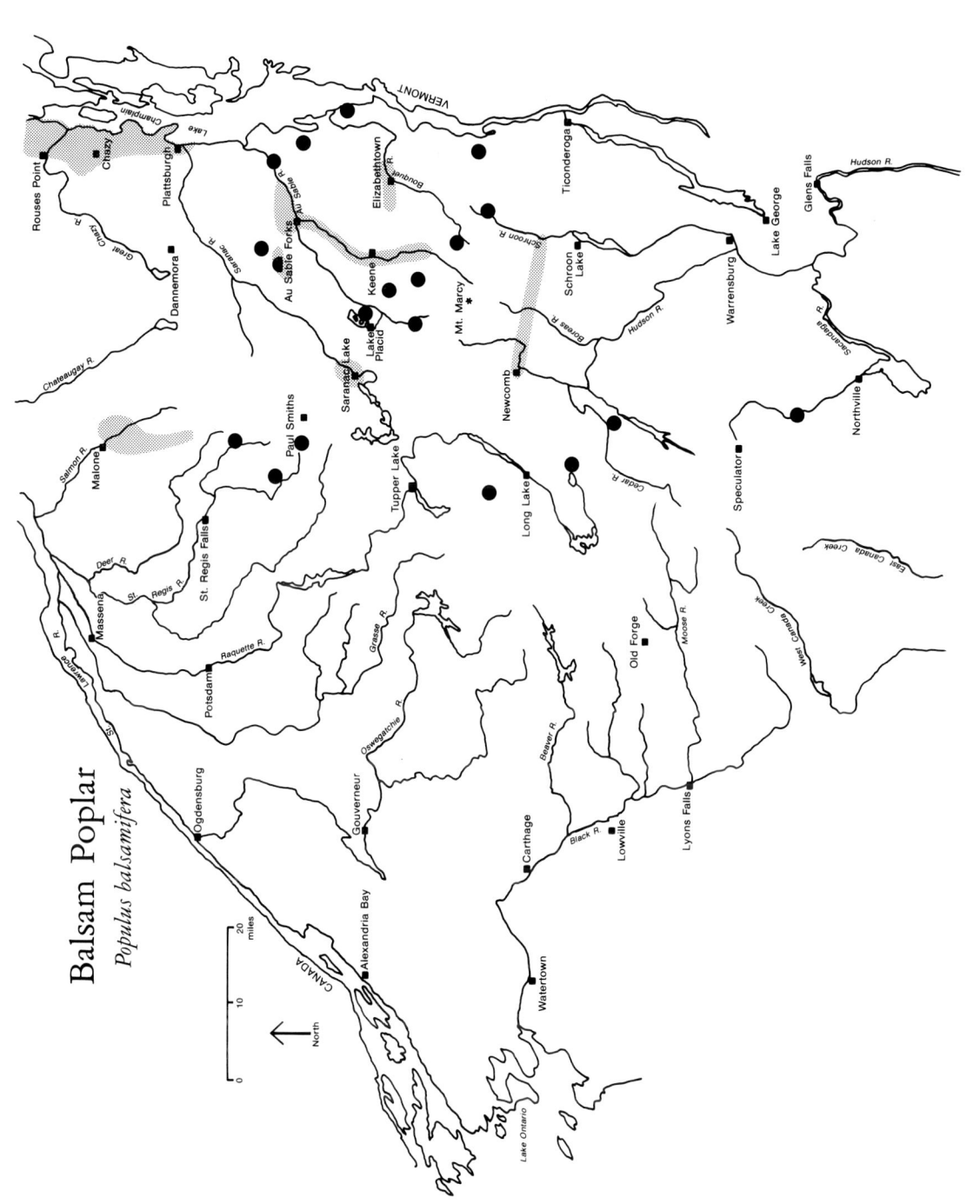

Elevation:	Catskills on shales and sandstones). From 100 feet along Lake Champlain to ca. 3300 feet on Whiteface Mountain.
Phenology:	Leafs out among the earliest trees beginning as early as April 21, but most often 1st week in May. It flowers about 2 to 3 weeks earlier than cousin Bigtooth and thus rarely hybridizes with it, although the two *can* set seed if artificially crossed. Fruits mature in June.
Frequency:	Common.
Miscellany:	The green bark on trunk and branches permits this species to manufacture sugar (photosynthesize) when the leaves are off - in late fall and early spring when the air temperature is above freezing and so is the soil (so that water is available); hence, the green bark can literally extend the growing season by several months and net the trees about 10% or so more sugar.
References:	Gadomski (1987) Schaedle & Foote (1971) Foote & Schaedle (1975).

SALIX (AS A GENUS)

Note that the "Weeping Willow" in Saranac Lake is not *S. babylonica*, but rather *S. alba* var *tristis*.

All our native Adirondack willows are shrubs; any tree-sized willows are either planted (*S. pentandra, S. alba* var *tristis*) or both planted and, sometimes, naturalized (*S. fragilis*). The maximum size of our largest native willows (*S. bebbiana*) is about 20 to 25 feet high and 4 inches in diameter.

References:	Ketchledge (1983, May). Ketchledge (1982, September).

Salix bebbiana Sarg. or *S. rostrata* Richards BEBB'S WILLOW

Site:	It is the only willow (with the uncommon exception of Pussy willow) found on both poorly and well-drained sites, provided there is full sun. Bebb's willow can be found in places where most people do not expect willows: dry, exposed hilltops and open places on deep, well-drained mineral soils. Soils: limestone soils along Lake Champlain to Adirondack gneisses and anorthosites.
Elevation:	From 100 feet along Lake Champlain to 4080 feet on Cascade Mountain.
Frequency:	Bebb's willow is perhaps the Flora area's most common species.
Miscellany:	Bebb's willow hybridizes with Pussy willow and Silky willow and intermediate forms result.

Salix discolor Muhl. PUSSY WILLOW
- Site: Pussy willow is much more common on poorly-drained sites such as roadside ditches and swamps; it seldom occurs on the well-drained. As all willows, shade-intolerant. Soils: limestone soils along Lake Champlain to Adirondack gneisses and anorthosite.
- Elevation: From 100 feet along Lake Champlain to 2878 feet on Haystack Mountain in Ray Brook.
- Frequency: The Flora area's second most common species.
- Miscellany: Sometimes crossing with Bebb's producing intermediates. Hybrid between this species and Bebb's willow at 3380 feet on Noonmark Mountain.

* *Salix fragilis* L. CRACK WILLOW
- Site: A canoe trip down the Saranac River below Saranac Lake Village revealed several medium-sized trees below the Sewage Treatment Plant where the River swells and floods very wide each spring.
- Miscellany: My feeling is that branches from ornamental trees break off in storms upstream in the Village and are carried downstream by the river to lodge in sandy banks. Hence, this may be the Flora area's only example of naturalization via vegetative (asexual) reproduction. All our other naturalized plants reproduce by seed. I have never seen *S. fragilis* set seed here. From Europe. If people fishing have taken willow branches from the Village and set them in the riverbank downstream (where they later root), then this species is not naturalized.

Salix lucida Muhl. SHINING WILLOW
- Site: In sunny swamps around Paul Smiths and Saranac Lake. Shade-intolerant.
- Frequency: Three stations.

Salix pedicellaris Pursh var. *hypoglauca* Fern. BOG WILLOW
- Site: First found in 1985 in a Paul Smiths bog which I have named after it: Bog Willow Bog. This willow is our most acid-tolerant; it grows in Sphagnum with a pH of upper 3s or low 4s. Intolerant of shade. Poorly-drained. The largest colony of it occurs in Spring Pond Bog west of Derrick in the Boreal Heritage Preserve.
- Phenology: In fruit at time of discovery June 27, 1985 in Bog Willow Bog.

Salix pyrifolia Anderss. BALSAM WILLOW
- Site: Poorly-drained areas in swamps and bogs. Shade-intolerant.
- Frequency: Five stations.

Miscellany: Along the Saranac River above Bloomingdale, the plants grade into *Salix rigida* Muhl., the Rigid willow, having features intermediate between the two species. At Paul Smiths the plants are typically *S. pyrifolia*.

Salix sericea **Marsh. SILKY WILLOW**
   Site: In swamps and boggy swamps and along streams. Shade-intolerant. Poorly-drained. One specimen high and dry on well-drained soil at the Asplin Fir plantations, Harrietstown Hill north of Saranac Lake.
   Elevation: Lake Champlain at 100 feet to Harrietstown at 1720 feet.
   Frequency: Seven stations but probably many more.
   Miscellany: Specimens at the NE corner of Heron Marsh at the beaver dam on the V.I.C. look like a hybrid, *S. sericea* X *S. bebbiana*. Both parent species are present and abundant. Specimens along the D & H railroad south of Plattsburgh station look like *S. sericea* X *S. rigida*.

42. BRASSICACEAE or CRUCIFERAE (MUSTARD FAMILY)

* *Barbarea vulgaris* **R. Br. YELLOW ROCKET**
   Site: Sometimes covers whole fields, the flowers turning them yellow.
   Phenology: In flower early June.
   Frequency: Locally abundant.
   Miscellany: European weed.
   Reference: Ives (1975) as an edible.

* *Berteroa incana* **(L.) DC. HOARY ALYSSUM**
   Phenology: In flower between June 9 and June 24.
   Miscellany: Garden escape. European weed.

* *Capsella bursa-pastoris* **(L.) Medic. SHEPHERD'S PURSE**
   Phenology: Flowers early May into summer.
   Frequency: Common.
   Miscellany: European weed.

Cardamine diphylla **(Michx.) Wood** or *Dentaria diphylla* **Michx. TOOTHWORT or PEPPERWORT**
   Site: Well-drained sites, shade-tolerant, with fairly high pH. Usually under Sugar maple, Ash, and Basswood.
   Elevation: From 200 feet at Plattsburgh High School to 2050 feet on Mount Baker.
   Phenology: Leafing April 30 through May 14.

                        In flower May 17 through 23.
    Frequency: 14 stations.

Cardamine pensylvanica **Muhl. ex Willd.** PENNSYLVANIA BITTERCRESS.
    Site: Poorly-drained soils at Barnum Pond Outlet at V.I.C. boardwalk
                        bridges, and Saranac River flood plain near the Brewster
                        Mountain quarry entrance.
    Elevation: 1510 feet to 1640 feet.
    Phenology: In flower June 12.

Cardamine pratensis **L.** var. *palustris* **Wimm. & Grab.** CUCKOO-FLOWER
    Site: Poorly-drained soils. Shade mid-tolerant. Saranac River flood
                        plain at Stevenson Lane substation. Peninsula Nature Trails, Lake
                        Placid.
    Elevation: 1510 feet to 1860 feet.
    Phenology: In flower late May to June 21.

* *Lepidium densiflorum* **Schrad.** or *L. apetalum* in **Peck.** PEPPERGRASS
    Site: European weed at Woodruff Street bridge over Saranac River,
                        Saranac Lake.
    Phenology: In flower and immature fruit July, 1974.

* *Nasturtium officinale* **R. Br. ex Ait.** WATERCRESS
    Site: European naturalized species, in flowing water. 3 stations
                        between Paul Smiths and Rainbow Lake.
    Phenology: Flower buds July 15, 1974 in Osgood River.

* *Raphanus raphanistrum* **L.** WILD RADISH
    Site: European weed. 3 stations around Saranac Lake.
    Phenology: In fruit September 15, 1974 on Mount Pisgah.
                        In flower and fruit August 12, 1986 at Hotel Saranac.

* *Sinapsis arvensis* **L.** or *Brassica kaber* **(DC.) L.C. Wheeler.** CHARLOCK,
CHARLOCK MUSTARD or WILD MUSTARD
    Phenology: In flower between June 2 and September 5 (a long season).
                        In fruit October 27.
    Miscellany: European weed.

43. ERICACEAE (HEATH FAMILY)

Andromeda polifolia **L.** var. *glaucophylla* **(Link) DC.** or *Andromeda glaucophylla*
Link. BOG ROSEMARY
    Site: In bogs.
    Elevation: From about 1550 feet to 1730 feet. Surely this range can be

extended in both directions.
Phenology: In flower between May 29 and June 12.
Frequency: Nine stations.

Arctostaphylos uva-ursi (L.) Spreng. BEARBERRY
Site: Open, sunny sites (shade-intolerant) and well to excessively-well drained, either on bedrock or on sand. Soils: anorthosite. Granitic gneiss on Coot Hill. (In the Catskills, on sandstone).
Elevation: From 1095 feet at Coot Hill to 2440 feet on Baxter Mountain.
Phenology: In flower May 8 in the Catskills, probably a little later in the Adirondacks.
Frequency: Three stations.

* *Calluna vulgaris* (L.) Hull HEATHER
Site: On the old golf course fairway at Paul Smiths.
Phenolgoy: In fruit November 22.
Miscellany: Either planted or naturalized from Europe.

Chamaedaphne calyculata (L.) Moench LEATHERLEAF
Site: A pioneer mostly in open bog waters, then eventually growing with *Sphagnum*, forming bog mats. Abundant. Shade-intolerant. One exceptional plant at the point on Horseshoe Pond SW of Saranac Inn was growing at least four feet above Pond level—high and dry—and none the worse off for it apparently. Plants atop Mount Marcy are also high and dry, not rooted in moist *Sphagnum*. Soils: largely Hemists (bog soils), and also Folists (alpine organic mats), but also on moderately well-drained outwash sand as at Horseshoe Pond.
Elevation: From 588 feet at Camp Pok-O-MacCready NW of Willsboro to 5300 feet on Mount Marcy.
Phenology: In flower from May 9 through 19 on the Upland. End flowering much later, June 14, on Algonquin summit.
In fruit August.
Frequency: Common.

Epigaea repens L. TRAILING ARBUTUS or MAYFLOWER
Site: On well-drained soils, usually in partial shade under semi-open pine forests on glacial outwash sands.
Elevation: From 750 feet at Altona to 1730 feet at Loon Lake.
Phenology: In flower, earliest date, April 19.
In flower, median date, May 4.
In flower, latest date, May 22.
Frequency: Protected by N.Y. State law, although not rare in our area. Ten stations.

Gaultheria hispidula (L.) Muhl. ex Bigel. or *Chiogenes hispidula* (L.) Torrey & Gray
CREEPING SNOWBERRY
    Site: On well to poorly-drained soils. Mostly under spruce and/or fir. Sometimes on mineral soils and sometimes on peat, but shade-tolerant always. To timberline.
    Elevation: From Paul Smiths at 1630 feet to Skylight 4926 feet.
    Frequency: Occasional.
    Miscellany: Fruits edible with wintergreen flavor.

Gaultheria procumbens L. WINTERGREEN, CHECKERBERRY or TEABERRY
    Site: Abundant on well-drained soils under semi-open mostly pine stands (shade mid-tolerant).
    Elevation: From 200 feet on the Ausable River Delta to about 2300 feet on Giant Mountain.
    Phenology: In flower August.
    Miscellany: Fruits tasty.
    Reference: Schottman (1987, May).

Gaylussacia baccata (Wang.) K. Koch HUCKLEBERRY
    Site: Locally abundant on well-drained sites under semi-open mostly pine stands, especially Red pine. Slightly shade-tolerant. Soils: gneisses and anorthosite. (In the Catskills on sandstones and conglomerates).
    Elevation: From 200 feet on the Ausable River Delta to 2040 feet on Prospect Mountain.
    Phenology: In flower June 15.
                    The best time of year to locate Huckleberries is mid-to late October when the leaves are scarlet. Look from a canoe for brilliantly-colored red shrubs about hip-high under Red pine groves on the east sides of lakes and on exposed peninsulas.
    Miscellany: Differs from the blueberries by its nearly black fruit with only 2 or 3 seeds which are considerably larger than those of the blueberries.

Kalmia angustifolia L. SHEEP LAUREL, SHEEP KILL or LAMB KILL
    Site: On poorly-drained to well-drained sites. Often in bogs. Usually shade-intolerant, but sometimes venturing into semi-open Red pine stands where it can also be mid-tolerant. To timberline. Soils: from Potsdam sandstone to Adirondack gneisses and anorthosite.
    Elevation: From 750 feet at Altona to 5076 feet on Algonquin.
    Phenology: In flower, earliest date, June 12.
                    In flower, median date, June 20.
                    In fruit, August through autumn.

Frequency: Protected in N.Y. State, but common in our area.
Miscellany: Poisonous to eat for sheep, other domesticated animals and people.
References: Jaynes (1968)

Kalmia polifolia Wang. BOG LAUREL or PALE LAUREL
Site: Poorly-drained peat soils in bogs. To timberline.
Elevation: From 1640 at Cooler Pond at Paul Smiths to 5250 feet on Mount Marcy.
Phenology: In flower May 29 through June 16.
In fruit August.
Frequency: Protected in N.Y. State.
Miscellany: Poisonous.
Reference: Doeffinger (1978).

Ledum groenlandicum Oeder. LABRADOR TEA
Site: Most abundant on bog mats, but also will grow on better-drained mineral soil provided that there is at least some sunlight (as under semi-open Red pine stands). Abundant in Labrador, the type locality for the common name. To timberline and on alpine summits.
Elevation: From 1620 feet at Paul Smiths to 5320 feet on Mount Marcy.
Phenology: In flower, earliest date, May 29.
In flower, median date, June 12.
In fruit, August through Autumn.
Frequency: Frequent.
Miscellany: Leaves make a good tea.
Reference: Riebesell (1981).

Vaccinium angustifolium Ait. or *V. pensylvanicum* var. *angustifolium* Gray BLUEBERRY, LOW SWEET BLUEBERRY or LATE SWEET BLUEBERRY
Site: Prolific on well-drained sites. Shade-intolerant and follows fires in profusion. To timberline. Soils: Potsdam sandstone at Plessis and Altona to Adirondack gneisses and anorthosite. Dominates the burned-over summit of Jenkins Mountain where it might inhibit tree reproduction by competition. Dominates the treeless bedrock knob summits SW of Jenkins Mountain where there have been no fires.
Elevation: 440 feet at Plessis to 4960 feet on Haystack.
Phenology: In flower May 22 through June 12.
In fruit August and September.
Frequency: The most common blueberry.
Miscellany: This species will cross with *V. myrtilloides*, producing intermediates.

Reference: To *all* Blueberries & Cranberries: Vander Kloet (1988).

Vaccinium macrocarpon **Ait.** LARGE-FRUITED CRANBERRY
- Site: On bog mats, seemingly less common than *V. oxycoccus*. Climbing the drier sandy or silty banks of the Raquette River between Axton & Raquette Falls.
- Phenology: Collect fruits after the first frosts in late September or October.
- Frequency: Four stations.
- Miscellany: Fruits edible.

Vaccinium myrtilloides **Michx.** or *V. canadense* **Kalm** SOUR-TOP BLUEBERRY or VELVETLEAF BLUEBERRY
- Site: Most often in poorly-drained sites as in bogs, swamps, and shorelines. Occasionally in open, well-drained sites. To timberline. Soils: anorthosite and Adirondack gneisses.
- Elevation: From 1620 feet in Paul Smiths area to 3830 feet on Lyon Mountain and to a possible hybrid with *V. angustifolium* atop Cascade Mountain at 4048 feet.
- Phenology: In fruit August.
- Frequency: Common.
- Miscellany: The two blueberries will cross and intermediates are common: *V. myrtilloides* X *V. angustifolium*.

Vaccinium oxycoccus **L.** SMALL-FRUITED CRANBERRY
- Site: Often growing on bog mats. To timberline.
- Elevation: Cooler Pond at Paul Smiths 1640 feet to 5076 feet on Algonquin.
- Phenology: In flower June 9 through 15 on the Upland.
  July 1 atop Wright Peak at 4550 feet.
  In mature fruit September and October.
- Frequency: The more common Cranberry.
- Miscellany: Collect fruits after the first frosts in late September and October.

Vaccinium uliginosum **L.** *ALPINE BILBERRY*
- Site: A mostly alpine species, often making up the largest fraction of the above-timberline biomass. It also occurs on lower summits lacking a true alpine zone such as Catamount Mountain (E of Franklin Falls) and Noonmark Mountain. Soils: organic mat (Folist) over anorthosite (syenite on Catamount).
- Elevation: From 3100 feet on Catamount and 3861 feet on McKenzie to 5330 feet on Mount Marcy.
- Phenology: Leaves turn purple September 5.
- Frequency: Rare in Mitchell.

44. PYROLACEAE (WINTERGREEN FAMILY)

Moneses uniflora (L.) Gray. One-flowered wintergreen. Originally, I had intended to include this species as a native plant in the Flora, but have since determined that it has been introduced by people into the Flora area. I had been informed in 1986 that *Moneses* was growing on trails to Oseetah Lake and Ampersand Mountain, and I did indeed find it in both places and in flower between June 19 and July 16. I thought it odd that the species was present only along the trails and not off them. Then, on July 18, 1990 I found *Moneses* in flower in the old quarry—a very heavily-used recreational spot—near the Mount Baker trail head in Saranac Lake. There is no question that someone has been planting *Moneses* at these three locations.

The species is common further north in Canada in spruce-fir forests, but only once have I seen it, apparently native, in the Adirondacks. This was well out of the Flora area on the floodplain of the Bouquet River above Wadhams, elevation 420 feet.

Chimaphila umbellata (L.) Bart. PIPSISSEWA or PRINCE'S PINE
- Site: On well-to imperfectly-drained sites, forming patches. Most often under pines on outwash soils, but also on tills with mixed woods. Shade midtolerant to tolerant, evergreen. Soils: mostly from anorthosite and Adirondack gneisses, but the Newcomb site may have some marble.
- Elevation: From 588 feet at Camp Pok-O-MacCready NW of Willsboro to 2450 feet on Pitchoff Mountain.
- Phenology: In fruit October 29.
- Frequency: Protected in N.Y. State. Occasional in the Flora area.

Orthilia secunda (L.) House or *Pyrola secunda* L. ONE-SIDED SHINLEAF
- Site: This is a northern Canadian species of bogs and coniferous woods, quite rare in the Adirondacks.
- Elevation: 1600 feet at Raquette Falls to 1630 feet at Bog Willow Bog and ca. 1800 feet on McKenzie Mountain trail.
- Phenology: In flower June 30 through July 16.
- Frequency: Three stations.

Pyrola elliptica Nutt. SHINLEAF
- Site: Shade-tolerant on well-drained soils, under hardwoods or conifers. pH of humus: n=8; min. 3.9; max. 5.8; median 4.75.
- Elevation: From ca. 450 feet on Tongue Mountain to 2100 feet on Crane Mountain.
- Phenology: In flower July 7 through 28.
  In fruit September 7.

Frequency: Occasional.
Miscellany: Evergreen.

45. MONOTROPACEAE (INDIAN PIPE FAMILY)

Monotropa hypopithys L. PINE SAP
- Site: Plants apparently dependent on mycorrhizal fungi as no chlorophyll is present. Under conifers on well-drained outwash or till.
- Elevation: From about 800 feet at Upper Jay to 1700 feet near Jenkins Mountain.
- Phenology: In flower June 30.
  In fruit August 9.
- Frequency: Three stations at Paul Smiths and Ray Brook.

Monotropa uniflora L. INDIAN PIPE
- Site: A non-green flowering plant dependent upon mycorrhizal fungi. Mostly well-drained sites. Shade tolerant hardly applies, as this plant could grow in total darkness if it had to. Under hardwoods or conifers.
- Elevation: From 200 feet at Chazy to 2670 feet along Loch Bonnie trail and 4050 feet on Porter Mountain.
- Phenology: In flower earliest date June 24.
  In flower normally late July into early August.
  Flowering in midsummer; the drooping flowers become erect and brown in fruit and overwinter.
- Frequency: Common.
- References: Schottman (1982, August)
  Hass (1970)
  Lutz & Sjolund (1973)

46. PRIMULACEAE (PRIMROSE FAMILY)

Lysimachia ciliata L. or *Steironema ciliatum* (L.) Raf. FRINGED LOOSESTRIFE
- Site: Often in poorly-drained sites, shade mid-tolerant.
- Elevation: From 100 feet along Lake Champlain at Point Au Roche to about 1700 feet.
- Phenology: In flower August 2.
- Frequency: Three stations in the Paul Smiths area.

Lysimachia quadrifolia L. WHORLED LOOSESTRIFE
- Site: Typically, this more southerly species is a shade-tolerant, on well-drained sites under primarily red and/or chestnut oaks. Locally,

> no oaks are present on the Paul Smiths area sites.
>
> Elevation: 500 feet on Prospect Mountain to 2330 feet on Buck Mountain.
> Phenology: In flower June 19 to July 15 and into August.
> Frequency: Four stations.
> Miscellany: I first discovered this species on the Paul Smith's College Campus on May 30, 1988. I had not noticed it in 1987 or earlier—a sudden new station.

Lysimachia terrestris (L.) BSP. SWAMP CANDLES, SWAMP LOOSESTRIFE or SPIKED LOOSESTRIFE

> Site: Occasionally in poorly-drained, mostly sunny situations.
> Elevation: From between 1000 and 1200 feet in Keene Valley to 2000 feet at Minnow Pond, Adirondack Museum.
> Phenology: Median flowering date August 12.
> In flower and immature fruit September 8, 1974.
> Frequency: Seven stations, but probably many more.

Lysimachia thrysiflora L. or *Naumbergia thyrsiflora* (L.) Duby TUFTED LOOSESTRIFE

> Site: Open, sunny, wet areas (i.e., marshes).
> Elevation: 1620 to 1640 feet.
> Phenology: In flower June 18 through 25.
> Frequency: Two stations in Heron Marsh area.

Trientalis borealis Raf. or *T. americana* Pursh STAR FLOWER

> Site: Shade-tolerant under northern hardwoods and conifers. Well to imperfectly drained. To timberline. pH of humus: n=16; min. 3.5; max. 5.0; median 4.55.
> Elevation: From 210 feet at Chazy Fir Swamp to 4900 feet on Skylight where dwarfed to 1 inch high.
> Phenology: In flower June 5 through 12 on the Upland.
> As late as June 30 atop McKenzie Mountain, elevation 3860 feet.
> References: Anderson & Loucks (1973)
> Cook, R.E. (1983)

47. GROSSULARIACEAE (CURRANT FAMILY)

Ribes cynos-bati L. PRICKLY GOOSEBERRY or DOGBERRY

> Site: Shade-tolerant, well-drained, usually under northern hardwoods. Soils: limestone along Lake Champlain and in the Saint Lawrence Valley to Adirondack gneisses and anorthosite. (Catskills shales and sandstones).

Elevation: From 200 feet at Chazy and at Barnhardt Island to Chimney Mountain, 2560 feet.
Frequency: Seven stations.
Miscellany: *Ribes* is an alternate host to the white pine blister rust fungus, *Cronartium ribicola*.

Ribes glandulosum Grauer SKUNK CURRANT
Site: The most common and boreal of the genus, reaching greatest abundance in the spruce-fir forest and even approaching timberline. Well-drained, shade-tolerant. Soils: Adirondack gneisses and anorthosite. (Catskills shales, sandstones and conglomerates).
Elevation: From 1100 feet on the Dickinson Esker to 5076 feet on Algonquin.
Phenology: Flower buds May 14; flowers late May; half-leaf May 7.
Frequency: 25 stations.
Miscellany: Easily identified by its skunk-like odor when crushed. This species, as other *Ribes*, is rare in white pine areas due to eradication programs for the control of the blister rust fungus, *Cronartium ribicola*, in the 1910s through 1930s. Locally, widely-scattered in northern hardwoods and mixed woods.

Ribes lacustre **(Pers.) Poir.** BRISTLY BLACK CURRANT
Site: Mostly in well-drained sites, shade-tolerant, under northern hardwoods usually, but occasionally in spruce-fir. Soils: Adirondack gneisses and anorthosite. (In the Catskills on shales and sandstones).
Elevation: From 1700 feet on Jenkins Mountain to 3650 feet on Cascade Mountain.
Frequency: Nine stations. Scarce because of eradication.
Reference: Phelps (1964) says that it attains timberline.

Ribes triste Pallas WILD RED CURRANT
Site: This may be the most demanding of the *Ribes* species, occurring on soils with the highest pH. Northern hardwoods with basswood, ash, elm, hop hornbeam. Well-drained to somewhat poorly drained, shade-tolerant.
Elevation: From 2040 feet at Owl's Head (Mountain View) and Cascade Lakes to 2200 feet near Heart Lake.
Frequency: Three stations.

48. SAXIFRAGACEAE (SAXIFRAGE FAMILY)

Chrysosplenium americanum Schwein. ex Hook. GOLDEN SAXIFRAGE

Site: In running water.
Elevation: From 588 feet at Pok-O-MacCready near Willsboro to about 1890 feet on Hill 1938 near Paul Smiths.
Phenology: In flower mid-May.
Frequency: Six stations.

Saxifraga virginiensis **Michx.** EARLY SAXIFRAGE
Site: Mostly on rock ledges under rich-site northern hardwoods (including basswood, white ash, elm, hop hornbeam) with a higher humus pH. Usually well-drained, but ledges sometimes dripping wet in early spring. Shade-tolerant.
Elevation: From 750 feet at Altona to about 2050 feet on Mount Baker.
Phenology: In flower May 12 through 17.
In fruit late June.
Frequency: Three stations.

Tiarella cordifolia **L.** FOAMFLOWER
Site: On higher-pH sites usually under northern hardwoods with basswood, white ash, elm, and hop hornbeam. However, this plant will grow in coniferous forests (spruce, fir, hemlock) along brooks where organic acids cannot accumulate readily and where the pH is about a unit higher (5 as opposed to 4) than that a short distance away in the forest. pH of humus: n=36; min. 3.7; max. 6.6; median 4.8.
Elevation: From 210 feet at Chazy Fir Swamp to 3650 feet on Cascade Mountain.
Phenology: In flower earliest date, May 17.
In flower median date, May 23.
In flower latest date, June 12.
Frequency: 64 stations. Common.

49. ROSACEAE (ROSE FAMILY)

Agrimonia **sp.** AGRIMONY
(possibly *A. gryposepala* **Wallr.** but not confirmed)
Frequency: Two stations in the High Peaks, elevation 1000 and 2250 feet.

Amelanchier arborea **(Michx. f.) Fernald** and var. *arborea* and var. *laevis* **(Wieg.) Ahles** (Both varieties often occur in the flora area on the same plant.) SERVICEBERRY, JUNEBERRY, SHADBUSH or SHADBLOW
Site: Fencerows, edges of woods, roadsides, and under open-canopy red pine stands. Well-drained. Shade-intolerant to slightly

Elevation: From 100 feet along Lake Champlain to 3800 feet on Lyon Mountain.
Phenology: Largely overlooked except in spring when in flower.
In flower, earliest date, May 12.
In flower, median date, May 21 just as the hardwoods leaf out.
In flower, latest date, May 29.
In mature fruit July; not June in our climate as one common name implies.
Frequency: Very abundant.
Miscellany: One unsolved problem regarding this species is why mature plants on Campus are smaller than those in the Saranac Lake Village area, especially those on Mount Baker. On Campus, a large plant is two or three inches in diameter at breast height, four and one half feet above ground, and twenty feet tall, while in Saranac Lake a large plant is up to six inches to one foot in diameter at breast height, and 30 to 45 feet tall. Could the explanation be better soil fertility in Saranac Lake? Fruits edible, sweet, and tasty.

Amelanchier bartramiana (Tausch) Roemer WILLIAM BARTRAM'S JUNEBERRY, SERVICEBERRY, SHADBUSH or SHADBLOW

Site: This is the high-elevation species, reaching well into the spruce-fir, to timberline. Locally, plants are scattered, isolated—never in groves—in the Paul Smiths area. Well-drained, occurring at edge of woods or in partial shade, never in full shade. Soils: outwash sands from Adirondack gneisses and anorthosite at Paul Smiths to high-elevation shallow tills and lithic borofolists (sub-alpine organic mats on bedrock). (Sandstone and conglomerate tills in Catskills).
Elevation: From about 1630 feet on the Paul Smith's College Campus to 4736 feet on Gothics.
Phenology: In flower between May 4 and May 17 on the Upland.
As late as June 5 on Algonquin at elevation 4450 feet.
Frequency: Less common than *A. arborea.*

Amelanchier stolonifera Wieg. BUSH JUNEBERRY or SHADBUSH
Site: Black Brook Sand plain.
Elevation: 1040 feet.
Phenology: In flower May 15.

Aronia melanocarpa (Michx.) Ell. or *Pyrus melanocarpa* (Michx.) Willd. BLACK CHOKEBERRY
- Site: From the wettest swamps and bog mats to the driest sandy outwash plains (such as at the Little Moose River Plains) to exposed bedrock ledges. Thus independent of water table depth but always shade-intolerant.
- Elevation: 750 feet at Altona to 2160 feet atop Pok-O-Moonshine Mountain. Certainly, someone can expand on this range from both ends.
- Phenology: In flower June 5 to 12.
- Frequency: Frequent shrub.
- Miscellany: Fruits on some plants tasty but more often puckery.

Crataegus sp. HAWTHORN or THORNAPPLE
- Site: Hawthorns are mostly old-field species, shade-intolerant, and largely on well-drained sites, although occasionally on imperfectly-drained. Much more common in the agricultural lowlands than on the Upland.
- Elevation: From 370 feet at the Rossie lead mine to about 2200 feet on the Owl's Head SE of Malone. Surely someone can find it below 370 feet.
- Miscellany: No attempt here is made to define species. This is the most difficult genus of vascular plants in the flora area to identify to species. Willows, sedges, grasses and rushes are much easier.

Dalibarda repens L. FALSE VIOLET
- Site: Shade-tolerant on well-to poorly-drained sites. Often under conifers, especially in fir and spruce swamps. pH of humus: n=7; min. 4.3; max. 5.0; median 4.7.
- Elevation: 1500 to 2200 feet at Heart Lake. To timberline?
- Phenology: This is one of the few shade-tolerant, woodland ground cover species which flowers in summer. The period is from June 16 through August 21. Evergreen.

Fragaria virginiana Mill. STRAWBERRY
- Site: In well-drained, open places. Shade-intolerant. Edges of woods, old fields, lawns etc.
- Elevation: From 210 feet at the Chazy Fir Swamp to 2873 feet on Saint Regis Mountain.
- Phenology: In flower, earliest date, May 1.
  In flower, median date, May 10.
  In flower, latest date, May 20.
  In fruit Mid-June.
- Frequency: Common.
- Reference: Wilhelm (1974)

Geum canadense Jacq. WHITE AVENS
    Site: Crest of Mount Pisgah, south of summit growing with *Pilea*. At old cistern on Campus between faculty housing and gymnasium.
    Elevation: 1900 and 1660 feet, respectively.
    Phenology: In flower early July.
                   In fruit August 4 through August 18.

Geum rivale L. PURPLE AVENS or WATER AVENS
    Site: Open or partly-shaded wetwoods, seepages, and swamps.
    Elevation: From 1800 feet on McKenzie Mountain to 2800 feet on Hurricane Mountain.
    Frequency: Abundant in wetlands further north in Canada, but only 2 stations in our Flora.

* *Malus pumila* Mill. or *Pyrus malus* of authors, not L. APPLE
    Site: Old fields and thickets. Cannot compete with native, taller trees in forest and succumbs to shade since it is intolerant.
    Phenology: In flower mid-May; in fruit September.
    Miscellany: Eurasian spread from cultivation and becoming naturalized. Seeds poisonous. Apparently self-pollinating, as widely-scattered individual trees, miles from other trees, bear fruits.
    Reference: Totemeier (1979)

Potentilla anserina L. SILVERWEED
    Site: Bloomingdale station site along old D & H railroad, in cinders.
    Elevation: 1550 feet.
    Phenology: In flower June 2.

* *Potentilla argentea* L. SILVERY CINQUEFOIL
    Site: European weed, low-growing, on lawns. On Campus and in Saranac Lake. Tolerates trampling. Leaves silvery on underside.
    Phenology: In flower in July.
    Frequency: Common.

Potentilla fruticosa L. SHRUBBY CINQUEFOIL
    Site: Right-of-way of the D & H railroad, in Bloomingdale Bog. Abundant further north in Canada where it grows to timberline and on tundra. Shade-intolerant. Mostly well-drained.
    Elevation: 1560 feet.
    Phenology: In flower August 1, 1984 atop Mont Albert, Gaspé.
    Miscellany: Planted more recently on Campus as an ornamental.

Potentilla norvegica L. NORWAY CINQUEFOIL
    Site: Native, but weedy.
    Elevation: 1630 to 2008 feet. Dr. Ketchledge found it on top of Mount Marcy.
    Phenology: On Campus, in flower July 18.
    Frequency: Common.
    Miscellany: Cosmopolitan—also native in Europe, apparently, hence the common name.

Potentilla palustris (L.) Scop. MARSH CINQUEFOIL
    Site: In shallow water or along edges of marshes or bogs.
    Elevation: From 1500 to 1700 feet.
    Phenology: In fruit August 20.
    Frequency: Eight stations.

* *Potentilla recta* L. ROUGH-FRUITED CINQUEFOIL
    Phenology: In flower June 19 through July.
                    In immature fruit August 4.
    Miscellany: European weed, often growing with the native *P. norvegica*.

Potentilla simplex Michx. OLD FIELD CINQUEFOIL or COMMON CINQUEFOIL
    Site: A native species. Weedy on lawns, roadsides, and old clearings. Shade-intolerant. Well-drained. Common on Campus.
    Elevation: 1500 feet to 2300 feet on old Algonquin Trail.
    Phenology: In flower between June 13 and 21.
    Frequency: Very common.

Potentilla tridentata Soland. ex Ait. THREE-TOOTHED CINQUEFOIL or ALPINE CINQUEFOIL
    Site: Intolerant of shade, well-drained. Common on alpine summits in the High Peaks, but also on exposed ledges in the Catskills and Adirondacks where there is no alpine zone. More abundant northward into Canada. Some along roadsides and others in old pastures.
    Elevation: 1560 feet on the Bigelow Road, Bloomingdale, to above 5103 feet on Algonquin and 5330 feet on Marcy.
    Phenology: In flower June 16 into early July.
    Frequency: Eleven stations in the Flora area.

Prunus pensylvanica L. f. RED CHERRY, PIN CHERRY, FIRE CHERRY or BIRD CHERRY
    Site: Shade-intolerant. Well-drained sites. Fast-growing; flowering and fruiting at an early age. Soils: limestone soils on Four Brothers Islands to Adirondack gneisses and anorthosite. (Sandstone,

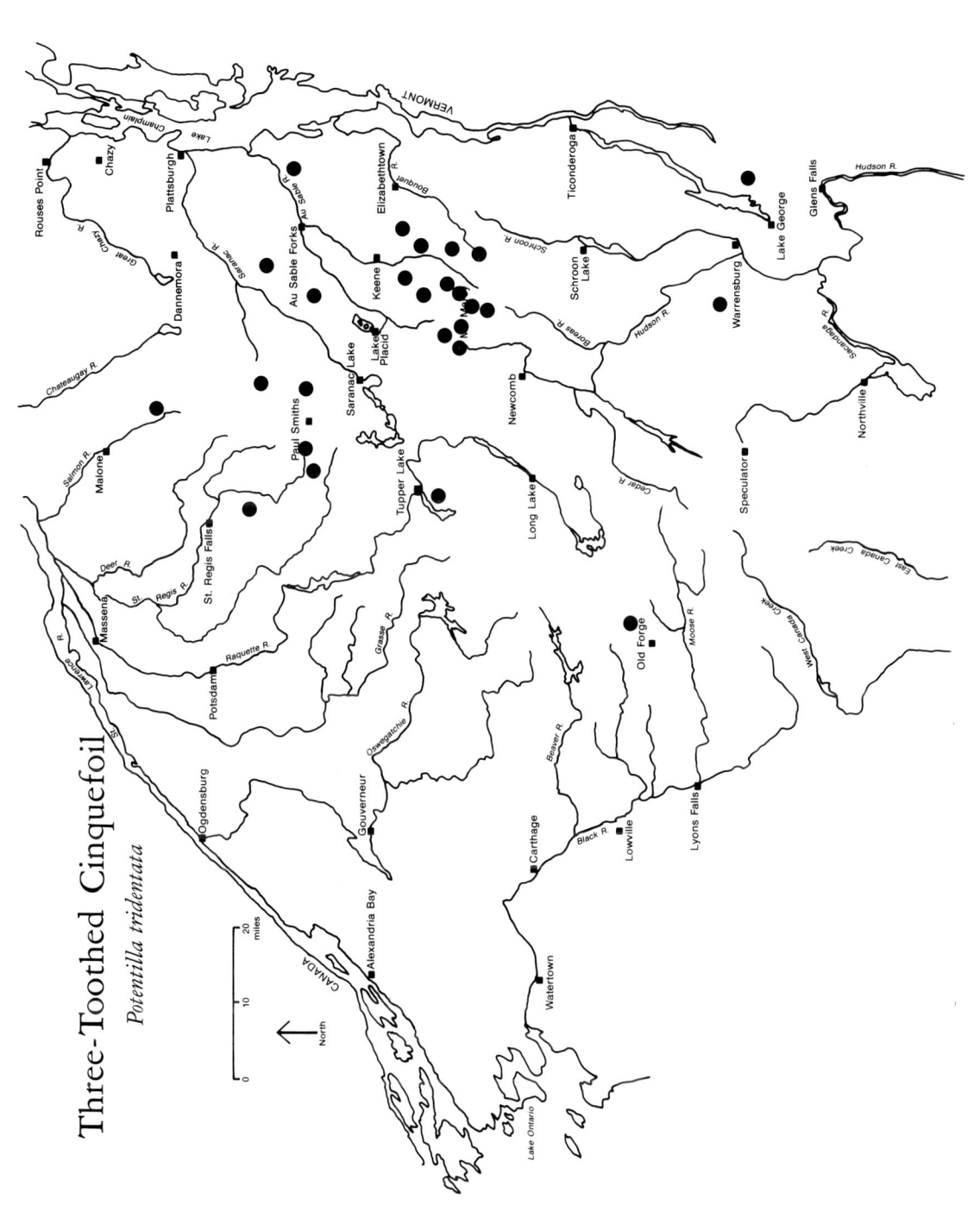

Three-Toothed Cinquefoil
Potentilla tridentata

                                shale, and conglomerate in the Catskills).
Elevation: 100 feet on Four Brothers Islands on Lake Champlain to 4080 feet on Cascade Mountain.
Phenology: The earliest of the cherries in the flora area, it flowers May 15 through 29.
Fruits in late July.
Frequency: Common.
Miscellany: Short-lived (to 25 or 30 years) pioneer. According to Marks (1974), seeds may lie dormant up to 50 years until the canopy is opened.
Reference: Marks (1974).

Prunus pumila L. var. *(pumila* SAND CHERRY
Site: Black Brook sand plain, in old gravel pit.
Elevation: 1040 feet.
Frequency: Rare according to Birmingham (1988), and to Mitchell (1986).

Prunus serotina Ehrh. BLACK CHERRY
Site: Mid-tolerant on well-drained soils. This species is less demanding than Sugar maple and will grow commonly on outwash sands although of poor form. Curran(1974) reports that it is frost-resistant and can survive in severe frost pockets and dry kettles. pH of humus: n=42; min. 3.5; max. 6.5; median 4.7. Soils: limestone soils along Lake Champlain to Adirondack gneisses and anorthosite. (Shales, sandstones, and conglomerates in the Catskills).
Elevation: From 100 feet along Lake Champlain to 3168 feet on Catamount Mountain. Uncommon above 2500 feet.
Frequency: Common.
Miscellany: Commonly contracts the Black knot fungus, *Dibotryum morbosum*. Infestations seem more severe on outwash soils than on till soils, but this must be checked before accepted as fact. Could the till soils be more fertile and the trees have more resistance to disease?
      Cherries have poisonous leaves (especially when wilted), twigs, and seeds—a form of cyanide. The pulpy fruits are edible, but are frequently very sour.

Prunus virginiana L. CHOKECHERRY
Site: Forming thickets from 3 to 20 feet high. Shade-tolerant, well-drained. Often along edges of woods and in old fields. Frequent on summits of lower hills. pH of humus: n=9; min. 4.4; max. 6.8; median 5.0.

    Elevation: From 100 feet along Lake Champlain to 2873 feet atop Saint Regis Mountain. Soils: limestone soils along Lake Champlain to Adirondack gneisses and anorthosite. (Catskills shales and sandstones).
    Phenology: In flower May 29 through June 5.
    Fruits late August.
    Also as an understory under Sugar maple and Black cherry where the Chokecherry leafs out about 1 to 2 weeks ahead of the overstory trees, and has full sun for photosynthesis at that time.
    Frequency: Locally abundant.
    Miscellany: Pulp of fruit (not seed) is edible, but often too sour.

Rosa palustris Marsh. SWAMP ROSE
    Elevation: From 1500 to 1770 feet, along edges of streams.
    Phenology: In flower July 17 through September 7.
    Frequency: Nine stations.

Rubus alleghaniensis Porter ex Bailey BLACKBERRY
    Site: Pioneer in old fields, following logging roads, logged-over areas, edges of woods, etc. Shade-intolerant. Well-drained.
    Elevation: From Pok-O-Moonshine Mountain at 1520 feet (and probably much lower) to 3903 feet on Snowy Mountain.
    Phenology: In flower June 16 through July 16.
    In fruit late August and September.
    Frequency: Abundant.
    Miscellany: Fruits tasty. Some authors consider this a species-group instead of a single species.

Rubus hispidus L. DEWBERRY
    Site: Shade-tolerant creeping low vine, often on poorly-drained soils as in spruce and fir swamps.
    Elevation: From 210 feet at the Chazy Fir Swamp to 3450 feet on Pitchoff Mountain.
    Phenology: In flower July 16.
    In fruit between July 28 and September 7.
    Frequency: Common.
    Miscellany: The fruits are small but sweeter than those of Blackberry and Raspberry, and well worth the effort of harvesting.
    Reference: Abrahamson (1975).

Rubus idaeus L. or *Rubus strigosus* Michx. RED RASPBERRY
    Site: Attaining timberline. Shade-intolerant and well-drained, rapidly filling logging roads and openings in woods, and along edges of

	roads and woods.
Elevation:	From 100 feet along Lake Champlain to Mount Marcy 5320 feet.
Phenology:	In flower June. Fruits—pick them from mid-July through mid-August.
Frequency:	Perhaps even more profuse than Blackberry but on similar sites.
Miscellany:	Fruits tasty, edible. Also considered by some authors to be a group of species rather than a single species.

Rubus odoratus L. PURPLE-FLOWERING RASPBERRY

Site:	Edges of woods, old log roads, etc. Shade-intolerant and well-drained sites. Not common in the Flora area but becoming more so in the eastern Adirondacks.
Elevation:	From 100 feet along Lake Champlain to 2100 feet at Cascade Lakes.
Phenology:	In flower June 19.
Frequency:	Two stations but probably many more.
Miscellany:	Fruits edible.

Sorbus americana Marsh. or *Pyrus americana* (Marsh.) DC MOUNTAIN ASH

Site:	Commonest in openings in the spruce-fir forest, 3000 feet elevation to timberline. Scattered on the Upland around Paul Smiths along edges of woods, roadsides, and under semi-open Red pine stands. Shade-intolerant. Well-drained to sometimes imperfectly drained.
Elevation:	From 1100 feet on the Dickinson Esker to 5280 feet on Mount Marcy.
Phenology:	In flower mid-June. In fruit late September.
Frequency:	Common.
Miscellany:	Fruits, non-poisonous but very sour, distributed by birds. The very similar and closely-related *S. decora* (Sarg.) Schneid. has not yet been found on Campus.

Spiraea latifolia (Ait.) Borkh. MEADOWSWEET

Site:	Growing on any site provided that there is nearly full sun. Well to poorly-drained. Often forming nearly impenetrable thickets to the exclusion of other species. Old fields and swamps, and especially on old burned-over outwash plains.
Elevation:	From 750 feet at Altona to ca. 4926 feet on Mount Marcy. Surely, someone can find it below 750 feet.
Phenology:	In flower July 16 to August 21.

Frequency: A most prolific shrub.

Spiraea tomentosa L. STEEPLEBUSH or HARDHACK
- Site: Most commonly found in poorly-drained areas such as along streams and ponds. Shade-intolerant, sometimes growing with its cousin, Meadowsweet. Occasionally on well-drained sites.
- Elevation: From 1630 feet around the Paul Smith's College Campus to 2873 feet atop Saint Regis Mountain.
- Phenology: In flower August 2 to 21.
- Frequency: Four stations, but probably many more.
- Miscellany: The name, hardhack, across the northernmost counties of New York and eastward into northern Vermont applies also to the tree, *Ostrya virginiana*.

50. FABACEAE or LEGUMINOSAE (LEGUME, PEA, or PULSE FAMILY)

There are no native legumes in the Adirondack Upland Flora area.

* *Lotus corniculata* L. BIRDSFOOT TREFOIL
- Site: Planted along roadsides to stabilize steep banks and in agricultural fields for crop rotation, but has become naturalized. European.
- Phenology: In flower June 26 through summer.
- Can turn whole fields yellow in summer. Locally abundant.
- Miscellany: As in many legumes, the roots of this species bear symbiotic bacteria which fix nitrogen.

* *Medicago lupulina* L. BLACK MEDICK
- Site: European weed occasional in Saranac Lake.
- Phenology: In flower July.

* *Medicago sativa* L. ALFALFA
- Site: European, planted to fix nitrogen in soils as part of crop rotations. Occasionally becomes naturalized. Paul Smiths.
- Phenology: In flower June 26 through July 1.

* *Melilotus alba* Desc. ex lam. WHITE MELILOT or WHITE SWEET CLOVER
- Phenology: In flower between July 14 and August 13.
- Frequency: Common.
- Miscellany: European naturalized weed.

* *Melilotus officinalis* (L.) Pallas YELLOW SWEET CLOVER or YELLOW MELILOT
    Site: Growing with its cousin, *M. alba,* at the site of the more southeasterly Bluebird Cabins on Campus. Both are introduced and naturalized from Europe.
    Phenology: In flower July 22 to August 13 at Bluebird Cabins.

* *Robinia pseudo-acacia* L. BLACK LOCUST
    Site: Occasionally planted in Saranac Lake and on Campus as an ornamental, but escaping and becoming naturalized.
    Phenology: Mature trees bear prolific, sweet-scented, white flowers in June.
    Miscellany: Seeds poisonous. Native of the southern Appalachian Mountains.

* *Trifolium arvense* L. RABBIT FOOT CLOVER
    Site: Naturalized European. Flowers gray, often on roadsides.
    Phenology: In flower July 17, Paul Smiths area.

* *Trifolium aureum* Pollich or *Trifolium agrarium* L., a nomen ambiguum YELLOW HOP CLOVER
    Site: Naturalized from Europe. Commonly growing with the other clovers.
    Phenology: In flower mid-July.

* *Trifolium pratense* L. RED CLOVER
    Site: Naturalized from Europe. Lawns, roadsides, old fields everywhere.
    Phenology: In flower June and July.
    Miscellany: Hybrids between this species and the next probably occur with pink flowers, intermediate: *T. pratense* × *repens.* = Alsike clover.

* *Trifolium repens* L. WHITE CLOVER
    Site: Naturalized from Europe. Profuse in lawns.
    Phenology: In flower June and July.
    Frequency: Common.

* *Vicia cracca* L. VETCH
    Site: Often trailing over ornamental hedges, and along roadsides and in old fields.
    Phenology: In flower, earliest date, June 9.
                  In flower through July and into October.
    Frequency: Very common.

Miscellany: European naturalized vine.

51. HALORAGACEAE (WATER-MILFOIL FAMILY)

Myriophyllum verticillatum L. var. *pectinatum* Wallr. WATER MILFOIL, WHORLED MILFOIL or GREEN MILFOIL
- Elevation: From 1500 to 1720 feet in the Flora area and at ca. 100 feet at Point Au Roche.
- Phenology: In flower August 5 through September 9.
- Frequency: Six stations.
- Miscellany: In 1979, plants in flower were found in Green Bay and in the Spitfire Narrows so that species determination was finally possible. Muenscher in 1930 had found *M. exalbescens* Fern. in the region, but all plants within the Flora are *M. verticillatum*.

52. LYTHRACEAE (LOOSESTRIFE FAMILY)

Decodon verticillatus (L.) Ell. WATER WILLOW or SWAMP LOOSESTRIFE
- Site: Locally abundant, forming dense stands in shallow water at edges of quiet rivers or lakes. Shade-intolerant.
- Elevation: 1500 to 1700 feet.
- Phenology: In June, leaves conspicuously pale yellow-green.
  In flower August 16.
  Leaves bright red on September 3, with abundant fruit.
- Frequency: Six stations.

* *Lythrum salicaria* L. PURPLE LOOSESTRIFE
- Site: In marshes and wet roadside ditches.
- Phenology: In flower summer.
  In flower, latest date, September 13.
- Frequency: A naturalized European plant, very abundant downstate, choking out other native vegetation. In the Upland, only three localities, but common in the Champlain Valley.
- References: Smith, S.J. (1962)
  Faber (1982)

53. THYMELAEACEAE (MEZEREUM FAMILY)

Dirca palustris L. LEATHERWOOD
- Site: Soils: metasedimentary rock sometimes including calcite (as off Brewster Mountain) to anorthosite on Baxter Mountain. Usually under sugar maple, ash and basswood.
- Elevation: From ca. 1500 feet on Mica Hill near Keene to about 2350 feet on Heaven Hill near Lake Placid.

Frequency: Four stations.

54. ONAGRACEAE (EVENING-PRIMROSE FAMILY)

Circaea alpina L. SMALL ENCHANTER'S NIGHTSHADE
- Site: Mostly in forests (shade-tolerant) where water percolates downslope and raises the pH, typically in springy spots in sugar maple stands.
- Elevation: From 1131 feet on Buck Mountain, Lake George, to 2540 feet on Cascade Mountain.
- Frequency: Twelve stations.

Circaea lutetiana L. ssp. *canadensis* (L.) Ascher & Magnus or *Circaea quadrisulcata* (Maxim.) Franch. & Sav. var. *canadensis* (L.) Hara ENCHANTER'S NIGHTSHADE
- Site: Under hardwoods, well to imperfectly drained.
- Elevation: From 100 feet at Wickham Marsh and on Four Brothers Islands along Lake Champlain to about 1500 feet on Pok-O-Moonshine Mountain.
- Frequency: In Eastern Adirondacks and Lake Champlain-Saint Lawrence Lowlands it is common. With only one station on the Upland at Keene Valley, elevation 1000 feet.

Epilobium angustifolium L. FIRE WEED
- Site: Along log roads, roadsides, edges of woods, and other open places. Shade-intolerant pioneer, following openings only seldom caused by fire. Well-drained.
- Elevation: From 1500 to 2200 feet at Heart Lake.
- Phenology: Flowers July 12 through summer.
- Frequency: Locally abundant.
- Miscellany: Transcontinental.
- Reference: Russell (1971).

Epilobium ciliatum Raf. ssp. *ciliatum* or *Epilobium adenocaulon* Hausskn. or *Epilobium glandulosum* var. *adenocaulon* (Hausskn.) Fernald NORTHERN WILLOW HERB
- Site: Poorly-drained sites, in the sun or in partial shade.
- Elevation: From 100 feet on Four Brothers Islands in Lake Champlain to 2000 feet at Adirondack Museum.
- Phenology: In flower July 28 through September 23.
- Frequency: Four stations at Paul Smiths and Saranac Lake.

Epilobium leptophyllum Raf or *E. lineare* in Peck, 1899. NARROW-LEAVED WILLOW HERB

Site: Along the railroad right-of-way south of Gabriels in a poorly-drained site, pole #346. Saranac River shore below the Stevenson Lane substation in Saranac Lake.
Elevation: 1500 to 1705 feet.
Phenology: In flower and immature fruit August 16.
Frequency: Two stations.

Oenothera biennis L. EVENING PRIMROSE
Site: Pioneer in open places and roadsides. A weedy native.
Elevation: From 100 feet on Four Brothers Islands in Lake Champlain to 2000 feet.
Phenology: In flower July 3 through August 28.
Frequency: Common.
Miscellany: Sweet, often lemon-scented flowers.
Reference: Schottman (1985, August)

Oenothera perennis L. SUNDROPS
Site: Native. Shade-intolerant.
Elevation: From about 800 feet above Upper Jay to 1720 feet at Gabriels.
Phenology: In flower June 19.
In fruit September 4.
Frequency: Two stations.

55. CORNACEAE (DOGWOOD FAMILY)

Cornus alternifolia L.f. ALTERNATE-LEAVED DOGWOOD or PAGODA DOGWOOD
Site: Scattered as individual shrubs, and thus rarely forming thickets in hardwood stands. Shade-tolerant. Well-drained. pH of humus: n=5; min. 4.8; max. 6.6; median 5.1. Soils: limestone along Lake Champlain to Adirondack gneisses and anorthosite. (In the Catskills, on sandstones and shales).
Elevation: From 100 feet at Wickham Marsh along Lake Champlain to 2875 feet on Mount Jo.
Phenology: In flower mid-June.
In fruit August.
Frequency: Occasional.
Miscellany: Blue fruits extremely sour. Dead branches bright orange; the only other woody plants with orange dead branches are big tooth aspen and white ash.

Cornus amomum Mill. SILKY DOGWOOD
Site: Poorly-drained areas in full sun, such as at edges of swamps,

	sluggish streams and ponds. Shade-intolerant.
Elevation:	From 100 feet along Lake Champlain to 1020 feet in Keene Valley, the latter our only station.
Phenology:	In fruit August 31 through October 2.
Miscellany:	Plants of this species and of *C. sericea* are not always clear-cut, because some hybridization probably has occurred.

Cornus canadensis L. DWARF DOGWOOD, BUNCHBERRY or DWARF CORNEL

Site:	Prolific on well-drained mixed woods sites where there is full sun or partial shade. Occurs also in shade on poorly-drained spruce and fir flats. To timberline and into the alpine zone. This species becomes more abundant as one travels north into Canada. pH of humus: n=18; min. 3.75; max. 5.0; median 4.7.
Elevation:	From 210 feet at the Chazy Fir Swamp to 5320 feet on Mount Marcy.
Phenology:	In flower, earliest date, May 26. In flower, median date, June 6. In fruit mid-August through September. In flower, last date, October 19.
Frequency:	Common.
Miscellany:	Fruits non-poisonous, as all Dogwoods, but tasteless.

Cornus foemina Mill. ssp. *racemosa* Lam. J. Wilson or *Cornus racemosa* Lam. GRAY DOGWOOD or PANICLED DOGWOOD

Site:	Common to occasional as an old-field species in the Saint Lawrence and Champlain Valleys. Shade-intolerant, mainly well-drained sites. Soils: largely on limestone and marble. (On sandstone, siltstones, and shales in the Finger Lakes). On anorthosite at Wilmington.
Elevation:	100 feet along Lake Champlain to 600 feet at Pierrepont to ca. 800 feet near Lake Bonaparte and 1300 feet at Wilmington Reservoir.
Phenology:	In fruit August 26 through October 2.
Frequency:	One station on the Upland.

Cornus rugosa Lam. ROUND-LEAVED DOGWOOD or ROUGH-LEAVED DOGWOOD

Site:	A species of the Eastern Adirondacks, Champlain and Saint Lawrence Valleys, occurring on exceedingly well-drained steep and rocky slopes, often on talus and on ledges. Moderately shade-tolerant. Further south of the Mohawk Valley, this species is often under Chestnut oak and Pitch pine. It seems to be migrating up the East Branch Ausable Valley also. Soils: limestone

Elevation: along Lake Champlain to Adirondack gneisses and anorthosite. (Sandstone at Hannawa Falls, and in Catskills with shales).
Elevation: 100 feet along Lake Champlain to Baxter Mountain summit, 2440 feet.
Frequency: Two stations in Keene Valley area.

Cornus sericea L. or *Cornus stolonifera* Michx. RED-STEMMED DOGWOOD or RED OSIER DOGWOOD
Site: In open, wet areas, such as along banks of slow-moving streams and roadside ditches. Shade-intolerant. Soils: limestone soils along Lake Champlain to Adirondack gneisses and anorthosite. (On sandstone and shale at Emmons Bog, SE of Oneonta, New York).
Elevation: 100 feet along Lake Champlain to 3700 feet at Indian Falls.
Frequency: Locally common.
Miscellany: Most conspicuous in winter when the bright red stems form dense thickets 3 to 6 feet high.

56. SANTALACEAE (SANDALWOOD FAMILY)

Comandra umbellata (L.) Nutt. BASTARD TOADFLAX
Site: Edge of Guideboard Road near Silver Lake Road, Town of Black Brook. It often grows with blueberries on burned-over lands.
Elevation: 1040 feet.
Miscellany: Outside the Flora area, *Comandra* occurs northwest of Clintonville, only a few miles away and, more distantly, on Coot Hill above Crown Point. Described in reference sources as a semi-parasite. Host plant/s not identified.

57. CELASTRACEAE (STAFF-TREE FAMILY)

Celastrus scandens L. BITTERSWEET
Phenology: In fruit August 31.
Frequency: Two stations, Saranac Lake and Keene Valley. Protected in N.Y. State.
Reference: Sculley (1987)

58. AQUIFOLIACEAE (HOLLY FAMILY)

Ilex verticillata (L.) Gray including forma *chrysocarpa* WINTERBERRY or BLACK ALDER HOLLY
Site: Locally, in poorly-drained situations. In full sun or partial shade. Mostly on river floodplains, but also in swamps.

Elevation: From 210 feet at the Chazy Fir Swamp to 1820 feet on the Chubb River.
Phenology: In flower late June.
Frequency: 19 stations.
Miscellany: Forma *chrysocarpa* has yellow fruits at Long Pond and Chubb River.

Nemopanthus mucronatus (L.) Loesener ex Koehne or *Nemopanthus mucronata* (L.) Trel. MOUNTAIN HOLLY
Site: Shrub in swamps and other poorly-drained soils, but also venturing up onto well-drained soils provided that the forest canopy is semi-open, as under Red pine. Intolerant to moderately-tolerant of shade. To timberline. Soils: Adirondack gneisses and anorthosite. (Potsdam sandstone at Altona and other sandstones in the Catskills).
Elevation: From 750 feet at Altona to 4857 feet on Dix Mountain.
Phenology: In flower late May into early June.
In fruit mid-August to September.
Frequency: Profusely abundant.

59. EUPHORBIACEAE (SPURGE FAMILY)

Chamaesyce maculata (L.) Small or *Euphorbia supina* Raf. MILK PURSLANE or SPOTTED SPURGE
Elevation: Saranac Lake railroad station, elevation 1560 feet.
Phenology: In flower August 27.
Miscellany: A native prostrate weed.

* *Euphorbia cyparissias* L. CYPRESS SPURGE
Site: Along sandy, well-drained roadsides and railroad grades.
Phenology: In flower between May 29 and June 16.
Frequency: Occasional.
Miscellany: Naturalized from Europe as a garden escape.

60. VITACEAE (VINE FAMILY)

Parthenocissus quinquefolia (L.) Planch. ex DC. VIRGINIA CREEPER
Site: Soils: limestone soils along Lake Champlain to Adirondack gneisses. (Sandstone and shales in the Catskills).
Elevation: From 100 feet along Lake Champlain to almost 1900 feet on Mount Baker.
Frequency: Common in the Champlain Valley and eastern Adirondacks, but uncommon in the interior Upland. Three stations on the Upland.

Miscellany: Escaped from cultivation in Saranac Lake on fences and building walls. Although it is totally harmless liana, or woody vine, with five leaflets it remotely resembles poison ivy. It can shade out host shrubs and trees that it grows on. Leaves turn brilliant scarlet in autumn.

Vitis riparia Michx. RIVER-BANK GRAPE or FROST GRAPE
Site: Soils: limestone soils along Lake Champlain to metasedimentary rocks at Pierrepont to Adirondack gneisses near Willsboro.
Elevation: 100 feet on Four Brothers Islands "C" and "D" in Lake Champlain to 600 feet at Bower's Farm in Pierrepont to 1020 feet at Keene Valley.
Phenology: Flower buds forming June 2, on Four Brothers Islands.
In fruit October 27, and August 31, on Four Brothers Islands.
Miscellany: Some grapes are cultivated in the Flora area and escapes are occasional. One example is at the Woodruff Street bridge in Saranac Lake over the Saranac River; the plant has not fruited and identification to species is thus difficult. Also in Keene Valley.

61. POLYGALACEAE (MILKWORT FAMILY)

Polygala paucifolia Willd. FRINGED POLYGALA, FLOWERING WINTERGREEN or GAY-WINGS
Site: Shade-tolerant, under northern hardwoods and/or oak and/or white pine-hemlock, well-drained.
Elevation: 420 feet above Wadhams to 1380 feet in the Town of Black Brook.
Phenology: In flower June 1 through June 19.
Frequency: Two stations. This species is infrequent in the eastern Adirondacks and in the Saint Lawrence Lowlands, but rare on the interior Upland.

62. ACERACEAE (MAPLE FAMILY)
References on maples in general:
Hudler (1984)
Ketchledge (1962, April-May)
Miller and Allen (1972)
Miller and Silverborg (1967)

* *Acer negundo* L. BOX ELDER or ASH-LEAVED MAPLE
  - Site: Introduced probably from the Midwestern United States. It is found only in the villages. Escaped and abundant in Saranac Lake, especially along the River as at Denny Park. Reproduces prolifically, with saplings arising on vacant lots, sidewalk cracks, etc. Many are planted as ornamentals. Shade-intolerant. Well to poorly-drained soils. Fast-growing.

Acer pensylvanicum L. STRIPED MAPLE, MOOSEWOOD, GOOSEFOOT MAPLE or WHISTLEWOOD
  - Site: Abundant, shade-tolerant understory tree mostly of the northern hardwood forest, but also reaching into the lower elevations of the spruce-fir. Well-drained. Often forming profuse thickets following logging when subject to sunlight. The first tree to show signs of drought stress during dry periods when the large, thin, membranous leaves readily wilt. *Aster acuminatus* and *Oxalis montana* are others giving early warnings of drought. On shallow-soiled bedrock knob summits, surrounded by northern hardwoods forest on deep soil, striped maple forms thickets often with Mountain maple, small red spruce, and shrubs.
  - Elevation: From 588 feet at Camp Pok-O-MacCready near Willsboro to about 3600 feet on Porter Mountain. Uncommon above 3000 feet.
  - Phenology: In flower late May into early June.
    In fruit September.
  - Frequency: Common.
  - Reference: Hibbs (1980).

* *Acer platanoides* L. NORWAY MAPLE
- Site: The largest station was on Woodruff Street in Saranac Lake at the railroad overpass, but most have been cut down. More on Marshall Street in Saranac Lake, and on Franklin Avenue at South Street. Shade tolerant.
- Phenology: The seeds germinate in large numbers in spring. Norway maple drops its leaves late in the autumn (almost November 1st, after our native species are bare) because the growing season in western Europe is longer than ours and it takes generations for most species to adjust. In other words, it is still on a European clock!
- Miscellany: European ornamental becoming naturalized.

Acer rubrum L. RED MAPLE, SOFT MAPLE or SWAMP MAPLE
- Site: Profusely common wherever Sugar maple cannot dominate, especially in poorly-drained, shallow, or exceedingly sandy soils, or in a combination of these conditions. Water table depth, soil depth, and soil texture thus seem immaterial to this moderately shade-tolerant tree. It will grow most anywhere except in dense shade (as under Sugar maple, Beech, and Hemlock). It will even grow, though not well and is short-lived, in bogs. pH of humus: n=56; min. 3.5; max. 5.05; median 4.5. Soils: limestone soils at Chazy to Adirondack gneisses and anorthosite. (Sandstones, shales, and conglomerates in the Catskills).
- Elevation: From 200 feet at Lake Alice near Chazy to 3320 feet on Loon Lake Mountain. Uncommon above 3000 feet. Surely one can find Red maple along Lake Champlain at 100 feet.
- Phenology: In flower first and second weeks of May.
  In fruit first week of June.
  Heavily-fruiting trees (as in 1982) leaf out later and only sparsely.
  This, and Silver maple are the Flora area's only spring-fruiting maples. All the others fruit in the autumn.
- Frequency: Common.
- Miscellany: The name Swamp maple is also used synonymously with Silver maple. This creates considerable confusion.

Acer saccharinum L. SILVER MAPLE or SWAMP MAPLE
- Site: Native to the Adirondacks. Always on alluvial soils (river flood plains) for reasons not understood. The need for a high water table appears unnecessary as trees planted on well-drained sites in the villages such as Saranac Lake do equally well. The trees *do* need, however, full sun, and cannot compete in the forest with other species above the flood plains. Silver maple appear to tolerate vernal and occasional aestival, or summer, flooding better than any other tree. Soils: limestone soils of Champlain Valley to outwash and alluvium from Adirondack gneisses.
- Elevation: From 100 feet along Lake Champlain at Four Brothers Islands (especially in swamps east of South Hero, Vermont) to 1600 feet above Raquette Falls.
- Phenology: Native trees in flower, early May; ornamentals in April.
  In fruit very late May through June.
  Silver maple, with Red, are the only vernally-fruiting maples. All the others fruit in the fall.
  Silver maple can be used to demonstrate that biological clocks do not change overnight. Native Silvers drop their leaves in early October, while those planted in the villages drop theirs several weeks to a full month later. The reason? The ornamentals have been brought in from places much further south where the growing season is longer and leaf drop is later. The ornamentals do not promptly adjust to the Adirondack growing season. In fact, their progeny in the villages do not either. How many generations are required for these maples to adjust to a short Adirondack growing season?
- Frequency: Six stations.

Acer saccharum Marsh. SUGAR MAPLE, HARD MAPLE or ROCK MAPLE
- Site: The dominant tree of the northern hardwood forest along with Beech. Probably also the single most important tree in the Adirondacks controlling the distribution of other tree species and ground cover plants. One of the most shade-tolerant trees. Mostly on deep well-drained soils, but it can grow on imperfectly drained sites provided that the slope is moderately steep and that the water is flowing, as at springs and seepages. Sugar maple cannot grow well where the soil is too poorly-drained (water not moving), too sandy (as on some outwash deposits), or too shallow (over bedrock ledges). Any one of these conditions can prevent Sugar maple from thriving and, certainly, two or more of them combined can eliminate it entirely. Other species, such as Red maple, Yellow birch, and the conifers will replace it on these marginal sites. 94% sand seems an upper limit for good Sugar

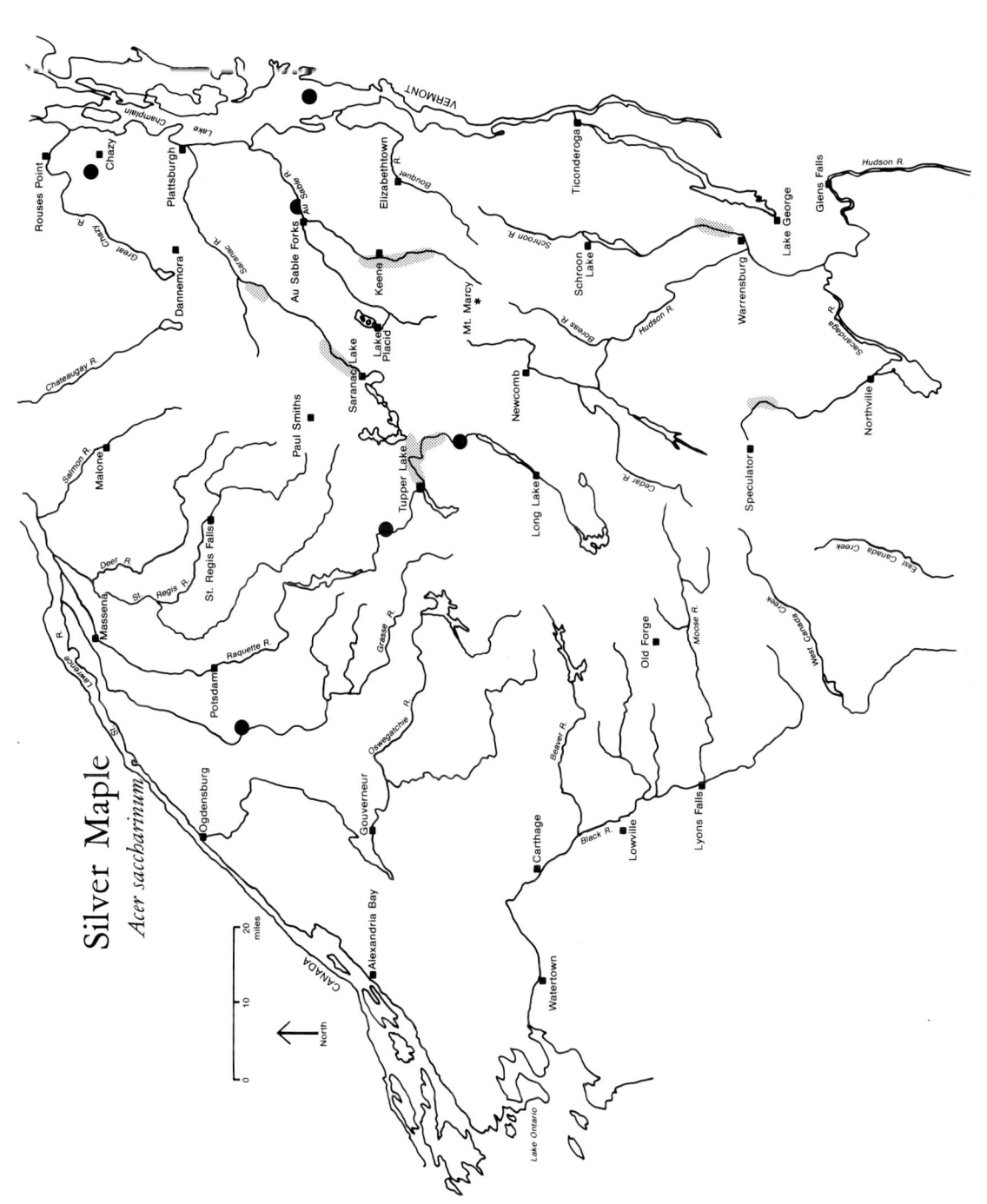

maple growth. A higher percentage will weaken the trees on that site.

Sugar maple makes richer humus which permits a number of ground cover species to exist where they could not under trees producing poorer, more acid humus (see chapter on soils).

SOIL TEXTURE:
    On till without Hop hornbeam, White ash, American elm, nor Basswood:
        % Silt
            B horizons: n=25; min. 3.2; max. 53.8; med. 22.6
            C horizons: n=46; min. 1.4; max. 29.1; med. 10.9
        % Clay
            B horizons: n=22; min. 1.4; max. 6.5; med. 2.35
            C horizons: n=37; min. 1.1; max. 5.3; med. 1.8
    On Outwash:
        % Silt
            B horizons: n=18; min. 3.8; max. 50.0; med. 18.5
            C horizons: n=50; min. 0.7; max. 30.0; med. 2.9
        % Clay
            B horizons: n=18; min. 0.7; max. 5.2; med. 2.7
            C horizons: n=36; min. 0.7; max. 3.0; med. 1.65

On till with Hop hornbeam, and/or White ash and/or American elm and/or Basswood. For comparison, see soil texture data for these other species.

pH OF HUMUS: on till where Hop hornbeam, White ash, American elm, and Basswood are all absent: n=30; min. 3.2; max. 5.0; med. 4.45.

    On outwash: n=24; min. 4.0; max. 5.0; med. 4.5.

On till with Hop hornbeam and/or White ash and/or American elm, and/or Basswood present. For comparison, see soil pH data for these other species. Generally, Sugar maple grows with these four species on humus pH up to 6.8.

    Soils: limestone soils along Lake Champlain to Adirondack gneisses and anorthosite. (Sandstone and shales in the Catskills).

        Elevation: From 100 feet along Lake Champlain to 3543 feet on S. slope of Porter Mountain in an old burn. Uncommon above 2500 feet in the northern Adirondacks, and above 2800 feet in the southern Adirondacks.

        Phenology: Sap rises mostly in March when it is tapped for the maple sugar industry.

                          In flower mid-May with leaf expansion.

                          In fruit September.

                          Lewis Staats states that every 4th year is a heavy fruiting year, as was 1988 for example.

        Frequency: Common.
        Miscellany: Official State Tree of New York.
        References: Borland (1976)
                          Drahos (1955)   Huber (1964)

Hibben (1961) King (1965)
Huber (1956) LaBastille (1974)

Acer spicatum Lam. MOUNTAIN MAPLE
    Site: The Flora area's smallest and least shade-tolerant maple. Shrubby, most common on talus, exposed ledges, on steep slopes where Yellow birch often dominates, and on tops of huge boulders, almost always in full sun. Very abundant between 2500 and 3000 feet in the transition zone between northern hardwoods and spruce-fir. Well-drained. Soils: limestone along Lake Champlain to Adirondack anorthosite and gneisses. (Catskills shales, sandstones and conglomerates).
    Elevation: From 100 feet along Lake Champlain in Plattsburgh to 3706 feet on Blue Mountain. Uncommon above 3000 feet.
    Phenology: In flower mid-June.
    In mature fruit September.
    Frequency: Common.

63. ANACARDIACEAE (CASHEW FAMILY)

Rhus typhina L. STAGHORN SUMAC
    Site: Scarce in the Flora area because of lack of fields and open places. Shade-intolerant, on well-drained sites; mostly commonly found along roadsides, fencerows and in old fields. Soils: limestone in Champlain Valley to Adirondack anorthosite, gneiss, and metasedimentary rocks. (Catskills shales and sandstones).
    Elevation: From 100 feet along Lake Champlain to about 2000 feet in Blue Mountain Lake. Common in Saint Lawrence and Champlain Valleys.
    Phenology: In flower June.
    In fruit September and well into winter.
    Frequency: Eight stations.
    Miscellany: Fruits are red-hairy and can be used to make a lemonade substitute. No part of the plant is poisonous. The largest local population was SE of the Mount Baker trailhead on an unpaved street near the power line. This population was still expanding in the early 1970s, and reached maximum size in the late 1970s. Since, it has been declining, mostly due to overshadowing by trees which outgrow it.
    Reference: Doust and Doust (1988).
    MacArthur (1958).

Toxicodendron radicans (L.) Kuntze or *Rhus radicans* L. POISON IVY
    Site: Poison ivy becomes more common at lower elevations and

	toward the periphery of the Adirondacks. It seems most abundant along river flats and flood plains, but can occur on well-drained areas as well. Moderately shade-tolerant. Not in the most acid sites. Soils: limestone soils along Lake Champlain to Adirondack gneisses. (Sandstones and shales in the Catskills).
Phenology:	White fruits persistent through December 26.
Frequency:	Eight stations.
References:	Crooks & Kephart (1945, 1951)
	Headstrom (1957–1958)
	Ketchledge (August-September, 1962)
	Kingsbury (1971)

64. OXALIDACEAE (WOOD-SORREL FAMILY)

Oxalis acetosella L. or *Oxalis montana* Raf. WOOD SORREL

Site:	The commonest herb of the spruce-fir forest since it is both shade-tolerant and will grow on very acid sites. Well to imperfectly drained places. Scattered in northern hardwood stands, most commonly under hemlock. pH of humus: n=30; min. 3.2; max. 5.1; median 4.5. Under direct sun and/or during times of drought, the leaflets fold up. To timberline.
Elevation:	From between 1000 and 1200 feet in Keene Valley to 4940 feet on Haystack Mountain.
Phenology:	In flower June 12 to July 15.
Frequency:	Common.
Miscellany:	Flowers pollinated by bees and other insects. The sour, edible clover-like leaves are good for salads, and unroll like fern fronds in spring, replacing the evergreen leaves of the previous year.

* *Oxalis corniculata* L. LADY'S SORREL

Site:	Weedy, often profuse in greenhouses.
Frequency:	Common.
Miscellany:	Naturalized from Europe.

65. GERANIACEAE (GERANIUM FAMILY)

Geranium robertianum L. HERB ROBERT

Site:	Moderately shade-tolerant, usually well-drained sites.
Elevation:	From 1294 feet on Pok-O-Moonshine Mountains to nearly 2000 feet on The Cobbles and 2080 feet on Mount Pisgah. Probably lower.
Frequency:	Two stations.

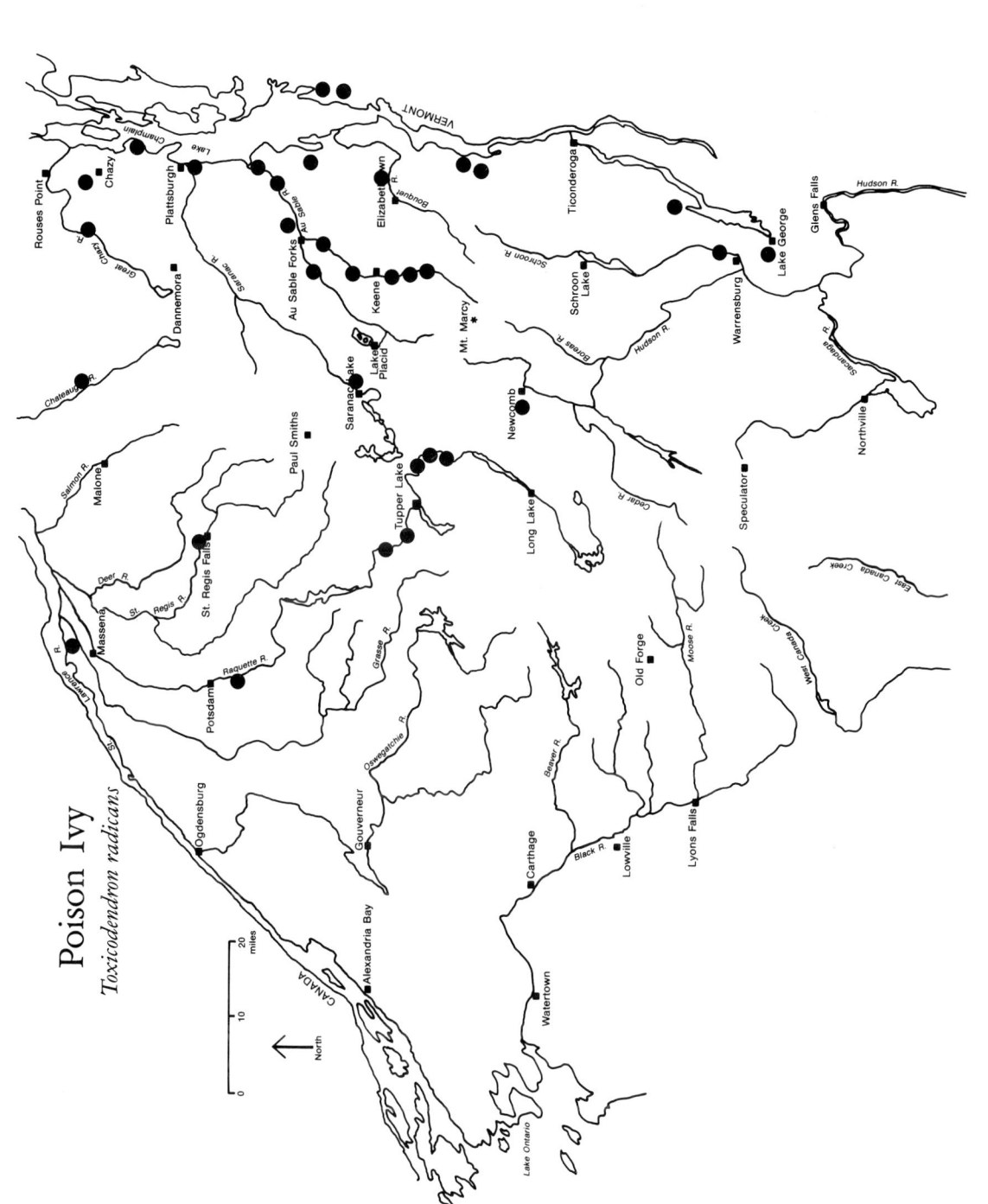

Poison Ivy
Toxicodendron radicans

66. BALSAMINACEAE (TOUCH-ME-NOT FAMILY)

Impatiens capensis **Meerb.** or *I. biflora* **Walt.** TOUCH-ME-NOT, JEWELWEED, SNAPWEED or SPOTTED TOUCH-ME-NOT

- Site: Shade-tolerant. This species most often has a high water table requirement and grows in poorly-drained sites and at springs. Where the water table is a foot from the surface, the plants rarely reach maturity and flower; on very well-drained soils, the seeds will not germinate.
- Elevation: From 100 feet along Lake Champlain to 2500 feet on Cascade Mountain.
- Phenology: In flower July through early September.
- Frequency: 23 stations but probably many more.
- Miscellany: A most unusual plant in many ways. One of the Flora area's few native annuals, lacking the overwintering rhizomes of the perennials; only the seeds of *Impatiens* overwinter. The name "Jewelweed" refers to the trapping of air bubbles (hence "Jewels" or "pearls") when the leaf is submerged. The name "Touch-me-not" refers to the fruits, which snap open abruptly when squeezed, shooting forth edible seeds. The sap is reported to be a soothing substance for poison ivy. Ruby-throated humming birds in New York City have been seen pollinating this species. A note in an advertisement for *Natural History* magazine states that the flowers start out as males, dispersing pollen; later, they change to female flowers to accept pollen. If the stamens and pistil matured at the same time, self-fertilization would be possible.
- References: Doeffinger (1978)
  Schemske (1978)
  Schottman (1986, August)

Impatiens pallida **Nutt.** PALE TOUCH-ME-NOT or SNAPWEED

- Site: High water table areas. Shade-tolerant. The more southerly, lower-elevation species of the two *Impatiens* in our region. Common in the Catskills and Finger Lakes, but not in the northern part of New York
- Elevation: About 1000 feet at Chateaugay Falls to 1900 feet on Baldface Mountain.
- Phenology: In flower August 8.
- Frequency: One Station.

67. ARALIACEAE (GINSENG FAMILY)

Aralia hispida **Vent.** BRISTLY SARSAPARILLA

- Site: Locally, on dry, bare outwash sands, and on exposed rocky ledges.

Elevation: Intolerant of shade, and capable of withstanding drought.
Elevation: From 1000 to 1200 feet in Keene Valley to 3950 feet on Cascade Mountain. Possibly to timberline.
Phenology: In flower bud June 16, and in flower July 15.
Frequency: Seven stations.
Reference: Thomson, McKenna & Cruzan (1989).

Aralia nudicaulis L. WILD SARSAPARILLA
Site: Under northern hardwoods and spruce-fir forests. Shade-tolerant, normally well-drained. To timberline. pH of humus: n=17; min. 3.75; max. 6.6; median 4.7.
Elevation: From 100 feet along Lake Champlain at Point Au Roche to Gothics 4736 feet.
Phenology: In flower May 25 through June 12.
              In fruit September.
              Leaves purple when expanding in spring, green in summer, and brilliant yellow in late September and into early October.
Frequency: Common.
Miscellany: Roots can be used for making a beverage, but my attempt failed; can one eat the roots raw? This is an unusual plant in that the flowers and fruits are borne on a separate stem beneath the leaves.
Reference: Schottman (June, 1990).

Aralia racemosa L. SPIKENARD, PETTY MORREL or LIFE-OF-MAN
Site: Well-drained, shade-tolerant herb of northern hardwood forests. Never in colonies, but always as widely isolated individuals.
Elevation: From 100 feet along Lake Champlain at Point Au Roche and Wickham Marsh to about 2000 feet above The Garden at Keene Valley.
Frequency: Six stations.
Miscellany: A large, spreading, tall herb.

Panax trifolius L. DWARF GINSENG
Site: Shade-tolerant, well-drained herb of northern hardwoods forest. Never abundant but, locally, in small colonies. Mostly under Sugar maple.
Elevation: From 1400 feet near Deer River to 1774 feet at Marsh Pond Mountain base.
Phenology: In flower May 14 through June 2.
Frequency: Ten stations.
Miscellany: Unlike its larger cousin, *P. quinquefolius* (which is nearly extinct due to human overpicking), this species fortunately is not attractive to the Chinese and others for medicinal purposes.

68. APIACEAE or UMBELLIFERAE (PARSLEY or CARROT FAMILY)

Angelica atropurpurea L. ALEXANDERS
    Site: Poorly-drained sites, in sun or under partial shade, often on flood plains.
    Elevation: From 550 feet below Ausable Forks to 1630 feet at Weller Brook.
    Phenology: In flower *bud* June 19.
                   In immature fruit July 28.
    Frequency: Three stations.

Cicuta bulbifera L. BULB-BEARING WATER HEMLOCK
    Site: Along edges of slow-moving streams and ponds. Poorly-drained, shade-intolerant.
    Elevation: From 100 feet at Point Au Roche along Lake Champlain to 1670 feet on the N Branch Saranac River.
    Phenology: In fruit August 19 through September 9.
    Frequency: Eight stations.
    Miscellany: Toxic?

* *Daucus carota* L. QUEEN ANNE'S LACE or WILD CARROT
    Site: European naturalized species scarce in the Paul Smiths area but abundant in the Saint Lawrence Valley. Well-drained, sunny disturbed places.

Heracleum lanatum Michx. or *H. maximum* Bartr. COW-PARSNIP
    Site: Open, sunny swamps, and edges of slow-moving streams. Poorly-drained and shade-intolerant.
    Elevation: Along E Branch of Saint Regis River above Everton Falls, elevation 1410 feet.
    Frequency: One station.

Hydrocotyle americana L. WATER PENNYWORT
    Site: Poorly-drained, often shaded, areas, especially at edges of brooks and in springs.
    Elevation: From 800 feet at Upper Jay to 1630 feet at Easy Street and Dan Brook near Paul Smiths.
    Phenology: In fruit September 13 through September 27.
    Frequency: Seven stations.

Osmorhiza claytonii (Michx.) C.B. Clarke SWEET CICELY or SWEET JARVIL
    Site: Shade-tolerant. Well-drained areas under rich-site northern hardwoods with sugar maple, and often basswood, white ash, elm, and

hop hornbeam. pH of humus: n=6; min. 5.0; max. 6.3; median 5.15.
- Elevation: From 100 feet at Point Au Roche to 2157 feet at Buck Hill.
- Phenology: In fruit August 8 to September 20.
- Frequency: 18 stations.

Sium suave **Walt.** WATER PARSNIP
- Site: Banks of meandering streams, poorly-drained and shade-intolerant.
- Elevation: From 100 feet at Point Au Roche to 1670 feet on N Branch Saranac River.
- Phenology: In immature fruit August 19 through September 8.
- Frequency: Three stations.

69. GENTIANACEAE (GENTIAN FAMILY)

Gentiana linearis **Froel.** NARROW-LEAVED GENTIAN, CLOSED GENTIAN or BOTTLE GENTIAN
- Site: Poorly-drained soils. Infrequent. To timberline. Shade-intolerant.
- Elevation: From about 1551 feet above Meacham Lake to 4736 feet on Gothics.
- Phenology: In flower August 10 through 24.
- Frequency: Nine stations. Protected in New York State.

70. *APOCYNACEAE (DOGBANE FAMILY)*

Apocynum androsaemifolium **L.** DOGBANE or SPREADING DOGBANE
- Site: Native, shade-intolerant species often along roadsides in old fields and open areas. Mostly well-drained.
- Elevation: From 210 feet at Chazy Fir Swamp to ca. 3450 feet on Pitchoff Mountain.
- Phenology: In flower June 19 through August 18. In fruit September.
- Frequency: Common.
- Miscellany: Probably poisonous.

71. ASCLEPIADACEAE (MILKWEED FAMILY)

Asclepias incarnata **L.** SWAMP MILKWEED
- Site: Infrequent in open, sunny, poorly-drained areas such as riverbanks.
- Elevation: From 100 feet along Lake Champlain to 1620 feet at Black Pond.

    Phenology: In flower July 17.
                         In fruit September 7.
    Frequency: Eight stations, but probably many more.
    Miscellany: Poisonous to eat.

Asclepias syriaca L. MILKWEED
    Site: Often growing with Eurasians in weedy areas such as roadsides, old fields, etc.
    Elevation: From 150 feet at Point Au Roche on Lake Champlain to perhaps 1700 feet.
    Phenology: In flower July 11.
                         In fruit with seed release October.
    Miscellany: Native, despite its specific name "of Syria". Young shoots edible when cooked.
    References: Doeffinger (1978)
                         Ives (1975)
                         Schottman (September 1988)

72. SOLANACEAE (NIGHTSHADE FAMILY)

* *Solanum dulcamara* L. NIGHTSHADE or BITTERSWEET
    Site: Poisonous vine, naturalized from Europe, twining over other vegetation. Shade-tolerant.
    Phenology: In flower June 19.
                         In fruit September.
                         The fruits, in various stages of ripening, look like red-amber-green railroad signals.

73. CONVOLVULACEAE (CONVOLVULUS or MORNING-GLORY FAMILY)

Calystegia sepium (L.) R. Br. or *Convolvulus sepium* L. BINDWEED, HEDGE BINDWEED or WILD MORNING-GLORY
    Site: Saranac River above Bloomingdale, along flood plain.
    Elevation: 1510 feet.

74. MENYANTHACEAE (BUCKBEAN FAMILY)

Menyanthes trifoliata L. BUCKBEAN or BOGBEAN
    Elevation: From 1630 feet in Bog Willow Bog to 1770 feet at Ferd's Bog.
    Phenology: In flower May 26–27.
    Frequency: Two stations around Paul Smiths.

75. HYDROPHYLLACEAE (WATERLEAF FAMILY)

Hydrophyllum virginianum L. WATERLEAF
- Site: It is a shade-tolerant of mostly imperfectly-drained sites, especially in springy areas and seeping banks under Sugar maple, often with White ash and Basswood.
- Elevation: From 200 feet at Chazy to about 2100 feet at Cascade Lakes.
- Frequency: One station.

76. BORAGINACEAE (BORAGE FAMILY)

* *Echium vulgare* L. BLUEWEED or VIPER'S BUGLOSS
- Site: European. Weedy, with the flowers an unusual combination of color: blue and pink-purple. Saranac Lake.

* *Myosotis sylvatica* Hoffm. FORGET-ME-NOTS
- Site: Often escaped from gardens and naturalized. Mostly in springy and poorlydrained sites. European. Shade-midtolerant. Paul Smiths and Saranac Lake.
- Phenology: In flower June 18 through July 4.

77. VERBENACEAE (VERVAIN FAMILY)

Verbena hastata L. BLUE VERVAIN
- Site: Riverbanks, wet ditches, and marshes.
- Elevation: From 100 feet along Lake Champlain to 1653 feet at Mud Pond, near Lake Kushaqua.
- Phenology: In immature fruit August 26.
  In fruit October 27.
- Frequency: One station.

78. LAMIACEAE or LABIATAE (MINT FAMILY)

* *Clinopodium vulgare* L. or *Satureja vulgaris (L.) Fritsch.* BASIL
- Site: European weed. Keese's Mills Quarry.

* *Galeopsis tetrahit* L. HEMP NETTLE
- Site: Often spreading into wooded areas and appearing with native species, as along log roads. Moderately shade-tolerant.
- Phenology: In fruit October 25.
- Frequency: Common European weed.

* *Glechoma hederacea* L. GILL-OVER-THE-GROUND

Site: In Saranac Lake, creeping through lawns.
Phenology: Flowers purple in late May to June 10.
Frequency: Common European weed.

Lycopus americanus Muhl. ex. Bart. BUGLEWEED or CUT-LEAVED WATER HOREHOUND
Site: Poorly-drained soils. Shade mid-tolerant.
Elevation: From 100 feet along Lake Champlain to about 1700 feet on Beech Hill.
Phenology: In fruit August 24 through October 27.
Frequency: One station.

Lycopus virginicus L. BUGLEWEED
Site: In wet sites, especially on stumps in ponds. Shade intolerant to midtolerant.
Elevation: From 100 feet along Lake Champlain to 2115 feet near Blue Mountain Lake.
Phenology: In flower August 18 through 21.
Frequency: Common.

Mentha arvensis L. MINT
Site: Circumboreal native, in poorly-drained, partly-shaded sites.
Elevation: From 100 feet along Lake Champlain to about 1700 feet.
Phenology: In flower August 29 to 31.
In fruit October 27.
Frequency: Five stations.

* *Prunella vulgaris* L. SELF HEAL or HEAL ALL
Site: In lawns, along roadsides, often penetrating deeply into forests only along log roads and trails. It cannot compete with the Flora area's native species. Tremendous ability to withstand trampling as the Path rush. Shade-intolerant.
Phenology: In flower July 16.

Scutellaria galericulata L. or *S. epilobiifolia* A. Hamilton. SKULLCAP
Site: Poorly-drained areas. Shade-midtolerant.
Elevation: 1500 to 2000 feet.
Phenology: In flower July 13 through August 28.
Frequency: Occasional.

Scutellaria lateriflora L. MAD DOG SKULLCAP
Site: Poorly drained areas.
Elevation: From 100 feet along Lake Champlain to 1800 feet on Jones Hill.

Phenology: In flower August 21.
In fruit October 27.
Frequency: Nine stations.

* *Thymus pulegioides* L. or *Thymus serpyllum* of American authors, not L. WILD THYME
Site: European garden escape on lawns in Saranac Lake.
Phenology: In flower July through October.

79. CALLITRICHACEAE (WATER-STARWORT FAMILY)

Callitriche palustris L. WATER STARWORT
Site: Aquatic.
Elevation: 1500 to 1700 feet.
Phenology: In fruit August 26.
Frequency: Two stations: Osgood River and Fish Creek.

80. PLANTAGINACEAE (PLANTAIN FAMILY)

* *Plantago lanceolata* L. NARROW-LEAVED PLANTAIN
Frequency: Abundant European weed.

* *Plantago major* L. COMMON PLANTAIN
Site: This is the species which often follows trails and log roads well into forests, being able to withstand much trampling.
Frequency: Commonest European weed.

Plantago rugelii Dcne. BROAD-LEAVED PLANTAIN
Elevation: 1630 feet.
Frequency: Two stations.
Miscellany: A native plant. Most of the broad-leaved plantains on Campus are *P. major*, a European weed, while a few are this native similar species. Often, both grow together.

81. OLEACEAE (OLIVE FAMILY)

Fraxinus americana L. WHITE ASH
Site: One of the Flora area's more demanding trees, usually on higher pH humus. Mostly well-drained sites, although infrequently in imperfect. Moderately shade-tolerant, and thus moving in on a site most often following disturbances. pH of humus: n=41; min. 3.8; max. 5.8; median 5.2. Soils: limestone along Lake Champlain to Adirondack gneisses and anorthosite. (Catskills shales and sandstones). Soil texture:

% silt:
> B horizons: n=51; min. 10.0; max. 57.4; med. 28.8
> C horizons: n=16; min. 3.2; max. 50.0; med. 20.6

% clay:
> B horizons: n=17; min. 0.5; max. 5.8; med. 2.5
> C horizons: n=14; min. 0.2; max. 4.3; med. 1.7

Elevation: 100 feet along Lake Champlain to 3200 feet on Noonmark Mountain. Uncommon above 2500 feet.
Phenology: In barely half-leaf on May 29, while other hardwoods are fully-leaved.
In fruit October 23.
Frequency: 67 stations.
Miscellany: Is the sap really good for soothing insect bites and stings?
References: On White ash specifically:
> Faust (1977)

On Ashes in general:
> Ketchledge (1964, Aug.-Sept.)
> Miller, G.N. (1955)
> Silverborg, Risely and Ross (1963)
> Zabel, Collins, and Yops (1963)

Fraxinus nigra Marsh. BLACK ASH
Site: Occasional in swamps and flood plains, often with Elm, Alder, Red maple and White cedar. One of the few trees that always indicates a high water table. Not tolerant of shade. Below Raquette Falls in swamp along road with swollen bases, like a Tupelo, at low water. Soils: limestone soils along Lake Champlain to Adirondack gneisses and anorthosite. (Shales and sandstones in Ulster County, N.Y.)
Elevation: 100 feet along Lake Champlain at Point Au Roche to greater than 2800 feet on Hurricane Mountain.
Phenology: Barely breaking dormancy (leafing) on May 25.
Frequency: 24 stations.

82. SCROPHULARIACEAE (FIGWORT FAMILY)

Chelone glabra L. TURTLEHEAD
Site: Poorly-drained sites, mostly in sun.
Elevation: From 800 feet at Upper Jay to 4735 feet on Gothics.
Phenology: In flower August.
Frequency: Nine stations.

* *Euphrasia stricta* Wolff ex Lehm. or *Euphrasia condensata* Jord. EYEBRIGHT
Site: Naturalized European in usually wet sand in open, disturbed areas.

Phenology: In flower on Campus August 23 through September 7.
Frequency: 4 stations.

* *Linaria vulgaris* Hill. BUTTER-AND-EGGS
Frequency: Prolific European weed.

Melampyrum lineare Desr. COW WHEAT
Site: Moderately shade-tolerant herb on well-drained sites, often under semi-open stands of red pine and bigtooth aspen.
Elevation: From 1095 feet on Coot Hill, above Bulwagga Bay, to 5300 feet on Mount Marcy, reports Dr. Ketchledge.
Phenology: In flower June 19 to August 18.
Frequency: Five stations, but probably many more.

Pedicularis canadensis L. WOOD BETONY or LOUSEWORT
Site: Well-drained, usually open, partially-shaded woods, under northern hardwoods.
Phenology: In flower May 7 in the Catskills.
Frequency: Baxter Mountain in Keene Valley.
Miscellany: A semi-parasitic plant.

Rhinanthus crista-galli L. YELLOW RATTLE
Site: Edge of Sabattis highway at Sperry Brook Bog.
Elevation: 1730 feet.
Phenology: In flower July 14.
Also at the summit of Mount Marcy at 5325 feet in immature and mature fruit on August 31.
Miscellany: This plant is common northward in Canada.

Scrophularia marilandica L. FIGWORT or CARPENTER'S-SQUARE.
Site: This is a native plant which often grows with Europeans in disturbed, open places.
Elevation: First observed on Paul Smith's College Campus at elevation 1625 feet. Also at Mount Pisgah, elevation 1620 feet.
Phenology: In immature fruit early to mid-July.
Frequency: Two stations.

* *Verbascum thapsus* L. GREAT MULLEIN
Site: Well-drained sites, intolerant of shade. Old fields, roadsides, disturbed places. (See page 18.)
Phenology: In flower, first of season, July 3 through August.
Frequency: Very common European.
Miscellany: Biennial (plant living two years), producing a rosette of velvety leaves on the ground the first year, and a 5 to 10–foot tall

                                spikelike inflorescence on the second.
        Reference: Schottman (1985, September).

Veronica americana (Raf.) Schwein, ex Bentham AMERICAN BROOKLIME or AMERICAN SPEEDWELL
        Site: Poorly-drained sites such as in small brooks (partly submerged), in springs and other places of seepage, often in wooded areas. Shade-tolerant. Often with *Hydrocotyle* and *Chrysosplenium.*
        Elevation: From about 1300 feet to 1750 feet.
        Phenology: In flower July 14.
                          In fruit October.
                          Evergreen and observed in a non-frozen pool in early March, 1983, SW of Buck Hill.
        Frequency: Three stations between Paul Smiths and Loon Lake.

Veronica officinalis L. COMMON SPEEDWELL
        Site: Native shade-tolerant on well-drained sites, often growing along trails and logging roads, and thus appearing as a naturalized species. It will grow also in old-growth areas distant from roads or any disturbance. Under northern hardwoods.
        Elevation: From ca. 450 feet on Tongue Mountain to from 3500 to 4000 feet along the Whiteface Mountain Highway.
        Phenology: In flower, first of season, June 7.

Veronica scutellata L. MARSH SPEEDWELL
        Site: Saranac River flood plain below the Stevenson Lane substation in Saranac Lake. Lake Placid lake near mouth of Minnow Brook.
        Elevation: 1510 and 1860 feet.
        Phenology: Flowers and immature fruit September 7.
        Frequency: Two stations.

* *Veronica serpyllifolia* L. THYME-LEAVED SPEEDWELL
        Site: On lawn at 72 Bloomindgale Avenue, Saranac Lake.
        Phenology: In flower June 10.
        Frequency: European common creeping weed.

83. OROBANCHACEAE (BROOMRAPE FAMILY)

Epifagus virginiana (L.) Bart. BEECH DROPS
        Site: A parasitic, non-green, flowering plant common on Beech roots and/or mycorrhiza.
        Elevation: same as for beech, from 100 feet along Lake Champlain to about 2500 feet. Uncommon above 2500 feet.

Phenology: In fruit October 28.
Reference: Schottman (1982, August).

84. LENTIBULARIACEAE (BLADDERWORT FAMILY)

Utricularia cornuta **Michx.** HORNED BLADDERWORT
   Site: Spurred yellow flowers on leafless stems arising from mud, muck, or peat. No leaves apparent. Poorly-drained. Shade-intolerant.
   Elevation: 1500 to 1700 feet.
   Phenology: In flower September 5.
   Frequency: Three stations: Paul Smiths and Middle Saranac Lake.
   Reference: Kondo (1972)

Utricularia gibba **L.** HUMPED BLADDERWORT or EASTERN BLADDERWORT
   Site: In bog NW of junction of Route 3 and Wawbeek Road, near Coreys and the Indian Carry in the Upper Saranac Lake area.
   Elevation: 1585 feet.
   Phenology: In flower September 10.
   Frequency: One station.

Utricularia intermedia **Hayne.** INTERMEDIATE BLADDERWORT
   Site: Growing prostrate on surface of mud, muck, or peat. Poorly-drained. Shade intolerant to midtolerant.
   Elevation: 1500 to 1700 feet.
   Phenology: In flower June 15.
   Frequency: Four stations around Paul Smiths.
   Miscellany: Carnivorous.

Utricularia vulgaris **L.** COMMON BLADDERWORT or GREAT BLADDERWORT
   Site: Floating carnivorous species.
   Elevation: From 1500 to 1720 feet.
   Phenology: In flower July 1 through August 27.
   Frequency: Six stations, but probably many more.
   Miscellany: Carnivorous.

85. CAMPANULACEAE (BLUEBELL FAMILY) including LOBELIACEAE (LOBELIA FAMILY)

Campanula aparinoides **Pursh** MARSH BELLFLOWER
   Site: In open wet spots, most often along streams.
   Elevation: From 1500 to 1700 feet.
   Phenology: In flower August 19 through September 7.
   Frequency: Five stations, Saranac Lake to Meacham Lake.

    Miscellany: The plants at Mud Pond Inlet may be *Campanula aparinoides* var. *uliginosa* (Rydb.) Gleason.

Campanula rotundifolia L. HAREBELL or BLUEBELL
    Site: Shade-intolerant. Well-drained. Occasional on mineral soil, but more often on cliffs and bedrock ledges. (In Canada, this is an alpine.) Dr. Ketchledge reports it at 4200 feet on Whiteface Mountain.
    Elevation: From 588 feet at Camp Pok-O-MacCready near Willsboro to 3168 feet on Catamount Mountain.
    Phenology: In flower June 24 through August 18.
    Frequency: Protected in New York State. Nine stations.

Lobelia cardinalis L. CARDINAL FLOWER
    Site: Swamps and river banks on poorly-drained, usually partly-sunny sites.
    Elevation: From 1500 to 1750 feet. At considerably lower elevations south of and outside the Adirondacks.
    Phenology: In flower August 4 into mid-September.
    Frequency: Protected in New York State. Six stations.
    Miscellany: Possibly poisonous to eat.
    Reference: Devlin (1988)

Lobelia dortmanna L. WATER LOBELIA
    Site: Aquatic, partly submerged and partly emergent, the flowers floating above the surface on flexible, long pedicels. Attached to pond bottoms.
    Elevation: From 1483 to 1700 feet.
    Phenology: In flower July 17 through August 24.
    Frequency: Eight stations.

Lobelia inflata L. INDIAN TOBACCO
    Site: Native, but weedy in old fields, roadsides, gravel pits, logging roads, trails, and other disturbed areas. A pioneer on mostly well-drained mineral soils, often sandy. Shade-intolerant.
    Elevation: From 1000 feet in Keene to about 1800 feet at Ermine Brook.
    Frequency: Two stations.
    Miscellany: Poisonous to eat.

86. RUBIACEAE (MADDER FAMILY)

Cephalanthus occidentalis L. BUTTONBUSH
    Site: Soils: limestone soils along Lake Champlain, to marble at Rich

Elevation: Lake, to anorthosite along Raquette River.
From 100 feet at Wickham Marsh and Point Au Roche to 1550 feet at both Raquette Falls and Rich Lake.
Phenology: Buds barely breaking dormancy (new shoots ½″ long) by May 26.
In fruit October 2.
Frequency: One station on Raquette River.

Galium asprellum **Michx. ROUGH BEDSTRAW**
Site: Open, sunny, wet areas such as streamsides, pond edges, and marshes.
Elevation: From 100 feet along Lake Champlain to about 2250 feet in the Pitchoff-Sentinel Notch.
Phenology: In flower between August 8 and September 11.
Frequency: Eight stations.

Galium circaezans **Michx. WILD LICORICE**
Site: Mostly in rich woods, shade-tolerant, well-drained to imperfectly drained.
Elevation: From 100 feet along Lake Champlain to 2215 feet on Crane Mountain.
Frequency: Three stations in Keene Valley and on Catamount Mountain, NE of Franklin Falls.

Galium palustre **L. SWAMP BEDSTRAW or MARSH BEDSTRAW**
Elevation: 1620 to 1630 feet at Paul Smiths.
Phenology: In fruit between September 5 and October 2.
Frequency: Two stations.
Miscellany: Intermediates do occur between *G. palustre* and *G. trifidum* so that some plants might be hybrids.

Galium trifidum **L. SMALL BEDSTRAW**
Site: In poorly-drained areas, often on mud. Shade intolerant to mid-tolerant.
Elevation: From 100 feet along Lake Champlain to 1705 feet at Gabriels.
Phenology: In flower July 4 to August 21.
Frequency: Seven stations.

Galium triflorum **Michx. SWEET-SCENTED BEDSTRAW or FRAGRANT BEDSTRAW**
Site: The Flora area's only shade-tolerant, forest-inhabiting species. Well-drained to imperfectly drained and on springy sites. More common on the higher pH sites from Sugar maple and richer.

> Two stations are under red maple, striped maple, fir, yellow birch and beech - no sugar maple.

Elevation: From 960 feet at Chateaugay Falls to 2940 feet on Cascade Mountain.
Phenology: In immature fruit July 13.
In leaf May 12.
In flower June.
Frequency: 25 stations.

* *Galium verum* L. YELLOW BEDSTRAW
Phenology: In flower and immature fruit August 18, Paul Smiths.
Miscellany: A European naturalization and the only *Galium* in the Flora area with yellow petals.

Hedyotis caerulea (L.) Hook. or *Houstonia caerulea* L. BLUETS or QUAKER LADIES
Site: In lawns or woods. Dr. Ketchledge reports it on Mt. Marcy summit.
Elevation: From 1550 feet near Raquette Falls to 2200 feet near Heart Lake.
Phenology: In flower May 21.
Frequency: Five stations, but probably many more.

Hedyotis longifolia (Gaertn.) Hook. or *Houstonia longifolia* Gaertn. PALE BLUETS or LONG-LEAVED BLUETS
Site: Edge of Guideboard Road, Black Brook, north of Silver Lake Road.
Elevation: 1040 feet.
Phenology: In flower July 7.

Mitchella repens L. PARTRIDGEBERRY
Site: Shade-tolerant under hardwoods, pines, or hemlock. Well-drained. pH of humus: n=16; min. 3.5; max. 5.7; median 4.5.
Elevation: From ca. 450 feet on Tongue Mountain, Lake George, to 1900 to 2000 feet on Pitchoff Mountain.
Phenology: In flower July 15.
In fruit mid-June persistent from previous year.
Frequency: Frequent.
Miscellany: Fruits paired and fused, edible. (Partridgeberry and apple jam is commercially prepared in Saint John's, Newfoundland).
Reference: Schottman (1987, May).

87. CAPRIFOLIACEAE (HONEYSUCKLE FAMILY)

Diervilla lonicera **Mill.** BUSH HONEYSUCKLE
- Site: Common on exposed, rocky sites and on well-drained soils where there is full sun or partial shade (mid-tolerant of shade). Commonly follows burns and the edges of woods. Soils: limestone soils of Champlain Valley to Adirondack gneisses and anorthosite. (In the Catskills sandstone and shale).
- Elevation: 100 feet along Lake Champlain at Plattsburgh to 4050 feet on Porter Mountain.
- Phenology: In flower June 19 through July 3.

Linnaea borealis **L. ssp. *americana* (Forbes) Hultén** TWINFLOWER
- Site: On mostly well-drained outwash sands (and occasionally on till) under partial shade (mid-tolerant). To timberline. Abundant ground cover in spruce-fir forests northward in Canada.
- Elevation: Below 1200 feet in Keene Valley (Porter Mountain Trail Head) to ca. 4500 feet on Skylight.
- Phenology: In flower June 16 and into July.
- Frequency: 20 stations.

Lonicera canadensis **Bartr.** FLY HONEYSUCKLE or CANADA HONEYSUCKLE
- Site: In hardwood stands, well-drained, shade-tolerant. Plants are usually isolated individuals rather than in vast colonies or stands. Soils: limestone soils of Champlain Valley to Adirondack gneisses and anorthosite. (Shales and sandstones in the Catskills).
- Elevation: 100 feet along Lake Champlain to 2732 feet on Algonquin.
- Phenology: In flower May 7 through 22.
  In fruit June 30 through July 6.
- Frequency: Frequent.
- Miscellany: Fruits edible.

Lonicera dioica **X *hirsuta*** A hybrid between MOUNTAIN and HAIRY HONEYSUCKLE
- Site: In open, sunny, well-drained brushy areas, climbing as a liana (woody vine) over shrubs.
- Phenology: In fruit August 30 through September 24.
- Frequency: One station in Bloomingdale Bog.
- Miscellany: This species is apparently a hybrid between the Mountain honeysuckle, *L. dioica* L. and *L. hirsuta* Eat., the Hairy honeysuckle. Another possibility is that there is but a single species involved here. The only difference between the two is the presence or absence of hairs and glaucous wax on the plants. The

Flora area's specimens lie somewhere in between the two extremes, normally hairy on the nodes only, or on parts of the leaves only.

Lonicera tatarica L. TARTARIAN HONEYSUCKLE
    Site: European ornamental commonly escaped about Saranac Lake and elsewhere.

Lonicera villosa (Michx.) R. & S. or *L. caerulea* MOUNTAIN FLY HONEYSUCKLE or NORTHERN HONEYSUCKLE
    Site: Northern species found in bogs, and well-drained, open outwash sands. This species is intolerant of shade but can grow over a wide range of water table depths. Acid sites.
    Elevation: 1550 feet at Bigelow Road in Bloomingdale to 1880 feet at Moose River Plains. Dr. Ketchledge reports it also atop Mount Marcy.
    Phenology: In fruit July 6 through 16.
    Frequency: Four stations.
    Miscellany: Fruits blue, and edible, but plant is so uncommon that it is not advisable to consume them.

Sambucus canadensis L. BLACK ELDERBERRY
    Site: Local along flood plains of major streams and in swamps. Shade-intolerant, high water table sites. The plants atop Saint Regis Mountain are a puzzle. This is a more southerly, low-elevation species reaching near its northern limits in the Adirondacks. Soils: limestone soils of Champlain Valley to Adirondack gneisses and anorthosites.
    Elevation: 100 feet at Wickham Marsh to the Chubb River at 1850 feet and Saint Regis Mountain at 2873 feet.
    Phenology: In flower July 17 through August 2.
    In immature (green) fruit September 10.
    In heavy ripe fruit September 26.
    Frequency: 12 stations.
    Miscellany: Further south in New York, where this species is abundant, the fruits can be made into wines and jellies.

Sambucus racemosa L. or *Sambucus pubens* Michx. RED ELDERBERRY
    Site: In open areas as well as shaded, with a wide range of shade tolerance. Well-drained sites, climbing well into the spruce-fir zone. The more northerly of the two species of elderberry and on more acid soils. Growth is fastest in full sun. This species is especially abundant on blown over trees with exposed vertical root systems; apparently, birds which distribute the seeds must

	commonly alight on the tops of the roots. pH of humus: n=23; min. 3.7; max. 6.8; median 4.7. Soils: limestone soils along Lake Champlain to Adirondack anorthosite and gneisses. (Catskills shale, sandstone and conglomerate).
Elevation:	100 feet on Four Brothers Islands in Lake Champlain to 3830 feet on Lyon Mountain.
Phenology:	In flower bud May 4 and 5. In flower May 17 through June 2. In fruit July 13 through August 8.
Frequency:	Common.
Miscellany:	Fruits too sour for comsumption. Both elderberries have poisonous roots and twigs.

Viburnum acerifolium L. MAPLE-LEAVED VIBURNUM or DOCKMACKIE
- Site: A shade-tolerant shrub of well-drained soils mostly under oak forests downstate, but venturing into the Adirondacks from the southeast most often following red oak. Soils: limestone soils of Champlain Valley to Adirondack gneisses and anorthosite. (Catskills sandstones and shales).
- Elevation: 100 feet at Wickham Marsh along Lake Champlain to 2100 feet on Crane Mountain.
- Frequency: Six stations, mostly around Keene.

Viburnum cassinoides L. WILD RAISIN or WITHE ROD
- Site: Abundant on mostly poorly-drained soils in swamps and bogs, but also occurs occasionally on well-drained soils provided that there is partial shade or full sun as under open red pine stands. This species can occur over a wide range of water table depths, but lies between shade intolerant and midtolerant. pH of humus: $n=19$; min. 3.8; max. 5.5; median 4.8. Soils: sandstone (Potsdam) and Adirondack gneisses and anorthosite. (Catskills shales and conglomerates). Limestone at Chazy.
- Elevation: From 210 feet at the Chazy Fir Swamp to 4736 feet on Gothics.
- Phenology: In flower June 16 through 18.
  In fruit September 26.
  Moist, blackened leaves emit butyric acid upon decay, from late August through late October, permeating portions of the Adirondacks with its putrid stench.
- Miscellany: Fruits edible, tasty, winey.

Viburnum edule (Michx.) Raf. SQUASHBERRY, MOOSEBERRY or PIMBINA
- Site: This shrub is common northward in Canada. In the Adirondacks, notes Dr. Ketchledge, it is the only shrub (maybe the only vascular plant species except *Streptopus amplexifolius*) which is confined only to streamsides in the spruce-fir forest. Scattered from Algonquin Peak through the Mount Colden, Marcy, and Haystack mountain areas. Well-drained to poorly-drained sites. Shade-tolerant.
- Elevation: From 3240 feet on Algonquin to 4170 feet on Haystack Mountain.
- Frequency: Rare in New York State; six stations in the Flora area.

Viburnum lantanoides Michx. or ***Viburnum alnifolium*** Marsh. WITCHHOBBLE or HOBBLEBUSH
- Site: Abundant on well-to imperfectly-drained soils under northern hardwoods or spruce-fir. One of the most common Adirondack

Maple-Leaved Viburnum
Viburnum acerifolium

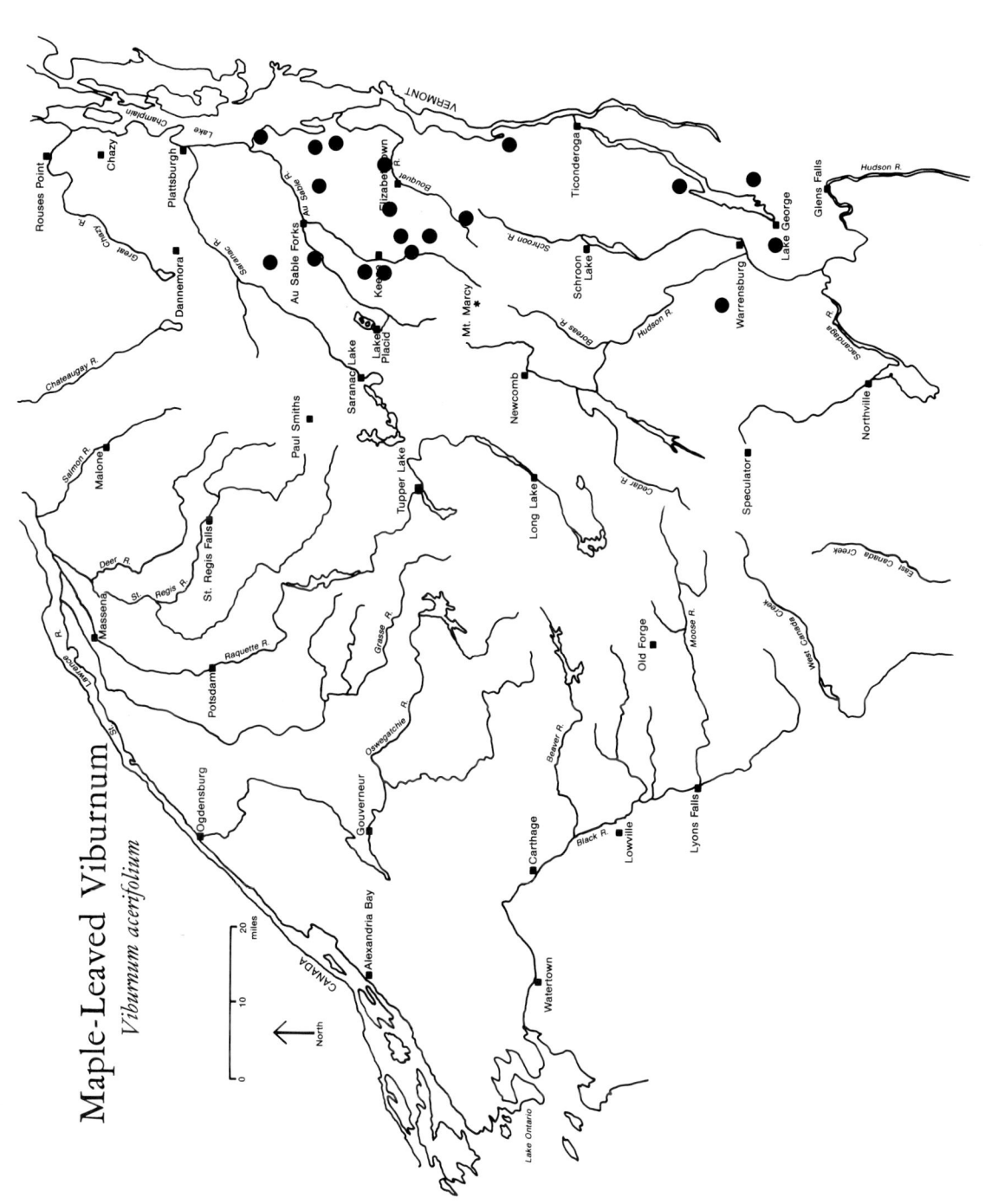

Elevation:	shrubs because it is so shade and acid tolerant, and the Adirondacks are chiefly forested. pH of humus: n=36; min. 3.5; max. 5.0; median 4.5. Soils: Adirondack gneisses and anorthosite. (In the Catskills, on shales, sandstones and conglomerates). From 1500 feet west of Duane to 3513 feet, and possibly as high as 3600 feet, on Whiteface. Uncommon above 3000 feet.
Phenology:	This shrub often flowers about mid-May, before the leaves are out and puts on quite a show. Some folks like to bring the shoots indoors about February and have them flower prematurely in a vase of water. In flower May 12 through June 2. In fruit September.
Miscellany:	Flowers *pink* on May 26, 1975, at junction of log roads NE of The Crevices. A favorite deer browse. Fruits edible but not as tasty as those of Wild raisin. Flowers dimorphic, the outer showy, sterile, and insect-attracting, the inner small and fertile. Observed with huge leaves, nearly a foot across, east southeast of Crane Mountain Pond along the trail.

Viburnum lentago L. NANNYBERRY

Site:	This shrub, similar to Wild raisin, is more southerly and more lowelevation in its distribution. It is invading the Adirondacks at the lower elevations from the Champlain Valley and from the Saint Lawrence. Soils: limestone soils along Lake Champlain to Adirondack gneisses and anorthosite. Marbles SW of DeKalb Junction.
Elevation:	100 feet along Lake Champlain to 1020 feet at Keene Valley.
Phenology:	In flower June 6 through 15.
Frequency:	One station at Keene Valley.

Viburnum recognitum Fern. ARROWWOOD

Site:	In open, poorly-drained sites, forming thickets with alders, red-stemmed dogwood, winterberry holly, black elderberry; etc. Also in open, well-drained areas in old fields, revealing a wide tolerance to water table depths. Shade-intolerant. pH probably higher than that of Wild raisin.
Elevation:	420 feet near Eben, SW of Potsdam, to 1600 feet on Mount Baker and ca. 1800 feet on Snowy Mountain trail. Ca. 350 feet at Indian Creek Nature Center at Rensselaer Falls.
Frequency:	Eight stations.
Miscellany:	This species is often considered a variety of *Viburnum dentatum*.

Viburnum trilobum Marsh. or *Viburnum opulus* var. *americanum* (Mill.) Ait.
HIGHBUSH CRANBERRY or CRANBERRY BUSH
    Site: Most often along river banks and along edges of woods, but sometimes in old fields. Shade-intolerant but accepting a wide range of water table depths.
    Elevation: From 1380 feet along Deer River near Mile Brook to 2100 feet on Crane Mountain.
    Phenology: In flower June 16.
    Frequency: Six stations, but probably many more.
    Miscellany: Some people consider the native Highbush cranberry only a variety of the European, hence *V. opulus* var. *americanum*. Fruits used for jellies and wines, but very sour eaten raw. These fruits, like those of European mountain ash, persist colorfully into winter, and provide food for birds. Flowers dimorphic as in Witchhobble.

88. VALERIANACEAE (VALERIAN FAMILY)

* *Valeriana officinalis* GARDEN HELIOTROPE
    Site: European garden escape, Saranac Lake.
    Phenology: In flower, June 19.

89. ASTERACEAE or COMPOSITAE (ASTER OR COMPOSITE FAMILY)

* *Achillea millefolium* L. YARROW or MILFOIL
    Site: Are all plants of European ancestry and thus introduced? Certain populations in areas remote from roadsides, gardens, and old fields (such as those at timberline) suggest that perhaps some plants are native. Do we have a cosmopolitan here?
    Phenology: In flower June 26 through July 29.
    Frequency: A most common weed.

Ambrosia artemisiifolia L. RAGWEED
    Site: Native, yet behaves as a European weed in Saranac Lake and on Campus. Most sterile and trampled soils.
    Elevation: Not recorded, but probably from Lake Champlain to about 2000 feet.
    Frequency: Frequent.
    References: Bazzaz (1974).
                      Egler (1971).

Anaphalis margaritacea (L.) Bentham & Hooker f. ex Clarke PEARLY EVERLASTING
    Site: An abundant native pioneer in well-drained forest openings. Shade-intolerant.
    Elevation: From 1500 feet around Saranac Lake to 3060 feet on Noonmark Mountain. Dr. Ketchledge reports it at 4000 feet on Whiteface Mountain.
    Phenology: In flower August and September.

Antennaria neglecta Greene var. *attenuata* (Fern.) Cron. or *A. neodioica* Greene PUSSY TOES
    Site: On well-drained soils on steep, rocky banks and roadsides. Moderately shade-tolerant to intolerant, and flowering before the hardwoods leaf out. Sometimes on lawns.
    Elevation: From 200 feet at Alice Lake to 2450 feet on Pitchoff Mountain.
    Phenology: In flower May 14 through June 16.
    Frequency: 11 stations.

* *Arctium lappa* L. BURDOCK
    Site: Common European weed in Saranac Lake.
    Miscellany: A classical example of a plant whose fruit (bur) goes for a ride on fur or human clothing.
    Reference: Davis (1973)

* *Artemisia vulgaris* L. COMMON MUGWORT
    Site: European. At the Saranac Lake railroad station.
    Phenology: Flower buds August 27.

Aster acuminatus Michx. SHARP-LEAVED ASTER or WHORLED WOOD ASTER
    Site: The most common aster of the spruce-fir forest, but also scattered in northern hardwoods. Moderately shade-tolerant, although best growth is in sunny spots. Well-drained usually, but on occasion even in Peat moss. To timberline. Abundant following disturbances such as blowdowns and landslides. This species is one of the first to wilt when a drought begins, with Striped maple not far behind to droop. pH of humus: n=8; min. 3.8; max. 5.6; median 4.3.
    Elevation: From 100 feet at Point Au Roche along Lake Champlain to 4857 feet on Dix Mountain.
    Phenology: In flower mid-August through October 3.
    Frequency: Common.
    Reference: Brown, Ashmun, and Pitelka (1985).

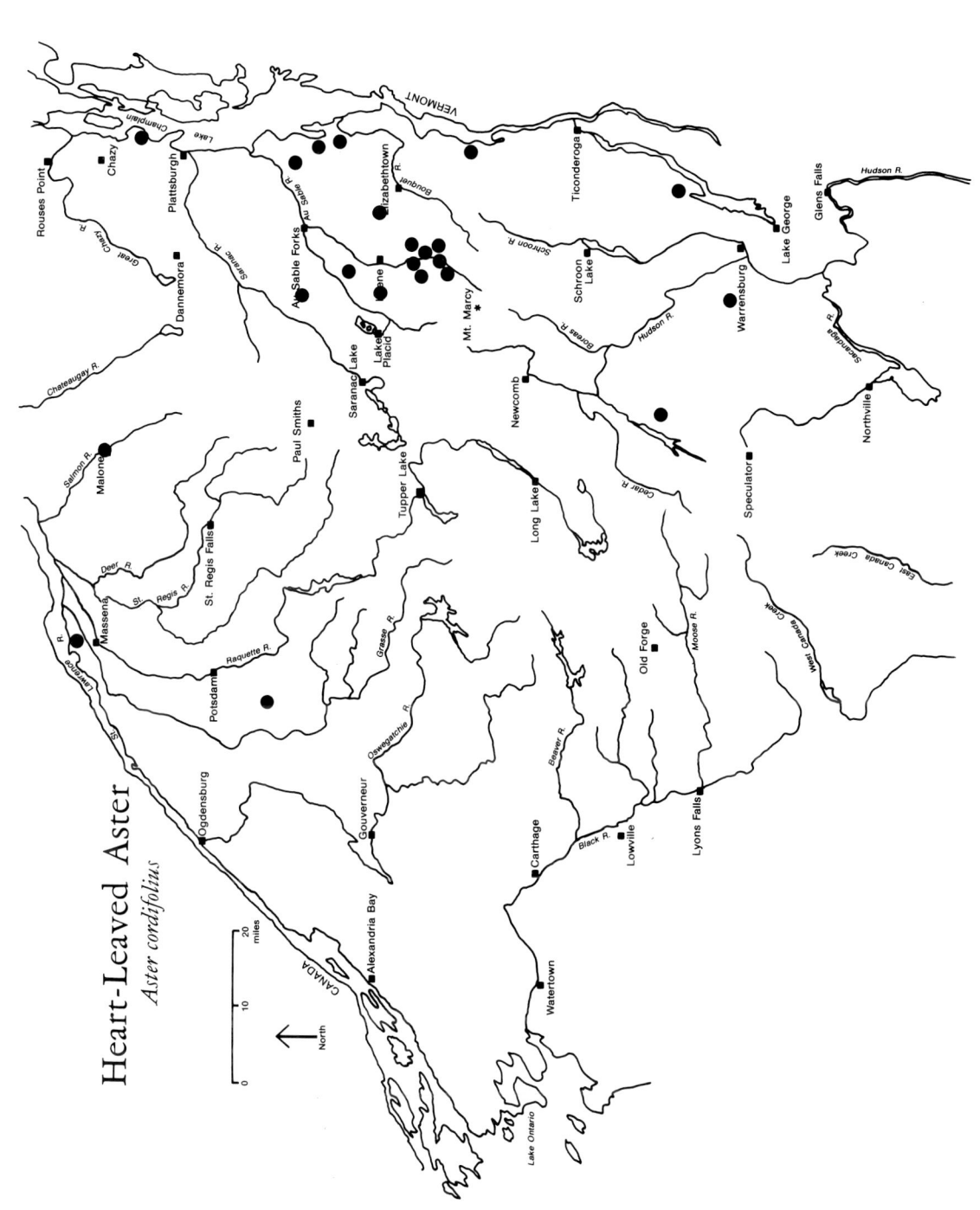

Aster cordifolius L. HEART-LEAVED ASTER
    Site: This species seems to be more abundant at the lower elevations in the Champlain and Saint Lawrence Valleys and in the eastern Adirondacks, not yet becoming fully established on the Upland around Paul Smiths. Usually under Sugar maple-White ash-Basswood-Red oak. Well-drained. Shade-tolerant.
    Elevation: From 100 feet at Point Au Roche on Lake Champlain to almost 3500 feet on Hurricane Mountain.
    Phenology: In flower September 27 through October 21.
    Frequency: Ten stations.
    Miscellany: On Campus at the intersection of the four roads below the gymnasia. Plant observed for the first time in September 1980, and number of stems was increasing annually. Removed summer 1989, by road shoulder reconstruction.

Aster divaricatus L. WHITE WOODLAND ASTER
    Site: A shade-tolerant species of lower elevations and more southerly affinities—usually under oaks. Well-drained, shade-tolerant.
    Elevation: From ca. 300 feet at Willsboro to 2100 feet on Crane Mountain, but lower in Vermont (ca. 200 feet).
    Phenology: In flower August and September.
    Frequency: One station on Porter Mountain, elevation ca. 2000 feet.

Aster lanceolatus Willd. or *Aster simplex* Willd. PANICLED ASTER or TALL WHITE ASTER
    Site: Along lake and stream shores. Intolerant of shade and in poorly-drained soils.
    Elevation: From 100 feet at Wickham Marsh on Lake Champlain to about 1700 feet on the Upland.
    Phenology: In flower September 8 through 14.
                    In fruit October.
    Frequency: Occasional.

Aster macrophyllus L. LARGE-LEAVED ASTER
    Site: Forming large colonies under hardwoods or conifers. Well-drained sites and moderately shade-tolerant. Often follows burns as on Pitchoff Mountain at 3430 feet and on the E slope of Jenkins Mountain at 2100 feet.
    Elevation: From 100 feet at Point Au Roche at Lake Champlain to 3430 feet on Pitchoff Mountain.
    Phenology: In flower August.
    Frequency: Occasional.

Aster novae-angliae **L.** NEW ENGLAND ASTER
- Site: This species, like *A. divaricatus*, is another peripheral species, having surrounded the Adirondacks but not quite crept into the interior. Unlike *A. divaricatus*, this species is shade-intolerant and occurs along roadsides and old fields.
- Elevation: From 100 feet along Lake Champlain at Point Au Roche to Keene Valley at about 1000 feet. It is not certain whether the Saranac Lake stations are naturalized or not, but these run 1500 to 1600 feet.

Aster prenanthoides **Muhl.** CROOKED-STEMMED ASTER
- Site: Moderately shade-tolerant. Well-drained.
- Elevation: Black Brook sand plain, elevation 1040 feet. Elevation down to ca. 300 feet at Willsboro.
- Frequency: One station.

Aster puniceus **L.** ROUGH-STEMMED ASTER
- Site: In swamps and roadside ditches in hardwood and coniferous forests. Poorly-drained, moderately tolerant of shade. Often growing with *A. umbellatus*.
- Elevation: From 210 feet at Chazy Fir Swamp to 2800 feet on Hurricane Mountain.
- Phenology: In flower August 12 through September.
- Frequency: Common.

Aster umbellatus **Mill.** FLAT-TOPPED WHITE ASTER
- Site: In poorly-drained open areas, (but occasionally on well-drained) especially in marshes and swamps. Intolerant of shade. Often with *A. puniceus*.
- Elevation: From ca. 450 feet on Tongue Mountain on Lake George to 2720 feet on Catamount Mountain.
- Phenology: In flower August 3 into September.
- Frequency: Common.

* *Bellis perennis* **L.** ENGLISH DAISY
- Site: European lawn weed. Johnson Road, off Moody Pond, Saranac Lake.

Bidens beckii **Torr. ex Spreng.** or *Meglalodonta beckii* (Torr. ex Spreng.) Greene WATER MARIGOLD
- Site: Native, aquatic. In Lower Saint Regis Lake along shore south of Bluebird Cabins site.
- Elevation: 1619 feet.
- Frequency: One station.

Bidens cernua L. NODDING BUR MARIGOLD
- Site: Poorly-drained sites and stream banks. Shade-intolerant.
- Elevation: From 100 feet along Lake Champlain to 1635 feet at Paul Smiths.
- Phenology: In flower August 21 through September 6.
- Frequency: Four stations, Paul Smiths and Saranac Lake.
- Miscellany: The Flora area's only native showy *Bidens* with ¼-inch-long yellow rays.

Bidens frondosa L. BEGGARS TICKS
- Site: Native weedy species. Shade-intolerant. Well to poorly-drained sites.
- Elevation: From 100 feet along Lake Champlain to 1630 feet at Paul Smiths.
- Phenology: In flower September 2.
  In fruit October 27.
- Frequency: Two stations, Paul Smiths and Saranac Lake, but probably many more.

Bidens tripartita L. SWAMP BEGGARS-TICKS or STICK-TIGHTS
- Site: Gravel pit, Saranac Lake.
- Phenology: In flower August 26 through September 2.
- Miscellany: The following plants had been classified as a distinct species: *Bidens connata* **Muhl. ex Willd.**, but now are considered part of *B. tripartita*. They occur at the east edge of Saint Regis Pond.

* *Centaurea maculosa* Lam. STAR THISTLE or SPOTTED KNAPWEED
- Site: In well-drained outwash sand, roadsides and cinders.
- Phenology: In flower July 19 through August 20.
- Frequency: Locally abundant.
- Miscellany: European.

* *Cichorium intybus* L. CHICORY
- Phenology: In flower July 11 through August 5.
- Frequency: Weedy and abundant outside the Flora area, but only occasional within it. European.

Cirsium muticum Michx. SWAMP THISTLE
- Site: Native. Marsh along inlet to Mud Pond near Lake Kushaqua, growing with *Verbena*.
- Elevation: 1653 feet at Mud Pond.
- Phenology: In fruit August 26.
- Miscellany: Reported in Averyville Marsh and in North Elba Flora by Peck (1899) as *Carduus muticus* or *Cnicus muticus*.

* *Cirsium vulgare* (Savi) Tenore. BULL THISTLE
    Site: Summit of Mount Pisgah, Saranac Lake.
    Phenology: In fruit September 14 through October 27.
    Miscellany: European weed.

Conyza canadensis (L.) Cronq. or *Erigeron canadensis* L. HORSEWEED
    Site: Native, but weedy, on lawns and wasteplaces. Around classroom buildings on Campus and in Saranac Lake Village.
    Elevation: From 1500 to 1700 feet but probably also much lower.
    Phenology: In flower August.
    Frequency: Common.

* *Crepis tectorum* L. HAWKSBEARD
    Site: Paul Smiths, Route 86.
    Phenology: In flower June 20 through July 23.
    Frequency: First observed by this author in 1981 and again in 1985, 1988 and 1990, but not in 1982, 1983, 1984, 1986, 1987 nor 1989. Could this species be introduced sporadically, say, by loads of passing hay on the highway? Fernald, p. 1559, says that these plants are annuals or biennials and do not long persist; our plants match this description.
    Miscellany: Naturalized from Europe.

Erigeron annuus (L.) Pers. DAISY FLEABANE
    Site: Native, but weedy species on roadsides and lawns at Paul Smiths and Saranac Lake.
    Elevation: From 1500 to 2200 feet at Heart Lake, but probably much lower also.
    Phenology: In flower July 13 through August 1.
    Frequency: Common.

Erigeron philadelphicus L. COMMON FLEABANE
    Site: At old water cistern site southwest of College gymnasia.
    Elevation: From 210 feet at Chazy to 1660 feet at Paul Smith's College.
    Phenology: In flower June 19 through 27.
    Frequency: One station.

Erigeron strigosus Muhl. ex Willd. FLEABANE
    Site: Route 30, south of Hoffman Road at Paul Smiths. Native species.
    Elevation: 1625 feet.
    Phenology: In flower July 15.
    Frequency: One station.

Eupatorium maculatum L. JOE PYE WEED
- Site: In open, poorly-drained areas. Shade-intolerant.
- Elevation: From 100 feet along Lake Champlain to 2480 feet on McKenzie Mountain.
- Phenology: In flower August 26.
- Frequency: Six stations but probably many more.

Eupatorium rugosum Houtt. or *E. urticaefolium* Reichard. WHITE SNAKEROOT
- Site: Shade-tolerant, often in somewhat poorly-drained or springy sites in forests under Sugar maple, White ash, Basswood, etc., with higher humus pH. Eastern Adirondacks only; not on the Central Upland.
- Elevation: 2100 feet at Cascade Lakes.
- Phenology: In flower July 23, 1986 at Cascade Lakes.
- Frequency: Five stations.
- Miscellany: Poisonous to eat. The poison may be transmitted by cow's milk.
- Reference: Cohn & Kucera (1969).

Euthamia graminifolia (L.) Nutt. ex Cass. or *Solidago graminifolia* (L.) Salisb. GRASS-LEAVED GOLDENROD
- Site: A third abundant old field species after *S. canadensis* and *S. rugosa*. Intolerant of shade and well-drained.
- Elevation: From 100 feet at Point Au Roche along Lake Champlain to atop Blue Mountain at 3759 feet.
- Phenology: In flower August and September.
- Frequency: Common.
- Reference: Gross & Werner (1983).

* *Galinsoga ciliata* (Raf.) Blake. GALINSOGA
- Site: Saranac Lake Village streets.
- Miscellany: The Flora area's only South American species which has become naturalized as a weed!

Gnaphalium uliginosum L. LOW CUDWEED
- Site: Native, but weedy. In a poorly-drained lawn on Campus. One of the Flora area's few native annuals. *Impatiens* and *Conyza canadensis* are others.
- Elevation: 1625 feet.
- Phenology: In flower September 7.
- Frequency: One station.

Gnaphalium viscosum **Kunth** or *Gnaphalium macounii* **Greene** or *G. decurrens* **Ives**
CLAMMY EVERLASTING
    Site: Native, but weedy.
    Elevation: 1630 to 1680 feet at Paul Smiths.
    Phenology: In flower August 30 through September 14.
    Frequency: Two stations.

* *Helianthus tuberosus* **L.** JERUSALEM ARTICHOKE
    Site: Native of U.S., but not the Flora area. Escaped from a garden and growing along Stevenson Lane, Saranac Lake.
    Phenology: In flower September 18.

* *Hieracium aurantiacum* **L.** ORANGE HAWKWEED or DEVIL'S PAINTBRUSH
    Site: Lawns, fields, roadsides.
    Elevation: To 3700 feet.
    Phenology: In flower June 9 into August.
    Frequency: Abundant.
    Miscellany: Introduced from Europe in the 1880s to New York, a relatively recent arrival.

* *Hieracium caespitosum* **Dumort.** or *Hieracium pratense* **Tausch.** YELLOW HAWKWEED or KING DEVIL
    Site: Lawns, fields, roadsides.
    Phenology: In flower June 9 through July 18.
    Frequency: Naturalized from Europe, often growing with *H. aurantiacum* and just as prolific.

Hieracium kalmii **L.** or *Hieracium canadense* **Michx.** CANADIAN HAWKWEED
    Site: A native hawkweed of well-drained, sunny sites.
    Elevation: From ca. 1250 feet at Roaring Brook to ca. 1600 feet in Saranac Lake.
    Phenology: In flower August 18.
    Frequency: Three stations.

* *Hieracium lachenalii* **Gmel.** or *Hieracium vulgatum* **Fries.** HAWKWEED
    Site: Keep an eye on this species. It is naturalized from Europe, but unlike most introductions, this one is partially tolerant of shade and can invade open, partly-sunny woods, competing with our native species. Only the European *Epipactis helleborine,* the orchid, is more shade-tolerant. (See page 19.)
    Phenology: In flower June 14 through October, nearly 4–month flowering period.

Hieracium scabrum **Michx.** ROUGH HAWKWEED
    Site: A native of sunny to partially-shaded roadsides.
    Elevation: From Willsboro, 588 feet at Pok-O-MacCready to 1700 feet at Gabriels.
    Phenology: In flower August 10 and into September.
    Frequency: Four stations, including Paul Smiths and Saranac Lake.

* *Hypochoeris radicata* **L.** CAT'S EAR
    Site: European lawn weed.
    Phenology: In flower June 22 and into July.
    Frequency: Common on Campus and Quadrangle.
    Miscellany: The genus is spelled *Hypochaeris* in Mitchell (1986).

Lactuca canadensis **L.** WILD LETTUCE
    Site: Moderately shade-tolerant native herb, reaching heights of 6 to 8 feet. Mostly well-drained sites.
    Elevation: From 1500 to 2020 feet near Heart Lake.
    Phenology: In immature fruit August 11.
    Frequency: Occasional at Paul Smiths.

* *Lapsana communis* **L.** NIPPLEWORT
    Site: European weed somewhat shade-tolerant. Paul Smiths.
    Phenology: In flower June 26 through July 20.

* *Leucanthemum vulgare* **Lam.** or *Chrysanthemum leucanthemum* **L.** DAISY, OXEYE DAISY or WHITE DAISY
    Phenology: In flower June 9 through July 20.
    Frequency: Commonest of all European weeds.

* *Matricaria matricarioides* **(Less.) Porter** PINEAPPLE WEED
    Site: Low-growing European lawn weed. Edges of sidewalks, etc., and capable of sustaining much trampling.
    Phenology: In flower July 14 through 19.
    Miscellany: Pineapple aroma when crushed.

Prenanthes altissima **L.** TALL RATTLESNAKEROOT
    Site: Shade-tolerant on mostly well-drained sites usually under hardwoods.
    Elevation: From 200 feet at Barnhardt Island and 210 feet at Chazy Fir Swamp to 2940 feet on Cascade Mountain.
    Phenology: In flower August 21.
                   In immature fruit September 18.
    Frequency: Five stations, but probably many more.

* *Rudbeckia hirta* L. var. *pulcherrima* Farw. or *Rudbeckia serotina* Nutt. BLACK-EYED SUSAN
- Phenology: In flower July 13 through August 1.
- Frequency: European weed. Common.
- Reference: Pangburn (1981, August).

Senecio schweinitzianus Nutt. or *Senecio robbinsii* Oakes ex Rusby ROBBINS' RAGWORT
- Site: Poorly-drained (most often), sunny areas in swamps and marshes. Shade-intolerant. Well-drained open fir plantation at Asplin's northwest of Saranac Lake; at least 30″ to water table, an exceptional site.
- Elevation: From 1550 feet at Bloomingdale to 1705 feet at Gabriels.
- Phenology: In flower June 16 through August 16.
- Frequency: Six stations.

SOLIDAGO (GOLDENROD)

Key to Goldenrod by sites:
- 1a. Alpine — *S. cutleri*
- 1b. Non-Alpine — 2
    - 2a. Bog — *S. uliginosa*
    - 2b. Well-drained — 3
        - 3a. Shade-intolerant in open fields, roadsides, etc. *S. rugosa, S. graminifolia, S. canadensis, S. nemoralis, S. puberula.*
        - 3b. Shade tolerant in woods — 4
            - 4a. Under Sugar maple, White ash, Basswood—*S. latifolia* & *S. caesia*
            - 4b. Under Spruce-fir-birch—*S. macrophylla*

Solidago bicolor L. SILVERROD
- Elevation: From 1550 feet in Saranac Lake to 2560 feet on Chimney Mountain.
- Phenology: In flower September 2 to October 2.
- Frequency: One station.
- Miscellany: The Flora area's only white-flowered goldenrod.

Solidago caesia L. BLUE-STEMMED GOLDENROD
- Site: Shade-tolerant, well-drained, restricted to hardwood forests with higher humus pH. One station, oddly, along Jenkins Mountain Road on an esker. Most often under Basswood, White ash, Elm, Hop hornbeam, sugar maple. pH of humus: n=17; min. 4.4; max. 6.8; median 5.7.

Elevation: From 100 feet at Point Au Roche along Lake Champlain to 2100 feet on Crane Mountain.
Phenology: In flower September and to October 2.
Frequency: 14 stations.

Solidago canadensis L. CANADA GOLDENROD
Site: One of the Flora area's most abundant old field species. Intolerant of shade. Well-drained. The effect of light on this goldenrod was observed along the Jenkins Mountain Road (now paved over as the V.I.C. entrance!) in October, 1972. Those plants at the edge of the road which had received full sun all summer grew to heights of 5 feet, and abundantly fruited. Those plants further away from the road in partial shade were about 3 feet tall with a few fruits. Those plants under the edge of the forest canopy had reached heights of 1 to 2 feet and had not flowered by October. No plants were observed in the deep shade.
Elevation: From 100 feet at Point Au Roche along Lake Champlain to 3759 feet atop Blue Mountain.
Phenology: In flower July 24 through August and September.
Frequency: Common.
References: Gross and Werner (1983).
Hartnett and Abrahamson (1979).

Solidago flexicaulis L. or *Solidago latifolia* L. WIDE-LEAVED GOLDENROD
Site: Shade-tolerant, well-drained species of higher humus pH hardwood sites, often with *S. caesia*. pH of humus: n=8; min. 4.8; max. 6.6; median 5.9.
Elevation: From 200 feet at Barnhardt Island to 2763 feet on Crane Mountain.
Phenology: In flower late August and September to October 18.
Frequency: 17 stations.

Solidago juncea Ait. EARLY GOLDENROD
Elevation: From 100 feet at Wickham Marsh along Lake Champlain to 2800 feet on Hurricane Mountain.
Phenology: This species flowers earlier than the others, beginning about July 24. In flower bud July 15.
Frequency: Two stations: Mount Baker and Raquette Falls.
References: Gross & Werner (1983).

Solidago macrophylla Pursh LARGE-LEAVED GOLDENROD
Site: The abundant high-elevation species of the spruce-fir forests, the companion Composite very often to *Aster acuminatus*. Occasional

	in hardwood stands. Tolerates some shade but is most vigorous in openings as along trails. Well-drained. To timberline.
Elevation:	From 1645 feet near Heron Marsh to Mount Marcy ca. 5250 feet. Uncommon below 2500 feet.
Phenology:	In flower August 8 through September 5. In fruit September 5 through October 3.
Frequency:	Frequent.

Solidago nemoralis Ait. GRAY GOLDENROD
- Site: Old fields, often on bare mineral soil as in gravel pits and on roadside banks. Intolerant of shade. Well-drained. A small, two-foot high species covered with very fine gray hairs.
- Elevation: From 100 feet at Point Au Roche along Lake Champlain to summit of Azure Mountain at 2500 feet.
- Phenology: In flower September 24. In fruit October 6.
- Frequency: Seven stations.
- Reference: Gross & Werner (1983).

Solidago puberula Nutt. DOWNY GOLDENROD or PUBERULENT GOLDENROD
- Site: Recorded in 1899 by Peck only around Rainbow Lake, but since has spread and has become common on well-drained outwash sands in full sun. Recognized by its cylindrical inflorescence.
- Elevation: From about 1500 feet to 1700 feet in the Flora area, but almost to 2057 feet on Lyon Mountain.
- Phenology: In flower late August until the end of September.
- Frequency: Eleven stations.

Solidago rugosa Ait. ROUGH-LEAVED GOLDENROD
- Site: This is one of the most abundant old field species, intolerant of shade and well-drained. Often growing with *S. canadensis* and *S. graminifolia*. It can reach moderately high elevations in clearings and openings. The old-field nature of many goldenrods would suggest European origin, but all the Flora area's species are native.
- Elevation: From 210 feet at Chazy to Mount Morris 3152 feet.
- Phenology: In flower August and September.
- Frequency: Common.

Solidago uliginosa Nutt. BOG GOLDENROD
- Site: Intolerant of shade and found in poorly-drained areas in Peat moss.
- Elevation: From 1500 to 1700 feet.

Phenology: In flower from August 10 through August 21.
Frequency: Five stations.

* *Sonchus oleraceus* L. SOW THISTLE
    Site: Church Street at the Saranac River, Saranac Lake.
    Lake Placid on Main Street - this species was found here and reported in the *Tri-Lakes Times,* August 16, 1972 as a Dandelion of unusually tall proportions.
    Phenology: In flower August 16 to 29.
    Miscellany: European weed.

* *Tanacetum vulgare* L. TANSY
    Phenology: In flower August.
    Frequency: In Saranac Lake and Paul Smiths, infrequent.
    Miscellany: European garden escape with aromatic leaves and stems.

* *Taraxacum officinale* **Weber ex Wiggers** DANDELION
    Site: It demands direct sun and well-drained soils. It thrives in places where other species cannot, such as after repeated tramplings and mowings. The rosette of prostrate leaves inhibits competition from neighboring plants, especially grasses.
    Phenology: In flower, earliest date, April 17.
    Median date of first flowering April 30. Continues to flower into summer and autumn.
    Frequency: Prolific European naturalized lawn weed.
    Miscellany: Leaves can be used for salads, and flowers for wine.
    References: Ives (1975).
    Solbrig (1971).

* *Tragopogon pratensis* L. YELLOW GOAT'S BEARD
    Site: Whitefathers near Onchiota, Saranac Lake, and Paul Smiths.
    Phenology: In flower June 9 and 10.
    Frequency: Occasional weed.
    Miscellany: European naturalized.

* *Tussilago farfara* L. COLT'S FOOT
    Site: Weedy, requiring full sun but varying from well-drained sites to springy banks.
    Phenology: In flower late April and May.
    Frequency: Two stations, but common in the Lowlands.
    Miscellany: Naturalized European.

MONOCOTYLEDONEAE (THE MONOCOTS)

90. ALISMATACEAE (WATER-PLANTAIN FAMILY)
Gleason (1963) spells this family as **ALISMACEAE**.

Sagittaria latifolia **Willd.** including forma *gracilis* **(Pursh)** Robinson ARROWHEAD, WAPATO or DUCK POTATO
- Site: Occasional in shallow waters.
- Elevation: From 100 feet at Point Au Roche to 1700 feet.
- Phenology: In flower August 7th.
  Median date of fruiting August 30th.
- Frequency: Eight stations.
- Miscellany: Forma *gracilis* has very narrow leaf blades, the basal lobes varying from two to one to none.
- Reference: Schottman (August, 1983).

Sagittaria graminea **Michx.** ARROWHEAD or GRASS-LEAF SAGITTARIA —
- Site: Aquatic.
- Elevation: 1500 to 1700 feet.
- Phenology: In flower August 18, 1985.
  In fruit August 13, 1986.
- Frequency: Two stations, at Paul Smiths and Middle Saranac Lake.

91. HYDROCHARITACEAE (FROG'S-BIT FAMILY)

Elodea canadensis **Rich. ex Michx.** or *Anacharis canadensis* **(Rich. ex Michx.) Planch.** ELODEA, WATERWEED or DITCH MOSS
- Site: Submerged aquatic.
- Elevation: 1600 to 1700 feet.
- Miscellany: Used commonly in aquaria. The species identified June 18 might be instead *E. nuttallii* **(Planch.)** St. John, the Narrowleaved waterweed.

Vallisneria americana **Michx.** EEL GRASS, WILD CELERY, TAPE GRASS or WATER CELERY
- Site: Floating aquatic.
- Elevation: 1500 to 1700 feet.
- Phenology: In flower August 16 to 20.
- Frequency: Common.

92. POTAMOGETONACEAE (PONDWEED FAMILY)
Reference: **Ogden** (1974)

Potamogeton amplifolius Tuckerm. LARGE-LEAVED PONDWEED
    Site: Aquatic. In Lower St. Regis Lake and Spitfire Narrows.
    Elevation: 588 feet at Willsboro to 1700 feet.
    Frequency: Four stations.

Potamogeton epihydrus Raf. PONDWEED
    Elevation: 1500 to 1866 feet at Bear Pond near Sabattis.
    Phenology: In flower between July 15 and August 24.
                   In fruit between August 16 and September 21.
                   A "bloom" of this species occurred in Cooler Pond at Paul Smiths in September and October 1977, but none was observed in 1978 nor 1979!
    Frequency: The Flora area's most common species of Pondweed.

Potamogeton gramineus L. PONDWEED
    Elevation: 1600 to 1700 feet.
    Frequency: Two stations, at Paul Smiths.

Potamogeton natans L. FLOATING PONDWEED
    Site: Aquatic.
    Elevation: 1500 to 1700 feet.
    Phenology: In fruit between August 28 and September 21.
    Frequency: Six stations.

Potamogeton perfoliatus L. or *P. bupleuroides* Fern. CLASPINGLEAF PONDWEED
    Site: This species requires a water pH of 6 or greater.
    Elevation: 1600 to 1700 feet at Paul Smiths and Lake Clear.
    Phenology: In flower between August 16 and August 28.
                   In fruit September 1.
    Frequency: Three stations.

Potamogeton pusillus L. PONDWEED
    Site: Aquatic.
    Elevation: 1651 feet. Osgood Pond.

Potamogeton richardsonii (Benn.) Rydb. PONDWEED or RED-HEAD PONDWEED
    Site: Aquatic.
    Elevation: 1500 to 1700 feet.
    Phenology: In flower August 19.
    Frequency: Probably the most common species after *P. epihydrus*. Eight stations.

Potamogeton robbinsii **Oaks. PONDWEED**
    Site: Aquatic.
    Elevation: 1600 to 1700 feet in the Paul Smiths area.
    Frequency: Three stations.

Potamogeton spirillus **Tuckerm. PONDWEED**
    Site: Aquatic.
    Elevation: 1600 to 1700 feet in the Paul Smiths area.
    Phenology: In fruit between August 16 and August 24.
    Frequency: Two stations.

93. NAJADACEAE (NAIAD FAMILY)

Najas flexilis **(Willd.) Rostk. & Schmidt NAIAD or WATER NAIAD**
    Site: Aquatic.
    Elevation: 1500 to 1700 feet in the Paul Smiths area.
    Phenology: Mid-August.
    Frequency: Two stations.

94. ARACEAE (ARUM FAMILY)

Arisaema triphyllum **(L.) Schott ex Schott & Endl. JACK-IN-THE-PULPIT or INDIAN TURNIP**
    Site: An indicator of higher humus pH, most often under Sugar maple and sometimes with Basswood, White ash and Elm. Common at Camp Canaras, Upper Saranac Lake, on outwash without any of these four tree species! Mostly on springy sites, but also on the well drained. Shade-tolerant. pH of humus: n=33; min. 3.6; max. 6.8; median 5.0.
    Elevation: 100 feet along Lake Champlain to 2200 feet on Pitchoff Mountain.
    Phenology: Earliest flowering date May 26.
                       Median flowering date June 1.
                       Latest flowering date June 6.
    Miscellany: Plants contain oxalate crystals which cause an intense burning sensation when eaten.
    References: Bierzychudek (1982)
                        Doeffinger (1978)
                        Doust (1982)
                        NY State *The Conservationist*, 1965
                        Schottman (1983, May)

Calla palustris **L. CALLA-LILY, WATER ARUM or WILD CALLA**
    Site: In marshes, sluggish streams, and bog mats.

Elevation: 1500 to 1900 feet.
Phenology: In flower June 25, 1974 at Cooler Pond Bog to August 21 at Heron Marsh.
Frequency: Locally common.

95. LEMNACEAE (DUCKWEED FAMILY)

Lemna minor L. DUCKWEED
Site: Floating dwarf vascular plant. Occasional in quiet waters.
Elevation: From 588 feet at Long Pond, Willsboro, to ca. 1750 feet.
Phenology: A "bloom" occurred in the Paul Smiths College Leach Field in November 1986.
Frequency: Six stations, but probably many more.
References: Keddy (1976)
Hillman & Culley (1978)

96. XYRIDACEAE (YELLOW-EYED GRASS FAMILY)

Xyris montana Ries YELLOW-EYED GRASS
Site: In the bog mat with *Drosera intermedia* in Bog Pond.
Elevation: Ca. 1630 feet.
Phenology: In flower August 26, 1974.

97. ERIOCAULACEAE (PIPEWORT FAMILY)

Eriocaulon aquaticum (Hill) Druce or *Eriocaulon septangulare* With. PIPEWORT, WHITE BUTTONS or HAT PINS
Site: Rooted in bottom of water several feet deep, the flower heads on long, unbranched stalks emerging above the water level.
Elevation: 1500 to 1750 feet.
Phenology: In flower July 17 to August 4.
Median date of fruiting August 30.
Frequency: Ten stations.

98. JUNCACEAE (RUSH FAMILY)
Reference: **Clemants (1990)**

Juncus articulatus L. RUSH or JOINTED RUSH
Site: Poorly-drained, sunny sites. Sandy outwash to bog mats.
Elevation: 1600 to 1700 feet.
Phenology: In fruit August 23, 1974.
Frequency: Two stations at Paul Smiths.

Juncus canadensis J. Gay ex LaHarpe RUSH or MARSH RUSH
    Site: Poorly-drained, sunny areas.
    Elevation: 1500 to 1700 feet.
    Phenology: Median fruiting date August 20.
    Frequency: Three stations: Paul Smiths, Lake Clear & Saranac Lake.

Juncus effusus L. RUSH, SOFT RUSH or CANDLE RUSH
    Site: Poorly-drained, sunny sites.
    Elevation: 1600 to 2000 feet.
    Phenology: In fruit September 4, 1973 on Campus.
    Frequency: Common.

Juncus nodosus L. RUSH or KNOT RUSH
    Site: Saranac River floodplain near Stevenson Lane substation, Saranac Lake.
    Elevation: 115 to 1510 feet.
    Phenology: In fruit September 12, 1987 at Point Au Roche State Park.

Juncus pelocarpus Mey. RUSH
    Site: Poorly-drained sites, shade-intolerant, often in wet sand. Saranac River floodplain and Heron Marsh.
    Elevation: 1500 to 1700 feet.
    Phenology: In fruit early September.

Juncus tenuis Willd. PATH RUSH or YARD RUSH
    Site: Surprising in its great ability to withstand trampling, this species is most abundant along trails and unpaved roads. From poorly to well-drained sites. This is the Flora area's only rush that will grow on a dry site.
    Elevation: 1500 to 2200 feet.
    Phenology: In fruit September 4 on Campus.
    Frequency: Common and weedy.

99. CYPERACEAE (SEDGE FAMILY)

Bulbostylis capillaris (L.) Clarke SAND-RUSH
    Site: Along cinders of railroad rights-of-way.
    Elevation: 1500 to 1700 feet.
    Phenology: In fruit August 11.
    Frequency: Three stations: one at Lake Clear Junction and two in Saranac Lake.

Carex arctata Boott ex Hook. SEDGE
    Site: Shade-tolerant and woodland-inhabiting. It is abundant along

with *C. communis* in the woods about Campus.
Elevation: From 1600 to 1900 feet.
Phenology: In flower and in immature fruit between May 14 and June 10.
Frequency: Three stations: Mount Baker and Mount Pisgah.

Carex argyrantha **Tuckerm. SEDGE**
Elevation: 1630 feet in Bay Pond area.
Phenology: In fruit August 1, 1974 on Campus.
Frequency: Rare in Mitchell.

Carex brunnescens **(Pers.) Poir. ex Lam. SEDGE**
Site: This is another abundant but delicate species of the forests, growing in well- or poorly-drained sites.
Elevation: From 1620 feet at Paul Smiths to over 5000 feet on Mount Marcy.
Phenology: In flower between June 12 and June 17.
In fruit between August 13 and September 8.
Frequency: Five stations, but probably many more not recorded.

Carex canescens **L. SEDGE**
Site: Poorly-drained bogs, brooks and ditches.
Elevation: From 1500 feet to over 5000 feet.
Phenology: In fruit late June through August.
Frequency: Three stations, two near Gabriels.

Carex communis **Bailey or** *C. pedicellata* **in Peck SEDGE**
Site: Perhaps the most common species of sedge in the woods around Campus.
Elevation: From 1620 feet to 2200 feet.
Phenology: In flower between May 18 and May 26.

Carex crinita **Lam. SEDGE**
Site: This is perhaps the most abundant species of sedge in wetlands, especially so along wet overgrown log roads in partial shade or full sun.
Elevation: From 1000 to 1200 feet in Keene Valley to 3700 feet at Indian Falls.
Phenology: In flower mid-June.
In fruit August & September.
Frequency: Common.

Carex debilis **Michx. var.** *rudgei* **Bailey or** *C. flexuosa* **Muhl. ex Willd. SEDGE**
Site: Northern hardwood and spruce-fir forests.

Elevation: From between 1000 and 1200 feet in Keene Valley to 3015 feet on Loon Lake Mountain, to almost 3500 feet.
Frequency: Common.

Carex exilis Dewey BOG SEDGE
Site: On the bog mats, Mud Lake north of Lake Kushaqua and Barnum Pond.
Elevation: 1653 feet in both cases.
Phenology: In fruit June 12 through August 20.

Carex flava L. SEDGE
Site: Bog along old railroad grade north of Lake Clear Junction.
Elevation: 1630 feet.
Phenology: August 16, 1974.

Carex intumescens Rudge SEDGE
Site: Primarily under northern hardwoods.
Elevation: From 210 feet at Chazy Fir Swamp to 3650 feet on Cascade Mountain.
Phenology: In immature fruit in June.
Frequency: This sedge is more common in the poorly-drained sites, but occasional on the well-drained.

Carex lasiocarpa Ehrh. or *C. filiformis* in Peck SEDGE
Site: This species was found along the Saranac River flood plain opposite Mount Pisgah. In Goose Pond just N of the Keese's Mills Road bridge.
Elevation: 1500 to 1700 feet.
Phenology: In fruit September 7, 1974.

Carex lurida Wahlenb. SEDGE
Site: This species occurs with *C. crinita* and *Scirpus* in marshes and in sunny wet spots in woods at the lower elevations.
Elevation: 1486 to ca. 2000 feet.
Phenology: In fruit late July through late August.
Frequency: Occasional.

Carex oligosperma Michx. SEDGE
Site: Bog mats.
Elevation: 1486 feet to 1700 feet.
Phenology: In fruit July 30 through September 8.
Frequency: Three stations at Paul Smiths.

Carex pallescens L. var. *neogaea* **Fern. SEDGE**
    Site: Roadside ditch, south side of Route 86, just west of entrance to Camp Gabriels.
    Elevation: 1720 feet.
    Phenology: June 22, 1980.

Carex pauciflora **Lightf. SEDGE.**
    Site: Bogs.
    Elevation: 1500 to 1700 feet.
    Phenology: In fruit July and August.
    Frequency: Three stations: Paul Smiths and Derrick.
    Miscellany: Common in bogs northward in Canada.

Carex paupercula **Michx.** or *C. magellanica* in **Peck SEDGE**
    Site: Bogs.
    Elevation: 1600 feet to 4888 feet on Mt. Marcy.
    Phenology: In fruit late June through early September. Spikelets brown in late summer, resembling Hemlock cones.
    Frequency: Three stations, Paul Smiths and Ray Brook.

Carex pedunculata **Muhl. ex Willd. SEDGE**
    Site: This species was found on Mt. Pisgah, Saranac Lake, on a near-neutral site harboring Sugar maple, White ash, and Basswood.
    Elevation: 1700 feet.
    Reference: Handel (1976).

Carex plantaginea **Lamb. PLANTAIN-LEAVED SEDGE**
    Site: This may be the Flora area's most demanding (i.e. high humus pH) *Carex* except for *C. pedunculata*. It grows most commonly under Sugar maple, often with White ash and Basswood. pH of humus: n=14; min. 4.5; max. 6.6; median 5.05.
    Elevation: From 1500 feet to 2100 feet on Brewster Mountain.
    Phenology: In flower, earliest date, April 13.
                      In flower, median date, early May.
                      In flower, latest date, May 12.
    Frequency: 19 stations.

Carex rostrata **Stokes ex With.** var *utriculata* **(Boott) Bailey SEDGE**
    Site: This is a species of marshes and open river flats.
    Elevation: 1486 to 1640 feet.
    Phenology: In fruit August and early September.
    Frequency: Two stations, Paul Smiths and Saranac Lake.

Carex scabrata **Schwein. SEDGE**
    Site: Wet swales in northern hardwoods; often in partial shade.
    Elevation: 1600 to 1900 feet.
    Frequency: Occasional.

Carex scoparia **Schkuhr ex Willd. SEDGE**
    Site: Common in wet spots, often in damp sand. Shade-intolerant.
    Elevation: 1500 to 4000 feet.
    Phenology: In fruit August.

Carex stricta **Lam. HUMMOCK SEDGE or TUSSOCK SEDGE**
    Site: Poorly-drained. Shade-intolerant. It forms foot-high hummocks above water table consisting of old leaves.
    Elevation: 1500 to 1700 feet (but attaining elevations near sea level in the greater New York City area).
    Phenology: In fruit August.
                        In flower late May to mid-June.
    Frequency: Locally profuse, covering acres in marshes and on river flats.

Carex trisperma **Dewey SEDGE**
    Site: Bogs.
    Elevation: 1486 to 2976 feet.
    Phenology: In fruit end of July through early September.
                        In flower June.
    Frequency: Two stations, Paul Smiths and Loch Bonnie northwest of Lake Placid.

Carex vulpinoidea **Michx. FOX SEDGE**
    Site: In damp sand behind the Science Building on campus.
    Elevation: 1630 feet.
    Miscellany: *Carex* is the genus with the largest number of species (24 identified to date in the Flora area). It is one of the most difficult genera for making a species identification.

Cyperus bipartitus **Torr. or** *Cyperus rivularis* **Kunth SEDGE**
    Site: Weedy on Paul Smith's College Campus.
    Elevation: 1620 to 1650 feet.
    Phenology: In fruit September 11.

Cyperus esculentus **L. NUT GRASS or YELLOW NUT GRASS**
    Site: Paul Smiths.
    Elevation: 1620 to 1700 feet.
    Phenology: In fruit August 2, 1974.
    Miscellany: Native weed. Swellings in the root system resemble nuts.

Cyperus filiculmis Vahl SEDGE
    Site: Near Saranac Lake Railroad Station.
    Elevation: 1560 feet.
    Phenology: In fruit August, 1974.

Dulichium arundinaceum (L.) Britt. THREE-WAY SEDGE
    Site: Poorly-drained. Shade-intolerant.
    Elevation: 150 to 1700 feet.
    Phenology: In fruit September 22, 1972.
    Frequency: Common along pond edges.

Eleocharis acicularis (L.) R. & S. NEEDLE SPIKE RUSH, HAIRGRASS or SLENDER SPIKERUSH
    Site: Plants about 4 inches high at most, appearing as a lawn on the bottoms of shallow lakes and slow streams. The roots bind sand.
    Elevation: 1500 to 1700 feet.
    Phenology: In fruit August 26, 1976.
    Frequency: Three stations at Paul Smiths and Saranac Lake.

Eleocharis robbinsii Oakes TRIANGLE SPIKE RUSH
    Site: Locally, common along river edge, partly submerged and partly emergent for about 6 inches. Raquette River near Tupper Lake Village.
    Elevation: 1550 feet.
    Phenology: In fruit August 23, 1987.
    Frequency: Rare in Mitchell.

Eleocharis smallii Britt. SPIKE RUSH
    Site: Shallow water and shorelines.
    Elevation: 1500 to 1700 feet.
    Phenology: In flower June 2.
                     In fruit between August 7th and September 10th.
    Frequency: Our most common *Eleocharis*.
    Miscellany: This and several other closely-related species are often combined into *E. palustris* (L.) R. & S.

Eriophorum vaginatum L. ssp. *spissum* (Fern.) Hulten or *Eriophorum spissum* Fern. COTTONGRASS or HARE'S TAIL
    Site: Bog mats. Two station examples are Hoffman Road Bog, E of junct. Route 30 and Hoffman Road. Alpine summits of the High Peaks.
    Elevation: 1600 to over 5000 feet.
    Phenology: In fruit between June 9 and July 6.

                Fruits already shed by September 29.
Frequency: Common.

Eriophorum virginicum L. COTTON GRASS or TAWNY COTTONGRASS
    Site: Bog mats.
    Elevation: 1600 to 1700 feet around Paul Smiths.
    Phenology: In fruit between August 20 and September 29.
    Frequency: Common.

Rhyncospora alba (L.) Vahl. BEAK RUSH or WHITE BEAK RUSH
    Site: Bog mats.
    Elevation: 1500 to 1710 feet.
    Phenology: In fruit between August 11 and September 18.
    Frequency: Six stations.

Rhyncospora capitellata (Michx.) Vahl. BEAK RUSH
    Site: In the Paul Smith's College axe-throw practice area near the Forestry Club Tool Room on Campus.
    Elevation: 1625 feet.
    Phenology: In fruit August 23, 1974.

Scirpus acutus Muhl. ex Bigel. or *S. occidentalis* (S. Wats.) Chase HARDSTEM BULRUSH or TULE
    Site: Grows in water, reaching heights of 5 feet.
    Elevation: 100 feet to 1700 feet.
    Phenology: In fruit Lake Clear September 1.
                     In fruit Osgood Pond August 24.
    Frequency: Three stations.

Scirpus atrocinctus Fern. WOOL GRASS or NORTHERN BULRUSH
    Site: Poorly-drained sites as in marshes and sunny wet spots along log roads and truck trails.
    Elevation: From 100 feet along Lake Champlain to Blue Mt. summit at 3759 feet.
    Phenology: In fruit Bluebird Road near Campus July 18.
    Frequency: Common.
    Miscellany: Often included within *S. cyperinus* (L.) Kunth.

Scirpus atrovirens Willd. BULRUSH
    Site: In poorly-drained sites such as in marshes and sunny wet spots along log roads and truck trails.
    Elevation: 1600 to 2000 feet.
    Phenology: In fruit Bluebird Road, in Paul Smiths, July 18.
    Frequency: Common.

Scirpus microcarpus Presl. or *Scirpus rubrotinctus* Fernald. BULRUSH
    Site: In sunny, poorly-drained marshy area along Bluebird Road near the Pole Barn on Campus. With cousins *S. atrovirens* and *S. atrocinctus.*
    Elevation: 1625 feet.
    Phenology: In fruit July 18.

Scirpus subterminalis Torr. WATER BULRUSH, SWAYING BULRUSH or CLUBRUSH
    Site: In shallow water.
    Elevation: 1600 to 1700 feet.
    Phenology: In fruit September 9 in Rainbow Lake.
                     In fruit between Osgood & Jones Ponds July 19.

100. POACEAE or GRAMINEAE (GRASS FAMILY)

References on grasses in general:
Brown (1979)
Hitchcock (1950)
Pohl (1954)
Smith, Stanley Jay (1965)

* *Agropyron repens* (L.) Beauv. QUACK GRASS
    Phenology: In fruit on Campus July 17 through August.
    Frequency: Common European weed.

* *Agrostis capillaris* L. or *Agrostis tenuis* Sibth. or *A. alba* L. var. *vulgaris* Gray RED TOP, COLONIAL BENT or RHODE ISLAND BENT
    Phenology: In fruit July 5 through August, coloring roadsides, fields, and the Paul Smith's Campus red in summer.
    Frequency: The commonest European weed.

Agrostis mertensii Trin. or *Agrostis borealis* Hartm. NORTHERN BENT GRASS
    Site: High Peaks.
    Elevation: From 3772 feet at Lake Arnold to over 5000 feet on Mount Marcy Summit.
    Phenology: July and August.

Agrostis perennans (Walt.) Tuckerm. UPLAND BENT GRASS or AUTUMN BENT GRASS
    Site: Two stations, at Saranac Lake and near Marcy Dam.
    Elevation: 1486 to 2366 feet.
    Phenology: In fruit August and early September.

Agrostis scabra **Willd.** or *A. hyemalis* **(Walt.) B.S.P.** HAIRGRASS, TICKLEGRASS or FLY-AWAY GRASS
    Site: Tolerates poorly-drained soils to the most dry, exposed rocky summits.
    Elevation: 1500 feet on the Saranac River to over 5000 feet on Mount Marcy.
    Phenology: In fruit between August 5 and 20.
    Frequency: Occasional: 5 stations.

* *Anthoxanthum odoratum* **L.** SWEET VERNAL GRASS
    Site: European. Paul Smiths.

Brachyelytrum erectum **(Schreb.) Beauv.** BRACHYELYTRUM GRASS
    Site: In northern hardwoods and mixed woods stands, equally as shade-tolerant as *Cinna* but not as common. Soils well-drained. pH of humus: n=6; min. 4.5; max. 6.6; median 4.95.
    Elevation: From 210 feet at Chazy Fir Swamp to 2450 feet at Catamount Mountain.
    Phenology: In fruit August 1 through September 20.
    Frequency: Sixteen stations but probably many more.

Bromus ciliatus **L.** FRINGED BROME
    Site: Our native woodland Brome grass. Shade-tolerant. Well-drained.
    Elevation: 1486 feet to over 2800 feet on Hurricane Mountain.
    Phenology: In fruit July and August.
    Frequency: Six stations.

* *Bromus inermis* **Leyss.** SMOOTH BROME
    Site: European naturalized grass in Saranac Lake.

Calamagrostis canadensis **(Michx.) Beauv.** BLUE-JOINT
    Site: Forming nearly pure stands of great density along slow-moving streams. Intolerant of shade. Also at pond margins and marshes. To timberline. From poorly-drained sites to open dry rocky summits, as atop Jenkins Mt. One stand on Campus near faculty housing is on well-drained outwash sand. (If the Blue Dot Reservoir had not been renamed Heron Marsh by the Visitors' Interpretive Center staff, I would have named it the Great *Calamagrostis* Meadows). Largest populations are along the Saranac River near Saranac Lake Village, and along Mountain Brook NE of Meno. Often with Speckled alder.
    Elevation: 100 feet at Wickham Marsh to Mount Marcy ca. 5250 feet.
    Frequency: One of the most abundant grasses.

Cinna latifolia (Trev. ex Goepp.) Griseb. ex Ledeb. DROOPING REED-GRASS
- Site: Well-drained under hardwoods and spruce-fir. To timberline. pH of humus: n=11; min. 4.0; max. 5.8; median 4.7.
- Elevation: 1486 feet at Cranberry Lake to Gothics 4736 feet.
- Phenology: In fruit August 1.
- Frequency: The Flora area's most common woodland shade-tolerant grass.

* *Dactylis glomerata* L. ORCHARD GRASS
- Frequency: Abundant.
- Miscellany: A European species naturalized and weedy, offering hayfever to many (including your author) via its prolific pollen.

Danthonia spicata (L.) Beauv. ex R. & S. WILDOAT GRASS, POVERTY GRASS or JUNEGRASS
- Site: Native, weedy, old-field species, often following old log roads and truck trails.
- Elevation: From 1486 feet to 3152 feet on Mount Morris.
- Phenology: In fruit July and August.
- Frequency: Common.

Deschampsia flexuosa (L.) Trin. HAIR GRASS or COMMON HAIR GRASS
- Site: To timberline and in alpine areas. Infrequent on lower-elevation exposed rocky dry summits, and on well-drained outwash sands. Shade-intolerant.
- Elevation: 1500 feet to ca. 5167 feet on Mount Marcy.
- Phenology: In fruit July and August.
- Frequency: Three stations in Paul Smiths area.

* *Digitaria ischaemum* (Schreb. ex Schweig.) Schreb. ex Muhl. SMOOTH CRABGRASS
- Site: European weed on Paul Smith's Campus.

* *Echinochloa crusgalli* (L.) Beauv. or *Echinochloa muricata* or *Echinochloa pungens* BARNYARD GRASS
- Site: European weed in poorly-drained sites, Paul Smiths.

Elymus hystrix L. or *Hystrix patula* Moench. BOTTLEBRUSH GRASS
- Site: Higher humus pH soils under Sugar maple, Basswood, and White ash. Well-drained. Shade-tolerant.
- Elevation: 1000 to 2300 feet.
- Phenology: In fruit July 10.
- Frequency: Three stations, including Keene Valley, Pitchoff Mountain, and Mount Baker.

*** Eragrostis minor** Host. or *E. poaeoides* Beauv. or *E. eragrostis* (L.) Karst
LOVEGRASS
    Site: In sand on Campus, exposed to full sun. Naturalized from Europe.
    Phenology: In fruit August 7 through 20.

Eragrostis spectabilis **(Pursh) Steud.** TUMBLE GRASS or PURPLE LOVE-GRASS
    Site: Shoulder of intersection of State Highways 86 and 30.
    Elevation: 1660 feet.
    Phenology: In fruit August 24.
    Miscellany: Native.

*** Festuca elatior** L. MEADOW FESCUE
    Site: Weedy European. In Saranac Lake and about the Classroom Building on Campus.
    Phenology: In fruit July 5, 1974.

Festuca obtusa **Biehler.** or *F. nutans* **Gray B & B.** FESCUE GRASS or NODDING FESCUE GRASS
    Site: The Flora area's native woodland fescue. Shade-tolerant. Well-drained sites on Mount Baker, under higher humus pH of Sugar maple, White ash, and Basswood.
    Elevation: 1900 feet.
    Phenology: In fruit July 10.
    Miscellany: Naturalized, according to Mitchell.

*** Festuca ovina** L. SHEEP FESCUE
    Site: On the Paul Smith's College Campus and at the Bay Pond Burn.
    Phenology: In fruit July 16.
    Frequency: Abundant European grass.

Glyceria canadensis **(Michx.) Trin.** RATTLESNAKE GRASS
    Site: Locally abundant along watercourses, as along the Saranac River.
    Elevation: 1486 to 1760 feet.
    Phenology: In fruit late July through September.
    Frequency: Four stations.

Glyceria maxima **(Hartm.) Holmb.** or *G. grandis* **S. Wats.** REED MEADOW GRASS
    Site: Poorly-drained soils. Shade-intolerant.
    Elevation: 1500 to 1700 feet.
    Phenology: In fruit July through September.

Frequency: Six stations.

Glyceria melicaria (Michx.) F.T. Hubbard. SLENDER MANNAGRASS
   Elevation: 1200 to 4000 feet.
   Phenology: In fruit late August.
   Frequency: Three stations, High Peaks.

Glyceria striata (Lam.) Hitchc. FOWL MANNAGRASS
   Site: In springs and along brooks in hardwood and spruce-fir forests. Shade-tolerant. Often in swales.
   Elevation: 1500 to 2200 feet at Heart Lake.
   Frequency: Common.

Leersia oryzoides (L.) Sw. RICE CUT GRASS
   Site: Poorly-drained areas in full sun. Edge of Church Pond at Paul Smiths and along floodplain of Saranac River above Bloomingdale.
   Elevation: 100 to 1770 feet.
   Phenology: In fruit August.
   Frequency: Two stations.

Milium effusum L. MILLETGRASS
   Site: On higher humus pH sites with Sugar maple, Ash, Basswood, *Elymus hystrix* and *Festuca obtusa* on Mount Baker. Also on Buck Hill summit near Rainbow Lake. Well-drained. Shade-tolerant.
   Elevation: 1600 feet to 3650 feet on Cascade Mountain.
   Phenology: In fruit September 26, 1974.
   Frequency: Seven stations.
   Miscellany: Recognized by its blue-green color.

Muhlenbergia frondosa (Poir.) Fern. WIRESTEM MUHLY
   Elevation: 100 feet on Four Brothers Islands in Lake Champlain to 1600 feet at Saranac Lake.
   Phenology: In fruit late August to late October.
   Miscellany: Native weed.

Oryzopsis asperifolia Michx. SPREADING RICEGRASS
   Site: Well-drained. Shade-tolerant. Along new V.I.C. Forest Ecology Trail under northern hardwoods.
   Elevation: 1650 feet.
   Phenology: In fruit June 15.
   Frequency: One station.

Panicum capillare L. WITCH GRASS
   Site: Native annual, forming tumbleweeds on Paul Smith's College

    Campus. There are other species of **Panicum** in the Flora area, but specimens collected never seem adequate to fit the keys in this most difficult genus.
Elevation: 100 to 1630 feet.
Phenology: In fruit late August through October.

Phalaris arundinacea **L.** REED CANARY GRASS
    Site: Flood plains of Saranac and Ausable Rivers.
    Elevation: 100 to 1700 feet.
    Phenology: In fruit late August through October.
    Frequency: Two stations.

* *Phleum pratense* **L.** TIMOTHY or HERD-GRASS
    Phenology: In fruit July 17 on Campus.
    Frequency: An abundant European weed.
    Miscellany: Originally introduced for forage, but now causing hayfever.

* *Poa compressa* **L.** CANADA BLUEGRASS
    Frequency: European. Less common than *P. pratensis* but also weedy.

Poa nemoralis **L.** WOOD BLUEGRASS
    Site: The Flora area's native woodland shade-tolerant *Poa*.
    Elevation: 1500 to 2200 feet.
    Phenology: In flower May.
                    In fruit early June.
    Frequency: Four stations Paul Smiths to Heart Lake.
    Miscellany: Peck lists only **P. alsodes** **Gray**. Possibly, they are two forms of the same species. Mitchell (1986) says naturalized.

* *Poa pratensis* **L.** KENTUCKY BLUE GRASS
    Phenology: In flower early June.
                    In fruit August through October.
    Frequency: Abundant and weedy. The commonest lawn grass.
    Miscellany: Native to the United States, but naturalized in the Flora area.

Schizachne purpurascens **(Torr.) Swallen.** FALSE MEDIC
    Site: McKenzie Pond Road near entrance to Saranac Lake Village Dump, along roadside.
    Elevation: 1560 feet.

* *Setaria glauca* **(L.) Beauv.** FOX TAIL GRASS or PIGEONGRASS
    Site: Saranac Lake.
    Frequency: Occasional European weed.

101. SPARGANIACEAE (BUR-REED FAMILY)

Sparganium americanum Nutt. BUR-REED or AMERICAN BUR-REED
    Site: In shallow water.
    Elevation: 1500 to 1700 feet in the Flora area.
    Phenology: In fruit September 8 & 12 at Heron Marsh.
                     In flower July 17 E end of Little Square Pond.
    Frequency: Locally profuse.

Sparganium emersum Rehm. or *Sparganium angustifolium* Michx. NARROW-LEAF BUR-REED
    Site: Leaves float on surface.
    Phenology: In fruit between July 22 and September 12.
    Frequency: Not nearly as common as *S. americanum*.
    Miscellany: Looks like a *Vallisneria* except for the distinctive Bur-reed fruit.

102. TYPHACEAE (CAT-TAIL FAMILY)

Typha latifolia L. CATTAIL, BROAD-LEAVED CATTAIL or COMMON CATTAIL
    Site: An indicator of poorly to very poorly drained sites, in full sun, but not acid bogs. Shade intolerant. Profuse in marshes.
    Elevation: 100 to 1700 feet.
    Frequency: Common.
    Miscellany: Roots, shoots, and heads of fruit edible.
    References: Brody (1979)
                     Hogg and Wien (1988)
                     Schottman (Sept./Oct. 1990).

103. PONTEDERIACEAE (PICKERELWEED FAMILY)

Pontederia cordata L. PICKEREL WEED
    Site: Forming extensive stands in shallow waters of marshes.
    Elevation: 1500 to 2008 feet.
    Phenology: In flower, earliest date, July 16.
                     In flower, median date, August 3.
                     In flower, latest date, September 16.
    Frequency: Common.
    Reference: Schottman (1983, August).

104. LILIACEAE (LILY FAMILY)

Mitchell (1986) has moved the genus *Smilax* into a separate family: the Smilacaceae, (Greenbrier Family).

Allium tricoccum Ait. WILD LEEK or RAMP
- Site: A shade-tolerant herb of primarily well-drained soils with higher humus pH, under Sugar maple, and often White ash, and Basswood.
- Elevation: 100 feet along Lake Champlain to 1700 feet.
- Phenology: In leaf May 14.
  In fruit August 14.
  Unique in the Flora area in that the leaves and flowers are not simultaneously visible; the leaves emerge soon after snow-melt and are withered by June when the flowers first emerge.
- Frequency: Three stations: two at Saranac Lake and one at Mile Brook.
- Miscellany: Leaves and bulb edible.

Clintonia borealis (Ait.) Raf. CLINTON'S LILY, BLUEBEAD LILY, WOOD LILY or CORN LILY
- Site: Occasionally in pure hardwood stands. Well-drained to imperfectly-drained soils. Shade-tolerant. pH of humus: n=14; min. 3.5; max. 5.8; median 4.25. This species persists under a powerline (NE of Substation on Route 30, near Campus) competing with dense Yellow birch saplings, Raspberries, and Bracken fern in full sun.
- Elevation: From 210 feet at Chazy Fir Swamp to Dix Mountain 4857 feet.
- Phenology: In flower, earliest date, May 25.
  In flower, median date, June 5.
  In fruit August.
- Frequency: One of the most common herbs of the spruce-fir forests and in mixed woods.
- Miscellany: Fruits poisonous, at least considered so by some authors.
- References: Schottman (1984)
  Brown, Ashmun & Pitelka (1985).

Erythronium americanum Ker. TROUT LILY, YELLOW ADDER'S TONGUE or YELLOW DOG-TOOTH VIOLET
- Site: Prolific in hardwood, conifer, or mixed woods, well-drained to imperfectly well-drained. pH of humus: n=23; min. 4.3; max. 5.6; median 4.6.
- Elevation: 200 feet at Chazy to ca. 2000 feet.
- Phenology: One of the first herbs to flower in the spring, the fruits maturing and the leaves withering by early June.

Leafing, earliest date, April 5; median date, April 13; latest date, May 1.

Full flower, earliest date, April 18; median date, April 27; latest date, May 17.

In fruit, earliest date, May 12; median date, May 25; latest date, May 29.

Last leaves wither June 30.

The strategy is to push the leaves out as soon as possible after snowmelt and get the year's photosynthesis done by the time the hardwoods are in mature leaf and block out much light. In June, the white rhizomes (underground stems) appear above ground for a short period.

Frequency: Common.
References: Muller & Bormann (1976)
Muller (1978)
Schottman (1984).

Maianthemum canadense Desf. CANADA MAYFLOWER, WILD LILY-OF-THE-VALLEY or FALSE LILY-OF-THE-VALLEY

Site: A most abundant herb under conifers, hardwoods, or mixed woods. Shade tolerant. Well-drained to imperfectly-drained. To timberline. pH of humus: n=36; min. 3.5; max. 6.8; median 4.6.
Elevation: 100 feet along Lake Champlain to 5330 feet on Mt. Marcy.
Phenology: In flower, earliest date, June 2.
In flower, median date, June 7.
In flower, latest date, June 16.
Frequency: Common.
Miscellany: Occasionally with such dense masses of flowers that the sweet scent can be smelled without one's bending over. Spreads by rhizomes just under the surface; the only above-ground stems bear 2 or 3 leaves with flowers; single leaves are attached to rhizomes.
Reference: Schottman (1984).

Medeola virginiana L. INDIAN CUCUMBERROOT

Site: In well-drained hardwoods or mixed woods. Shade-tolerant. pH of humus: n=9; min. 3.7; max. 4.9; median 4.1.
Elevation: From 210 feet at Chazy Fir Swamp to 2500 feet on Loch Bonnie trail.
Phenology: In flower, earliest date, June 5.
In flower, median date, June 12.
Frequency: Ten stations.
Miscellany: Eating the root destroys the plant, so consume garden cucumbers instead.

Polygonatum pubescens **(Willd.) Pursh** SOLOMON'S SEALS
- Site: On well-drained sites under mostly northern hardwoods. Shade-tolerant.
- Elevation: 100 feet along Lake Champlain to 2250 feet on Pitchoff Mt.
- Phenology: In flower bud April 30 to May 4.
  In flower late May to early June.
- Frequency: Occasional.
- Miscellany: Peck (1899) lists *P. biflorum.*

Smilacina racemosa **(L.) Desf.** SOLOMON'S PLUMES, FALSE SOLOMON'S SEALS or SPIKENARD
- Site: On well-drained sites, mostly under hardwoods. Shade-tolerant. Leaves horizontally-oriented to maximize the catching of incident light on the forest floor. pH of humus: n=12; min. 3.5; max. 6.6; median 5.4.
- Elevation: 100 feet along Lake Champlain to 2400 feet on Debar Mountain.
- Phenology: In flower, median date, June 1.
- Frequency: Common locally.

Smilacina trifolia **(L.) Desf.** THREE-LEAVED FALSE SOLOMON'S SEAL or FALSE SOLOMON'S SEAL
- Site: Bogs.
- Elevation: 1500 to 1800 feet.
- Phenology: In flower, median date, June 10.
  In fruit September 2.
- Frequency: Common.

Streptopus amplexifolius **(L.) DC** LARGE-LEAVED TWISTED STALK, WHITE MANDARIN or LIVERBERRY
- Site: In spruce-fir forest.
- Elevation: In the High Peaks from elevation 3600 to 4300 feet.
- Frequency: Four stations.
- Miscellany: This is a more common plant northward in Canada under spruce-fir forests.

Streptopus roseus **Michx.** ROSY TWISTED STALK or ROSE MANDARIN
- Site: In well-drained woods mostly under hardwoods or mixed stands. Shadetolerant. To timberline.
- Elevation: From about 1600 feet at Paul Smiths to Gothics Mt. 4736 feet.
- Phenology: In flower May 22 NW corner of Long Pond near E base of Jenkins Mountain.
- Frequency: Frequent.

Trillium erectum **L. RED TRILLIUM, PURPLE TRILLIUM, WAKE ROBIN** or **STINKING BENJAMIN**
- Site: Mostly under hardwoods. Well-drained. Shade-tolerant. Pollinated by carrion insects, and therefore producing the foul stench of decaying animals and color of dried blood. pH of humus: n=31; min. 3.5; max. 6.6; median 4.6.
- Elevation: 200 feet at Chazy and on Barnhardt Island to 3650 feet on Cascade Mountain.
- Phenology: Flower buds, earliest date, April 19.
  Flower buds, median date, April 26.
  Flower buds, latest date, May 5.
  Full flower, earliest date, April 30.
  Full flower, median date, May 12.
  Full flower, latest date, June 7.
- Frequency: Protected in N.Y. State, yet common in our area.
- Miscellany: Forma *luteum* has yellow petals and is not uncommon: Buck Hill, Saint Regis Mountain, lean-to at NW corner of Long Pond near E base of Jenkins Mountain, and Campus faculty housing.
- References: Pangburn (1981, June)
  Schottman (1987, June)

Trillium undulatum **Willd. PAINTED TRILLIUM**
- Site: Grows under northern hardwoods, but more under spruce-fir than *T. erectum*. Tolerant of shade. Well-drained. pH of humus: n=15; min. 3.7; max. 5.0; median 4.5.
- Elevation: From 750 feet at Altona to 3450 feet on Algonquin.
- Phenology: In flower, earliest date, April 26.
  In flower, median date, May 9.
  In flower, latest date, June 3.
  Dr. Ketchledge reports it at elevation 4400 feet on Nippletop Mountain where it flowers three weeks later.
- Frequency: Protected in N.Y. State, yet common in this Flora's area.
- Reference: Schottman (1987, June)

Uvularia sessilifolia **L.** or ***Oakesia sessilifolia*** **in Heimburger. SESSILE-LEAVED BELLWORT** or **WILD OATS**
- Site: Mostly under northern hardwoods and often under Sugar maple, White ash, Basswood. Shade-tolerant. Well drained. pH of humus: n=7; min. 3.7; max. 6.6; median 4.7.
- Elevation: 100 feet along Lake Champlain to 2350 feet on Algonquin.
- Phenology: In flower May 13 through 24.
- Frequency: 14 stations.

Veratrum viride Ait. FALSE HELLEBORE, WHITE HELLEBORE or INDIAN POKE
- Site: Poorly-drained and imperfectly-drained soils. Shade-tolerant.
- Elevation: From 420 feet along the Boquet River to ca. 5250 feet to 5325 feet on Mount Marcy.
- Phenology: Leaves appear April 20 to 23 into early May.
  In flower June.
  Plant withers July.
  Plant still in fruit August 31 on Mount Marcy tundra.
- Frequency: 17 stations.
- Miscellany: This is sometimes confused with Skunk-cabbage. *Veratrum*, unfortunately, is a poisonous plant to eat, so that this confusion can be serious.

105. IRIDACEAE (IRIS FAMILY)

Iris versicolor L. BLUE FLAG or WILD IRIS
- Site: Poorly-drained soils and marshes. Shade-intolerant.
- Elevation: 115 feet at Duck Pond, Point Au Roche State Park, to 1700 feet at Paul Smiths.
- Phenology: In flower, earliest date, June 16.
  In flower, median date, June 27.
  In fruit, September.
- Frequency: 11 stations. Occasional.
- Miscellany: "Flagge" in Middle English means "reed."

Sisyrinchium angustifolium Mill. BLUE-EYED GRASS
- Site: On usually poorly-drained sites. Shade-intolerant. Occasionally on lawns.
- Elevation: From 210 feet at Chazy Fir Swamp to 1700 feet at Paul Smiths.
- Phenology: In flower, earliest date, May 29.
  In flower, median date, June 15.
- Frequency: Six stations.

106. SMILACACEAE (GREENBRIER FAMILY)

Smilax herbacea L. CARRION FLOWER or JACOB'S LADDER
- Site: Saranac River floodplain south of Bloomingdale.
  Edge of Stony Creek Ponds at Coreys, near Upper Saranac Lake.
- Elevation: 100 feet at Lake Champlain to 1550 feet.

107. ORCHIDACEAE (ORCHIS FAMILY OR ORCHID FAMILY)
References for Orchids in general (not individual species):
Knaus (1982)
Luer (1975)
All orchids are protected in New York State.

Arethusa bulbosa L. SWAMP PINK, ARETHUSA or DRAGON'S MOUTH
    Site: Upper Saint Regis Bog.
    Phenology: End of flowering July 1, 1979.

Calopogon tuberosus (L.) BSP or *Calopogon pulchellus* (Salis.) R.Br. ex Ait. GRASS PINK
    Site: Bogs.
    Elevation: 1600 to 1700 feet in the Paul Smiths area.
    Phenology: In flower, earliest date, July 1.
                     In flower, median date, July 13.
    Frequency: Four stations.
    Miscellany: Several flowers per culm.

Corallorhiza maculata (Raf.) Raf. SPOTTED CORALROOT
    Site: A non-green orchid of well-drained forested sites. Shade or lack of does not affect this plant.
    Elevation: 1600 to 1900 feet.
    Frequency: Four stations Paul Smiths to Franklin Falls areas.

Cypripedium acaule Ait. MOCCASIN FLOWER or PINK LADY'S SLIPPER
    Site: Well-drained sites most often under partial shade in coniferous forests, but also sometimes under hardwoods, as at the east base of Saint Regis Mountain.
    Elevation: From 1630 feet at Paul Smiths to 3630 feet on Cascade Mountain. Probably to much lower elevations as well.
    Phenology: In flower, earliest date, May 30.
                     In flower, median date, June 13.
    Frequency: The Flora area's most common Orchid, yet still not abundant.
    References: Doeffinger (1978)
                    Phelps (1976)
                    Schottman (1982, May)
                    Smith, Ralph (1965, June-July)

* *Epipactis helleborine* (L.) Crantz. HELLEBORINE or WEED ORCHID
    Site: A European species naturalized in the U.S. and unusual in that it is one of the very few that is shade-tolerant and appears in forests with native species. Well-drained sites, often under hardwoods,

>   and along sanded roadsides.
> Elevation: From 100 feet along Lake Champlain to about 2000 feet.
> Phenology: In flower August.
> Frequency: Frequent.
> Miscellany: Flowers green.
> Reference: Doeffinger (1978)

Goodyera repens (L.) R. Br. DWARF RATTLESNAKE-PLANTAIN
>   Site: A shade-tolerant species in well-drained to imperfectly-drained soils.
> Elevation: 1600 to 1800 feet.
> Phenology: In flower bud July south of Upper Saint Regis Lean-to.
>   End of flowering end of August south of Upper Saint Regis Lean-to.
> Frequency: Five stations.

Platanthera blephariglottis (Willd.) Lindl. or *Habenaria blephariglottis* (Willd.) Hook. WHITE FRINGED ORCHID
>   Site: Bogs at Bog Pond and near Mount Meenahga, off Rainbow Lake.
> Elevation: 1683 feet.
> Phenology: In flower July 2.
> Frequency: Two stations.

Platanthera clavellata (Michx.) Luer or *Habenaria clavellata* (Michx.) Spreng. GREEN WOODLAND ORCHID
>   Site: Mostly in bogs.
> Elevation: 1600 to 1720 feet.
> Phenology: In flower, earliest date, July 23.
>   In flower, median date, August 1.
>   In immature fruit August 10.
> Frequency: Four stations at Paul Smiths.

Platanthera dilatata (Pursh) Lindl. ex Beck or *Habenaria dilatata* (Pursh) Hook. LEAFY WHITE ORCHIS, BOG CANDLE, SCENT-BOTTLE or WHITE BOG ORCHID
>   Site: Swamps and bogs.
> Elevation: 1620 to 1650 feet.
> Phenology: In flower between June 17 and July 23.
> Frequency: Two stations at Paul Smiths.

Platanthera lacera (Michx.) G. Don. or *Habenaria lacera* (Michx.) R. Brown. RAGGED FRINGED ORCHID
>   Site: Bogs.

Bunchberry (*Cornus canadensis*)

This is a dwarf cousin of Flowering dogwood, hence the strong resemblance of the flowers, fruits, and leaves. The four large white structures are flower bud scales. The true petals, four to each tiny flower, are inconspicuous and yellow.
© Jim Kraus 1992.

Elevation: 1620 to 1650 feet.
Phenology: In flower between July 1 and July 23.
Frequency: Two stations, both near Paul Smiths.

Platanthera orbiculata (Pursh) Lindl. or *Habenaria orbiculata* (Pursh). Torr. LARGE ROUND-LEAF ORCHID
Site: Shade-tolerant. Well to imperfectly drained. Under northern hardwoods, hemlock, or spruce-fir.
Elevation: 1360 to 2100 feet.
Phenology: In flower between June 19 and July 19.
Frequency: Three stations: near Wilmington, near Bloomingdale and near Heart Lake.

Pogonia ophioglossoides (L.) Juss. not Ker ROSE POGONIA or SNAKE MOUTH
Site: In bogs.
Elevation: 1620 to 1650 feet in the Paul Smiths area.
Phenology: In flower between July 13 and July 23.
Frequency: Three stations.

Spiranthes cernua (L.) Richard AUTUMN LADIES' TRESSES or NODDING LADIES' TRESSES
Site: Mostly on poorly-drained sandy outwash. Shade-intolerant.
Elevation: 800 to ca. 2050 feet.
Phenology: In flower August 30 through September 28.
Frequency: Six stations.

Sugar maple (*Acer saccharum*)
Male Sugar maple flowers are yellow and appear as the leaves expand and produce pollen. Female flowers typically are borne on separate trees. © Jim Kraus 1992.

Chapter 6

PLANTS IN JEOPARDY

Obviously, the forces of exterpation, whatever form they take, are no respectors of species abundance status.
Kudish

The first attempt to protect plants in New York State was in 1950. Section 1425, Subdivision 2, of the New York State Penal Law was under the general heading "Malicious Injury to and Destruction of Property". The plants listed as protected then do not necessarily occur in the Adirondack Upland. They are:

Epigaea repens, trailing arbutus.
Nelumbo lutea, yellow lotus.
Cornus florida, flowering dogwood.
Kalmia latifolia, mountain laurel.
Gentiana crinita and *G. andrewsii,* fringed and closed gentians.
Cypripedium spp., lady slippers, including *C. acaule, C. parviflorum, C. reginae,* and *C. candidum.*
Scolopendrium vulgare, the Hart's tongue fern, and all other ferns.

This law protected plants on public lands and on private lands where collecting was done without permission of the owner. Public lands included State and municipal (town, village, county, city, etc.) lands.

In 1974, a new wildflower law (Section 193.3) protected native plants. Statutory Authority: Environmental Conservation Law S 9-1503 was passed in New York State. The protected native plants list was quite lengthy; of this, only those plants which occur on the Adirondack Upland are listed here:

Campanula rotundifolia, harebell.
Celastrus scandens, bittersweet.
Chimaphila spp., pipsissewas.
Drosera spp., sundews.
Epigaea repens, trailing arbutus.
Ilex spp., hollies.
Kalmia spp., laurels.
Lilium spp., lilies.
Lobelia cardinalis, cardinal flower.
Lycopodium spp., clubmosses.
Orchidaceae, all orchids.
Rhododendron spp., rhododendrons.
Sanguinaria canadensis, bloodroot.
Sarracenia purpurea, pitcher plant.
Trillium spp., trilliums.
All ferns except:
  Pteridium, bracken,
  Dennstaedtia, hay-scented, and
  Onoclea, sensitive.

Note that some of these plants are protected but not rare. Others are rare but not protected because, probably, they are inconspicuous and not likely to be collected. An example of the former is *Dryopteris intermedia,* woodfern, the most abundant fern in the Adirondacks. An example of the latter is *Betula glandulosa,* dwarf birch, an Arctic-alpine (now protected, but not in 1974).

A section of the most recent (ca. 1990) Protected Native Plants booklet

Milkweed (*Asclepias syriaca*)

The dry fruit of the Milkweed, called a follicle, splits open along one seam and releases myriad seeds, each bearing long silky hairs to be carried by the wind. © Jim Kraus 1992.

published by the New York State Department of Environmental Conservation states the protected plants law as follows:

> It is a violation for any person, anywhere in the state, to pick, pluck, sever, remove, or damage by the application of herbicides or defoliants, or carry away, without consent of the owner, any protected plant. Each protected plant so picked, plucked, severed, removed, damaged, or carried away shall constitute a separate violation.

ENDANGERED, THREATENED, RARE, AND EXPLOITABLY VULNERABLE PLANTS

Beginning in 1979, all vascular plant species in New York State were evaluated in terms of their abundance status; as more and more data became available a series of publications ensued (1980, 1981, 1986, 1988, 1989, 1990). Each was assigned one of the ranks listed in the above heading.

In New York State, an endangered species occurs in only one to five stations over the whole state. Clemants (1989) assigns endangered plants an S1 ranking, where S denotes New York State and 1 the highest priority in locating, monitoring, and protecting. The three other criteria for determining whether or not a species is endangered are: there must be fewer than 1000 individuals in the State; stations are restricted to fewer than four United State Geological Survey 7½-minute topographic quadrangles Statewide; and the species is listed as endangered by the United States Department of the Interior. An example of an endangered species in the Adirondack Mountains is *Betula glandulosa,* an Arctic-alpine, which grows above the 4000–foot limit of this Upland Flora.

Threatened species occur in six to twenty stations over the whole State. These are assigned an S2 ranking, with 2 indicating the second highest priority. The three other criteria for defining a threatened species are: there must be between 1000 and 3000 individuals Statewide; stations are restricted to from four to seven U.S.G.S. 7½-minute topographic quadrangles; and the species is listed as threatened by the U.S. Department of the Interior. The following examples of threatened plants are all Arcticalpines, growing above the 4000–foot ceiling of this Upland Flora: *Diapensia lapponica,* mountain bride; *Juncus trifidus,* Arctic rush; *Rhododendron lapponicum,* Lapland rosebay; and *Scirpus cespitosus,* tufted bulrush.

Rare plants occur in 21 to 100 stations throughout New York State and are given the rank S3. There must be 3000 to 5000 individuals Statewide. Examples of rare plants in the Arctic-alpine zone above the Adirondack Upland are: *Carex bigelowii,* Bigelow's sedge; *Empetrum nigrum,* black crowberry; *Minuartia groenlandica,* mountain sandwort; *Salix uva-ursi,* bearberry willow; and *Vaccinium boreale,* high-mountain blueberry. Examples of rare plants within the Adirondack Upland are: *Arethusa bulbosa,* swamp pink; *Betula pumila,* swamp birch; *Carex houghtonii,* Houghton's sedge; *Pinus banksiana,* jack pine; *Prunus pumila* var. *pumila,* sand-cherry; *Vaccinium uliginosum,* alpine or bog bilberry; and *Viburnum edule,* squashberry.

Exploitably vulnerable species may become rare, threatened, or endangered in the future if collection and/or development continues. These species are also protected despite the fact that they are locally common in portions of the State. Examples in the Upland Flora area include:

Campanula rotundifolia, harebell.
Celastrus scandens, bittersweet.
Chimaphila spp., pipsissewa.
Epigaea repens, trailing arbutus.
All ferns except:
    Pteridium, bracken,
    Dennstaedtia, hay-scented, and
    Onoclea, sensitive.
Gentiana spp., gentians.
Ilex spp., hollies.
Kalmia spp., laurels.
Lobelia cardinalis, cardinal flower.
Orchidaceae, all orchids.
Sarracenia purpurea, pitcher plant.
Trillium spp., trilliums.

Clemants (1989) also presents global ranks for the species in his list. The symbol "G" is used for the world-wide distribution and is followed by one of the following numbers 1 through 5. G1 is the most critically imperiled worldwide and G5 the least. Some plants, though rare, threatened or endangered in New York State, e.g. many Arctic-alpines, are common in other portions of the world, e.g. northern Canada and even northern Eurasia.

The following lists of publications reflect the increased universal interest in endangered, threatened, rare, exploitably vulnerable, and protected species in recent years. The first is arranged chronologically to draw attention to the "burst" of activity after 1974 in New York State, four years after Earth Day in 1970. Both lists are included here as examples of the beginning of a major new phenomenon in the study of flora emerging in the second half of the Twentieth Century, namely the awareness of the fragile nature of vegetation.

NEW YORK STATE:

1950: E.W. Littlefield. *To Pick or Not To Pick - Fact and Fancy About the New York Wildflower Law.* The Conservationist, April-May, 1950, page 24.

1974: J.W. Aldrich. *Protected Plants—Official List Protects Rare, Unique Species.* NYS Environment. November 1, 1974, page 12.

1975: (No author, but likely J.W. Aldrich.) *Section 193.3, Protected Native Plants.* A list published by the D.E.C. in January, 1975.

1975: J.W. Aldrich. *Our Wildflowers and a Program to Protect Them.* The Conservationist, April-May, 1975, pp. 23-29.

1979: Richard S. Mitchell. *Preliminary Lists of Rare, Endangered and Threatened Plant Species in New York State.* New York State Museum Leaflet number 21, 18 pp.

1980: Richard S. Mitchell, Charles J. Sheviak, and J. Kenneth Dean. *Rare and Endangered Vascular Plant Species in New York State.* The State Botanist's

Office, N.Y. State Museum, in cooperation with the U.S. Fish & Wildlife Service, 38 pp.

1981: Richard S. Mitchell and Charles J. Sheviak. *Rare Plants of New York State.* New York State Museum Bulletin no. 445, 96 pp.

1986: Steven E. Clemants. *New York State Rare Plants—Spring 1986 Status Report.* N.Y. State D.E.C. and The Nature Conservancy, 26 pp.

1987: Frank Knight. *Rare and Protected - Plants That Need Help.* The Conservationist, May-June, 1987, pp. 3–9.

1988: Michael J. Birmingham. *Draft Regulations on Protected Native Plants.* News Release from the N.Y. State Department of Environmental Conservation, Albany, October 17, 1988. This News Release was followed by a booklet, also published by the D.E.C., entitled *Protected Native Plants,* unauthored in 1990.

1989: Steven E. Clemants. *New York Natural Heritage Program—New York Rare Plant Status List,* February, 1989.

GENERAL (Beyond New York State):

The Nature Conservancy News. March/April 1979 (features rare plants.)

The Nature Conservancy Magazine. November/December, 1987 (features rare plants.)

The Smithsonian Magazine, November, 1978, pp. 122–129. Article by Edward S. Ayensi: *Calling the roll of the world's vanishing plants.*

The New York Times: December 8, 1984; February 9, 1975, p. D39; November 19, 1978, p. 32LL++; September 7, 1980.

EXTIRPATED PLANTS

Obviously, the forces of exterpation, whatever form they take, are no respectors of species' abundance status. Specifically, two species discussed in *Paul Smiths Flora II* (1981) have since been removed from the Paul Smiths area. (I have not found either one anywhere else on the Adirondack Upland since.) *Prunus nigra,* wild plum, had been removed by crews clearing roadside brush along State Highway 30 in 1987. A second plant, *Carex houghtonii,* Houghton's sedge, had been removed during the demolition of an old coal-loading facility on the Campus about 1985.

Betula pumila, swamp birch, was not totally extirpated from the Upland, but a fifty percent loss occurred in the spring of 1990 when the population along the Keese's Mills Road was extirpated by brush cutting to widen the highway.

Cardinal flower (*Lobelia cardinalis*)
Unlike the radially symmetrical arrangement of flowers in most plant families, that of lobelia is bilateral. © Jim Kraus 1992.

Appendix I
AGES AND DIAMETERS OF NATIVE TREES

A recent sampling of tree ages in the Paul Smiths area gives a rough idea how large and how old local species commonly become. Note that the statistics for any tree in the following table do not indicate record breakers. Ages have been determined either by means of an increment borer or by ring counts on stumps and logs.

SPECIES	YEARS	DIAMETER, IN INCHES	LOCALITY
Hemlock	300	33	The Tongue
White pine	275	49	The Tongue
Red spruce	250	22	The Tongue
Yellow birch	235	36	Fish Pond Truck Trail
Red pine	230	15	Paul Smiths Campus
Sugar maple	200	24	Johnson Hill
Beech	200	26	Upper St. Regis Post Office
American elm	175	24	Buck Hill
Northern white cedar	105	6.5	Paul Smiths Campus
Black spruce	100	2	Ferd's Bog
Black spruce	54	13.5 (note contrast)	Moose River Plains
Bigtooth aspen	90	19.7	Paul Smith's College Campus
White spruce	70	15.5	Bloomingdale
Eastern larch	41	10	Bluebird Road, Campus
Trembling aspen	33	7	Mountain Pond Ridge
Striped maple	30	2.5	Jenkins Mountain

Appendix II
PLANT STRATEGIES: REPRODUCTION

Pollination may be by insects which are attracted to bright colors and sweet, pleasant aromas to gather nectar from flowers. An example is bees and goldenrods. Pollination may also be by carrion insects such as certain flies and beetles which are attracted to flowers by "trickery." Why does the flower of Purple trillium have a dark red color and unpleasant stench? To attract insects which think that the flower is a chunk of rotting meat derived from a dead vertebrate. While the insects climb about the flower looking for the "meat" in vain, they inadvertently pollinate! Pollination may also be by wind (Poplars, Hickories, Walnuts, Oaks, Birch family, Sweet gale, Beech); here the flowers need not be showy nor aromatic, but large quantities of pollen must be shed due to a great percentage loss. Think of the huge volumes of White pine pollen in June in puddles and on vehicles; conifers are wind-pollinated also.

Dimorphic flowers—Most insect-pollinated plants have one type of flower which has two jobs: to attract the insect and to produce the pollen and seed. Witchhobblebush, in contrast, has a division of labor among its flowers; the large, showy ones are sterile and serve only to attract insects. The small, numerous ones are not very showy, but are fertile, producing both pollen and seeds.

Dimorphic fronds—Most ferns have one kind of frond (or leaf) which both photosynthesizes and produces spores (reproduces). However, a few ferns bear two kinds of leaves, one kind to photosynthesize and the second to produce spores. The fronds are quite different. Examples are Cinnamon fern and Sensitive fern. The Field horsetail also has two kinds of shoots, one photosynthetic and the other reproductive.

Vegetative reproduction—Some plants reproduce vegetatively (asexually) in addition to sexually (by seeds or spores). The advantage of vegetative reproduction is speed, but the disadvantage is that the offspring are genetically identical to the parents (evolutionarily not sound). Examples of such reproduction are:

    Shining clubmoss: bulblets
    Bulblet bladder fern: bulblets
    Willow branches: fragmentation

Dioecism vs. Monoecism—Most plants are monoecious, that is, both male and female flowers are borne on the same plant. But a small number of plants are dioecious (male plants and female plants as in higher animals); examples are Willows, Poplars, Ashes, Yews, Maples (mostly), Sweet gale, Staghorn sumac.

Long-lived seeds—Several plants' seeds can outlive by far the plant. See Marks

Bluets (*Hedyotis caerulea*) Most dicots have flower parts in fives. But several families, including the one which contains Bluets, have four: four sepals, four petals, and four stamens; and one ovary consisting of two fused leaves or carpels. © Jim Kraus 1992.

(1974) for details. Examples are Red cherry, Blackberry, Raspberry. The plants are short-lived pioneers following disturbances in which the forest canopy is removed, permitting sunlight entry. The cherry is our shortest-lived tree, only 20 to 30 years, but the seeds can persist in the soils for 50 to 75 years and still germinate when the soil is disturbed the next time due to logging, clearing, blowdown, fire, etc. The advantage is that seeds not be brought in each time by birds and other animals when a disturbance occurs; the seeds can wait out until the next disturbance.

Appendix III

\# = outside Flora area

LOCALITY	LATITUDE NORTH	LONGITUDE WEST
Adirondack Museum #	43°52'	74°27'
Algonquin Mountain	44°08½'	73°59'
Alice Lake #	44°52'	73°29'
Altona #	44°53'	73°39½'
Ampersand Mountain	44°14'	74°12½'
Ausable Chasm #	44°32'	73°28'
Ausable Delta #	44°33½'	73°27½'
Averyville Marsh	44°14'	74°04'
Axton	44°12'	74°20'
Azure Mountain	44°32½'	74°30½'
Baldface Mountain	44°37'	74°11'
Barnhardt Island #	45°00'	74°50'
Barnum Pond	44°27½'	74°15½'
Bartlett Carry	44°15'	74°18'
Bartlett Pond	44°20'	74°01'
Baxter Mountain	44°13'	73°46'
Beech Hill	44°27'	74°15'
Black Brook	44°28'	73°43½'
Black Pond	44°26½'	74°17½'
Black Ridge, Piercefield	44°15½'	74°36'
Black Spruce Mountain #	43°25'	73°45½'
Bloomingdale	44°24'	74°05'
Bloomingdale Bog	44°23'	74°08'
Blue Hill	44°27'	74°09'
Blue Mountain #	43°51½'	74°24½'
Blue Mountain Lake #	43°52'	74°28'
Bog Pond	44°24'	74°17'
Bog Willow Bog	44°26½'	74°14½'
Brasher Center #	44°53'	74°46'
Brewster Mountain	44°22½'	74°07'
Brighton Town Hall	44°26'	74°14'
Buck Hill	44°28'	74°12'
Buck Mountain #	43°31'	73°35'
Bulwagga Bay #	44°01'	73°27'
Camp Canaras	44°19½'	74°20'
Camp Gabriels	44°26'	74°11'
Camp Pok-O-MacCready #	44°22½'	73°27½'
Canton #	44°36'	75°10'
Cascade Lakes	44°13½'	73°52'
Cascade Mountain	44°13'	73°51½'
Cascade Notch	44°13½'	73°52'
Cat Mountain #	44°06'	74°52'
Catamount Mountain	44°28'	73°52½'
Chapel Pond #	44°08'	73°45'

LOCALITY	LATITUDE NORTH	LONGITUDE WEST
Chateaugay Chasm #	44°55'	74°05½'
Chateaguay Falls #	44°55'	74°05½'
Chazy Fir Swamp #	44°53½'	73°29½'
Chestertown #	43°39'	73°48'
Chimney Mountain #	43°41½'	74°13'
Church Pond	44°26'	74°15'
Clintonville #	44°28'	73°36
Cobble Hill	44°17½'	73°58'
Cooler Pond & Bog	44°26'	74°15'
Coot Hill #	44°00'	73°28'
Copperas Pond	44°19'	74°23'
Coreys	44°13'	74°19'
Cranberry Lake #	44°10'	74°50'
Crane Mountain #	43°32½'	73°57½'
Crown Point #	44°01'	73°25'
Cumberland Head #	44°43'	73°24'
Currier Forest	44°25	73°56'
Dan Brook	44°31'	74°16'
Dannemora #	44°43'	73°44'
Debar Mountain	44°36	74°13½'
Deer River	44°41'	74°23'
DeKalb Junction #	44°30'	75°16½'
Derrick	44°22'	74°29'
Dewey Mountain	44°19'	74°09'
Dickinson Esker	44°42'	74°31½'
Dix Mountain	44°05'	73°47'
East Hill, Keene	44°15'	73°45'
Easy Street	44°26	74°14'
Essex #	44°18'	73°21½'
Everton Falls	44°40'	74°25'
Ferd's Bog #	43°47½'	74°45'
Fish Creek & Ponds	44°18'	74°22'
Follensby Jr. Pond	44°27½'	74°21'
Fort Ticonderoga #	43°50½'	73°24'
Four Brothers Islands #	44°25½'	73°20'
Franklin Falls	44°26'	73°58½'
Gabriels	44°26'	74°11'
Giant Mountain #	44°10'	73°43'
Glens Falls #	43°18'	73°38'
Goldiana Pond	44°24'15"	74°18'20"
Goodnow Mountain #	43°57½'	74°13'
Goose Pond	44°26'	74°16'
Gothics Mountain	44°07½'	73°52'
Gouverneur #	44°20'	75°28'
Green Bay	44°26'	74°17'
Hannawa Falls #	44°36½'	74°59'
Haystack Mountain	44°06'	73°54½'
Haystack Mountain, Ray Brook	44°19'	74°03½'
Heart Lake	44°11'	73°58½'
Heaven Hill	44°15'	74°00'

Wild raisin (*Viburnum cassinoides*) These, initially green fruits, pass through a magenta phase in late summer before turning a deep purple. When ripe, they are edible and tasty, but each contains a large seed. © Michael Kudish 1992.

LOCALITY	LATITUDE NORTH	LONGITUDE WEST
Helena #	44°55½'	74°44'
Heron Marsh	44°27'	74°16'
Hidden Pond, Lot 15	44°29½'	74°15'
High Peaks	44°-44°15'	73°45'-74°15'
Hoel Pond	44°21'	74°21'
Hoffman Road Bog	44°25'	74°00'
Horseshoe Pond, Saranac Inn	44°19'	74°22'
Hurricane Mountain #	44°14'	73°42½'
Indian Falls	44°07½'	73°56'
Indian Lake #	43°41'	74°20'
Jenkins Mountain	44°27'	74°18'
Jenks Swamp #	43°42'	73°48'
Johns Brook Loj	44°09'	73°52'
Jones Hill	44°27'	74°10'
Jones Pond	44°27'	74°11½'
Juniper Hill	44°22'	73°49'
Kate Mountain	44°29½'	74°04'
Keene	44°15'	73°47½'
Keene Valley	44°11'	73°47½'
Keese's Mills Quarry	44°26'	74°01'
Keese's Mills Road	44°26'	74°15' to 22'
Keeseville #	44°30'	73°29'
Lake Arnold	44°07'	73°57'
Lake Bonaparte #	44°08'	75°23'
Lake Clear	44°22½'	74°15'
Lake Kushaqua	44°31'	74°07'
Lake Placid	44°19'	73°59'
Limekiln Lake #	43°43'	74°48½'
Lincoln Brook	44°25'	73°56½'
Little Moose River Plains #	43°41½'	74°45½'
Little Square Pond	44°19'	74°23½'
Loch Bonnie	44°21'	73°59'
Long Lake #	43°58'	74°26'
Long Pond, Willsboro #	44°23'	73°27'
Long Pond, Paul Smiths	44°27'	74°17½'
Long Sault Dam #	45°00'	74°52'
Loon Lake	44°33'	74°04'
Loon Lake Mountain	44°33½'	74°09'
Lower Saint Regis Lake	44°26'	74°15'
Lyon Mountain	44°42½'	73°51½'
MacIntyre Range	44°08'	73°59'
Madrid #	44°45'	75°08'
Marsh Pond Mountain	41°31'	74°04'
McKenzie Mountain	44°20'	74°02'
Meacham Lake	44°34'	74°17½'
Meno	44°31½'	74°24'
Merwin Hill #	43°52½'	74°26½'
Mica Hill	44°16'	73°49½'
Middle Saranac Lake	44°16'	74°16'
Mile Brook	44°40'	74°23'

LOCALITY	LATITUDE NORTH	LONGITUDE WEST
Minnow Pond #	43°53'	74°26½'
Mooers #	44°58'	73°35'
Moose Mountain	44°21'	74°00½'
Moose Pond	44°03'	74°11½'
Mount Arab	44°12'	74°35'
Mount Baker	44°20'	74°06'
Mount Colden	44°07½'	73°58'
Mount Jo	44°11½'	73°58'
Mount Marcy	44°07'	73°56'
Mount Meenhaga	44°29½'	74°09½'
Mount Morris	44°09½'	74°28½'
Mount Pisgah	44°20½'	74°08'
Mount Whitney	44°18'	73°57½'
Mountain Pond	44°28½	74°16½'
Mountain Pond Ridge	44°28'	74°16'
Mountain Brook, Meno	44°32'	74°22'
Mud Pond, Kushaqua	44°32'	74°06'
Mud Pond Inlet, Kushaqua	44°32'	74°06'
Natural Bridge #	44°04'	75°29½'
Noonmark Mountain #	44°07½'	73°46'
Old Forge #	43°43'	74°59'
Onchiota	44°29'	74°07'
Oseetah Lake	44°17'	74°08'
Osgood Pond	44°27'	74°15'
Osgood River	44°30'	74°17'
Owl's Head, Mountain View	44°45'	74°09½'
Owl's Head, Keene	44°15'	73°49½'
Palmer Hill #	44°29'	73°40'
Parishville #	44°38'	74°49'
Paul Smith's College	44°26'	74°15'
Phelps Mountain	44°09½'	73°55½'
Pierrepont #	44°32'	75°01'
Pitchoff Mountain	44°14'	73°15'
Plattsburgh #	44°41½'	73°27'
Plessis #	44°16'	75°51'
Point Au Roche #	44°46½'	73°22½'
Pok-O-Moonshine Mountain #	44°24'	73°31'
Porter Mountain	44°13'	73°51½'
Prospect Mountain #	43°25'	73°44½'
Rainbow Lake	44°28'	74°10'
Raquette Falls	44°08½'	74°20'
Ray Brook	44°18'	74°05'
Raymondville #	44°50½'	74°59'
Rensselaer Falls #	44°35'	75°18'
Rich Lake #	43°58½'	74°18'
Roakdale Bog	44°30'	74°05'
Roaring Brook #	44°09'	73°46'
Rossie Lead Mine #	44°21'	75°41'
Rouses Point #	44°59'	73°22'
Saint Regis Mountain	44°24½'	74°20'

LOCALITY	LATITUDE NORTH	LONGITUDE WEST
Saint Regis Pond	44°23'	74°19'
Saranac	44°39'	73°45'
Saranac Inn	44°20½'	74°19½'
Saranac Lake Village	44°20'	74°08'
Sawyer Mountain #	43°48'	74°20½'
Scarface Mountain	44°16'	74°04½'
Schuyler Falls #	44°38'	73°34'
Sentinel Notch	44°15'	73°52½'
Skylight Mountain	44°06'	73°56'
Snowy Mountain #	43°42'	74°24'
Sperry Pond Bog	44°04½'	74°32'
Spitfire Narrows	44°25'	73°15½'
Spring Pond Bog	44°22'	74°30'
Star Mountain Pond	44°32'	74°16'
Stony Creek Ponds	44°13'	74°19'
Sugar Hill (Bay Pond)	44°25'	74°22'
The Cobbles	44°26½'	74°00½'
The Crows	44°16½'	73°44½'
The Crevices	44°26'	74°17½'
The Tongue	44°25'	74°17'
Thurman #	43°32'	73°55'
Tongue Mountain #	43°38'	73°35'
Tupper Lake Village	44°13'	74°28'
Upper Jay	44°20'	73°46½'
Upper Saranac Lake	44°17'	74°20'
Valcour #	44°36'	73°27'
Visitors Interpretive Center	44°27'	74°16'
Wadhams #	44°14'	73°28'
Wawbeek	44°15'	74°20½'
Wallface Ponds	44°08½'	74°03'
Weller Brook	44°26'	74°14'
Westport #	44°11'	73°26'
Whey Pond	44°18'	74°24'
Whisker Hill	44°15'	74°23½'
Whiteface Mountain	44°22'	73°54'
Whitefathers	44°31'	74°07½'
White's Pine Camp	44°28'	74°14'
Wickham Marsh #	44°32'	73°25½'
Willsboro #	44°22'	73°24'
Wilmington	44°23'	73°49'
Wilmington Reservoir	44°23½'	73°51'
Wright Peak	44°09'	73°59'

Bibliography

Bibliographic Inclusions:
Trees—Articles on tree species included in this Flora are either state-wide or Adirondack regional.
Herbs and shrubs—since herb and shrub references are, for the most part, limited in number, articles about species included in this Flora but the studies done in other states as well as in New York are included.
All plants—References have been selected only if they deal with plants and related topics included in this Flora:
    carnivorous/insectivorous plants;
    bog plants;
    orchids;
    alpine plants;
    edible, toxic, and medicinal plants;
    protected, endangered, threatened, rare, and exploitably vulnerable plants;
    and largest trees.
Related topics:
    plant disease and forest decline,
    geographic distribution,
    ecology,
    taxonomy, and nomenclature.

Symbols for frequency-listed publications:
    NYSC = New York State *The Conservationist*
    EC = *Ecology*
    EM = *Ecological Monographs*
    AS = *American Scientist*
    ADK = *Adirondac,* Magazine of the Adirondack Mountain Club
    NH = *Natural History*
    AJB = *American Journal of Botany*

A

Abrahamson, Warren G. 1975. *Reproductive Strategies in Dewberries.* EC 56 (3): pp. 721-726. (*Rubus hispidus* in Massachusetts.)

Adams, C.C., G.P. Burns, T.L. Hankinson, B. Moore and N. Taylor. 1920. *Plants and Animals of Mount Marcy, New York.* EC 1: pp. 71-94, 204-233, 274-288.

Aldrich, J.W. 1974. *Protected Plants—Official List Protects Rare, Unique Species.* NY State Environment, November 1, 1974, page 12.

Aldrich, J.W. 1975. *Our Wildflowers and a Program to Protect Them.* NYSC April-May, 1975, pp. 23–29.

Anderson, Roger C. and Orie L. Loucks. 1973. *Aspects of the Biology of Trientalis borealis* Raf. EC 54 (4): pp. 798–808. (Starflower in Wisconsin.)

Argo, Virgil N. 1964. *Insect-Trapping Plants.* NH, Volume 73, March 1964, pp. 28–33.

Ashley, LuAnne. 1987. A series of unpublished floras written for The Adirondack Nature Conservancy on the Boreal Heritage Preserve, near Derrick, Northeast of Tupper Lake: "Plants at Spring Pond Bog," "Plants at the Jordan River," "Plants at Willis Brook Bog," "Plants at Black Brook Bog."

Ashley, Terry and Joseph F. Gennarro, Jr. 1971. *Fly in the Sundew.* NH, Dec. 1971, pp. 80–85, 102.

B

Baar, C.F. 1962. *Some Outstanding Big Trees of New York State.* NYSC Feb.-March, 1962, p. 46.

Baker, Sarah. *The Saranac Valley.* Published by the author.
 Vol. I. 1970. *The Pioneers.* 54 pp.
 Vol. II. 1970. *The Boom Days.* 58 pp.
 Vol. III. 1974. *The Plank Road Gazette.* 42 pp.

Bakuzis, E.V. and H.L. Hansen. 1965. *Balsam Fir—A Monographic Review.* University of Minnesota Press, Minneapolis. 445 pp.

Baum, Werner. 1984. *Trailside: Of Horsetails and Scouring Rushes and Equisetum.* ADK, July, 1984, pp. 24 & 25.

Bazzaz, F.A. 1974. Ecophysiology of *Ambrosia artemisiifolia,* a Successful Dominant. EC 55 (1): pp. 112-119. (Ragweed in Illinois.)

Beattie, A.J. and N. Lyons. 1975. *Seed Dispersal in Viola (Violaceae): Adaptations and Strategies.* AJB 62,7: pp. 714–722.

Bierzychudek, Paulette. 1982. *The Demography of Jack-in-the-pulpit, A Forest Perennial That Changes Sex.* EM 52 (4): 335–351.

Birmingham, Michael J. 1988. *Draft Regulations on Protected Native Plants in New York State.* New York State Department of Environmental Conservation publication dated October 17, 1988, ca. 26 pp. (This publication was shortly followed, in March 1990, by *Protected Native Plants,* a booklet published by the Division of Lands and Forests by the New York State Department of Environmental Conservation. Booklet without author listed, but probably Birmingham.)

Bishop, Jean. 1970. *New Pest Control Promising.* Syracuse Herald-American, Sept. 6, 1970. (On Dutch Elm disease.)

Borland, Hal and Clyde H. Smith. 1976. Year of the Maple. Audubon 78: 22–27, November, 1976 (on Sugar maple).

Bray, William L. 1915. *The Development of the Vegetation of New York State.* NY State College of Forestry at Syracuse. Technical Publication #3, Volume 16, Number 2, November 1915, pp. 1–186.

Brody, Jane. 1979. (on cattails edibility) New York Times, Sept. 4, 1979.

Brown, Lauren. 1979. *Grasses: An Identification Guide.* The Peterson Nature Library, Houghton Mifflin Company. 240 pp.

Brown, R.L., J.W. Ashmun, and L.F. Pitelka. 1985. *Within-and-Between Species Variation in Vegetation Phenology in Two Forest Herbs.* EC 66 (1): 251–258. (Involves *Aster acuminatus* and *Clintonia borealis* at Hubbard Brook, N.H.)

Brumstead, Harlan B. 1956. *Pond Plants.* NYSC Aug.-Sept. 1956, p. 23–26.

Buck, Ellsworth G. 1985. *The Beauty of Tamaracks.* NYSC Nov.-Dec. 1985, pp. 20–23.

Buttrick, Steven C. 1980. *Predatory Plants.* Nature Conservancy News, March-April, 1980, pp. 24–25.

Buzzard, William H., Jr. 1970. *Department Fights Beech Disease and Decline.* NYSC, June-July, 1970, pp. 10–13. (See also NYSC Dec.-Jan. 1965–1966, p. 41.)

C

Canham, Charles D. 1988. *Growth and Canopy Architecture of Shade-Tolerant Trees: Response to Canopy Gaps. Ecology* 69 (3): pp. 786–795. (The studies were done in the Adirondacks at Five Ponds Wilderness Area and at the Huntington Forest in Newcomb.)

Cate, David. *A Fragile World.* Adirondack Life, July-Aug., 1979, pp. 28–33. (On alpine plants.)

Clemants, Steven E. 1986. *New York State Rare Plants—Spring 1986 Status Report.* NY State Department of Environmental Conservation and The Nature Conservacy, 26 pp.

Clemants, Steven E. 1989. *New York Natural Heritage Program: New York State Rare Plant Status List.* February, 1989. 26 pp. Published by The Nature Conservancy.

Clemants, Steven E. 1990. *Juncaceae (Rush Family) of New York State.* NY State Museum Bulletin 475. 67 pp.

Cline, M.G. and R.L. Marshall. 1977. *Soils of New York Landscapes.* Information Bulletin 119, Physical Sciences, Agronomy 6. Extension publication of the New York State College of Agriculture and Life Sciences at Cornell University, Ithaca. 61 pp.

Cobb, Boughton. 1956. *A Field Guide to the Ferns.* The Peterson Field Guide Series. Houghton Mifflin Company, Boston. 281 pp.

Cohn, Robert J. and C.L. Kucera. 1969. *Photoperiodic Adaptations in Eupatorium rugosum.* AJB 56 (5): 571–574.

Cook, David B. 1959. *Fern Picking.* NYSC Aug.-Sept., 1959, pp. 6–7.

Cook, David B., Ralph H. Smith, and Earl L. Stone. 1952. *The Natural Distribution of Red Pine in New York.* EC 33 (4): pp. 500–512.

Cook, David B. and Neil J. Stout. 1959–1960. *The White Spruce of Fulmer Valley.* NYSC Dec.-Jan. 1959–1960, pp. 4–6.

Cook, David B., Ralph H. Smith, and Earl L. Stone. 1973. *The Natural Distribution of White Spruce in New York.* New York Forester, Aug. 31, 1973, 12 pp.

Cook, John C. 1974. *From Farmland to Forest.* NYSC Oct.-Nov., 1974. pp. 10–13

Cook, Robert E. 1983. *Clonal Plant Populations.* AS 71 (3): pp. 244–253. (Includes Starflower, Canada violet, Bracken fern, Ground cedar clubmoss, etc.)

Cox, Donald D. 1959. *Some Postglacial Forests in Central and Eastern New York State as Determined by the Method of Pollen Analysis.* New York State Museum Bulletin #377, December 1959. 52 pp. (includes Chestertown Bog and Perch Lake Bog near Watertown.)

Crooks, Donald M. and Leonard W. Kephart. 1945, revised 1951 *Poison-ivy, Poison-oak and Poison sumac.* Farmer's Bulletin No. 1972. U.S. Dept. of Agriculture. 30 pp.

Curran, Raymond. 1974. *Vegetational Development of the Plains of the Oswegatchie.* MS Thesis, State University of New York College of Environmental Science and Forestry, Syracuse.

Cutler, Lewis M. and Cathy Murray. 1976. *Vegetational Survey.* pp. 1–28, A-1, A-2. Chapter in "Preserve Master Plan for the Silver Lake Camp Preserve," Adirondack Conservancy Committee of The Nature Conservancy. Elizabethtown, New York.

D

Dady, Peter. 1974. *Adirondack Loj Tree Trail.* Adirondack Mountain Club, Glens Falls, NY 5 pp. unbound.

Davis, George D. 1988. *Vision 2020--Fulfilling The Promise of the Adirondack Park.* Volume 1 is Biological Diversity: Saving All The Pieces. The Adirondack Council, Elizabethtown, New York, 64 pp.

Davis, Millard C. 1973. *Burdock.* NYSC, Aug.-Sept. 1973, pp. 11 and 40.

Dean, J. Kenneth and Robert F. Trozzo, compilers. 1990. *Preliminary Vouchered Atlas of New York State Flora.* New York Flora Association of The New York State Museum Institute, Albany. 496 pp.

Delcourt, Hazel R. and Paul A. Delcourt. 1984. *Ice Age Haven for Hardwoods. Natural History,* September, 1984, pp. 22, 24, 26, 28.

DeMent, James A. and Earl L. Stone. 1968. *Influence of Soil and Site on Red Pine Plantations in New York.* Bulletin 1020, July 1968. Cornell University Agriculture Experiment Station, NY State College of Agriculture, Ithaca. Part II.

Dethier, B.E. 1966. *Precipitation in New York State.* Cornell University Agricultural Experiment Station Bulletin #1009. Ithaca. 78 pp.

Dethier, B.E. and M.T. Vittum. 1967. *Growing Degree Days in New York.* Cornell University Agricultural Experiment Station Bulletin #1017. Ithaca. 50 pp.

deVlaming, Victor and Vernon W. Proctor. 1968. *Dispersal of Aquatic Organisms: Viability of Seeds from the Droppings of Captive Killdeer and Mallard Ducks.* AJB 55 (1): pp. 20–26.

Devlin, B. 1988. *The Effects of Stress on Reproductive Characters of Lobelia cardinalis. Ecology* 69 (6): pp. 1716–1720. (On cardinal flowers.)

DiNunzio, Michael. 1972. *A Vegetational Survey of the Alpine Zone of the Adirondack Mountains, New York.* M.S. Thesis. State University of New York College of Environmental Science and Forestry, Syracuse.

DiNunzio, Michael. 1984. *Adirondack Wildguide—A Natural History of the Adirondack Park.* Adirondack Conservancy and The Adirondack Council. 160 pp.

Doeffinger, Derek. 1978. *Adirondack Booby-traps.* Adirondack Life, Sept.-Oct. 1978, pp. 36–38. (Includes Milkweed, Helleborine orchid, Lady's slipper, Mountain and sheep laurel, Touch-me-not, Jack-in-the-Pulpit.)

Doust, Jon Lovett and Paul B. Cavers. 1982. *Sex and Gender Dynamics in Jack-in-the-pulpit, Arisaema triphyllum (Araceae).* EC 63 (3): pp. 797–808.

Doust, Jon Lovett and Lesley Lovett Doust. 1988. *Modules of Production and Reproduction in a Dioecious Clonal Shrub, Rhus typhina.* Ecology 69 (3): pp. 741–750. (On staghorn sumac.)

Drahos, Nick. 1953. *Fall Foods for Wildlife.* NYSC Oct.Nov., 1953, pp. 6–10.

Drahos, Nick. 1955. *Sugaring Off.* NYSC Feb.-March, 1955, p.4.

Drahos, Nick. 1958–1959. *Cedar Oil.* NYSC Dec.-Jan., 1958–1959, pp. 6, 7.

Dreby, Edward. 1982. *Cursory Studies of Silver Lake, St. Lawrence County, N.Y.* Division of Environmental Research of the Academy of Natural Sciences of Philadelphia, PA. 12 pp. (Contains aquatic vascular plant list of this pond adjacent to Cranberry Lake.)

E

Egler, Frank. 1971. *Ragweed.* NYSC Aug.-Sept., 1971, p. 27.

Eisner, Thomas. 1967. *Life on the Sticky Sundew.* NH, June-July, 1967, pp. 32–35.

Eldblom, Nancy. 1988. *Newsletter to Native Plant Lovers of Saint Lawrence County/Valley.* Newsletter No. 2, Spring 1988. Potsdam, NY. (Lists phenology and flowering dates for a number of species in the Saint Lawrence Valley.)

Ellison, Patricia and John M. Kingsbury. 1965. *Common Wild Flowers of New York State.* Cornell Extension Bulletin 900. NY State College of Agriculture, 32 pp.

E.S.F. 1987. New York State College of Environmental Science and Forestry (Syracuse) Alumni Bulletin, Dec. 1987, page 3 has an article entitled *Grandmother Tree Named to DEC Historic Tree Register.* (The tree is a white pine at Pack Forest, Warrensburg.)

F

Faber, Harold. 1980. *New York Expands Quarantine on Pine.* New York Times, Dec. 14, 1980. (*Scleroderris* Canker on Red and Scotch pines.)

Faber, Harold. 1982. *Spread of Wildflower Said to Peril Plants in the Northeast's Wetlands.* New York Times, circa June 11, 1982. (On Purple loosestrife.)

Faber, Harold. 1985. *Insect That Threatens Red Pine Spotted in Upstate New York.* New York Times, April 21, 1985.

Fassett, Norman C. 1940, 1975. *A Manual of Aquatic Plants.* University of Wisconsin Press, Madison. 405 pp.

Faust, Joan Lee. 1977. *Maladies That Linger.* New York Times, Nov. 13, 1977. (White ash cankers.)

Faust, Joan Lee. 1978. *Wasps to the Rescue in Hemlock Attacks.* New York Times, June 4, 1978.

Faust, Joan Lee. 1984. *Is There Hope For a Comeback of Two Native Trees?* New York Times, Aug. 19, 1984. page 24h. (American chestnut and elm.)

Fernald, Merritt Lyndon. 1950. *Gray's Manual of Botany.* American Book Company, New York. 1632 pp. Eighth Edition.

Fischer, Richard N. 1964. *Some Plant Galls of New York.* NYSC Feb.-March, 1964, pp. 22-27.

Fisher, Donald W., Yngvar W. Isachsen, and Philip R. Whitney. 1980. *New Mountains from Old Rocks: The Adirondacks.* Pamphlet #80-6685 of the New York State Geological Survey of the New York State Museum and Science Service. (First printing December 1979, revised July 1980.)

Flint, Harrison L. 1972. *Cold Hardiness of Twigs of Quercus rubra as a Function of Geographic Origin.* EC 52 (6): pp. 1163-1170.

Flynn, Robert. 1967-1968. *Winter Fruit.* NYSC Dec.,-Jan., 1967-1968, pp. 2 & 3. (Some shrubs and subshrubs.)

Follos, Alison M.G., 1986. *Adirondack Forest Decline: A Price of Acid Rain?* ADK, Aug., 1986, pp. 20-23.

Foote, Knowlton and Michail Schaedle. 1975. *Bark photosynthesis.* NYSC Dec., 1975, p. 38. (Involves aspens, but also mentions willows, cottonwood, and basswood.)

Ford, Mary S. (Jesse). 1990. *A 10,000-Year History of Natural Ecosystem Acidification.* Ecological Monographs 60 (1), pp. 57-89. (In Vermont and New Hampshire.)

Fowells, H.A. 1965. *Silvics of Forest Trees of the United States.* U.S. Department of Agriculture Handbook No. 271. Forest Service, Washington, D.C. 762 pp.

Frahn, M. Winnifred. 1987. *Wondering as I Wandered Along an Adirondack Lake.* Published by the author, Duane, NY, 163 pp. Printed by the Industrial Press, Malone, NY. (Involves plant phenology at Deer River Flow, Duane.)

Frederick, R.H., E.C. Johnson, and H.A. MacDonald. 1959. *Spring and Fall Freezing Temperatures in New York State.* New York State College of Agriculture, Cornell Miscellaneous Bulletin #33. 16 pp.

G

Gadomski, Michael P. 1987. *Aspens.* NYSC Jan.-Feb., 1987, pp. 25, 26.

Gaffney, Jeffrey S. 1984. *Are Air Pollutants Damaging Our Forests?* ADK, May, 1984, pp. 8-10.

Gleason, Henry Allan. 1963. *The New Britton and Brown Illustrated Flora.* New York Botanical Garden and Hafner Publishing Company, Inc., New York. Three volumes.

Greason, Michael C. 1986. *Butternut—A Good Friend to a Small Woodland Owner.* NYSC Nov.-Dec., 1986, pp. 42-45.

Greason, Michael C. 1986. *The White Pine in New York.* NYSC, Jan.-Feb., 1986, p. 17.

Gross, Ronald S. and Patricia A. Werner. 1983. *Relationships Among Flowering Phenology, Insect Visitors, and Seed-set of Individuals: Experimental Studies on Four Co-occurring*

Species of Goldenrod . . . EM 53 (1): pp. 95–117. (Includes *Solidago juncea, S. graminifolia, S. canadensis* and *S. nemoralis,* all four occurring in the Adirondacks.)

H

Handel, Steven N. 1976. *Dispersal Ecology of Carex pedunculata (Cyperaceae), A New North American Myrmecochore.* AJB 63 (8): pp. 1071–1079. (This species is dispersed by ants.)

Hartnett, David C. and Warren G. Abrahamson. 1979. *The Effects of Gall Insects on Life History Patterns in Solidago canadensis.* EC 60 (5): pp. 910–917. (In Pennsylvania. See also the pages following, 918–926, on Canada goldenrod.)

Hass, Eric. 1970. *Nature's Botanical Oddity.* New York Times, Aug. 9, 1970. (On Indian pipe.)

Headstrom, Richard. 1957–1958. *Poison Ivy and Wildlife.* NYSC Dec.-Jan., 1957–1958, p. 6.

Headstrom, Richard. 1958. *Acorns.* NYSC Oct.-Nov., 1958, p. 29.

Heady, Harold F. 1940. *Annotated List of the Ferns and Flowering Plants of the Huntington Wildlife Station.* Roosevelt Wildlife Bulletin Volume 7, Number 3, pp. 234–369. (In Newcomb.)

Heimburger, Carl C. 1934. *Forest Type Studies in the Adirondack Region.* Cornell University Agriculture Expt. Station Memoir 165. 122 pp.

Heinrich, Bernd. 1976. *Flowering Phenologies: Bog, Woodland and Disturbed Habitats.* EC 57 (5): 890–899. (In Maine.)

Heslop-Harrison, Yolande. 1978. *Carnivorous Plants.* Scientific American, Feb. 1978, pp. 104–115.

Hibben, C. 1961. *Decline of the Sugar Maple in the Northeast.* NYSC June-July, 1961, pp. 7 and 8.

Hibbs, David E. and Brayton F. Wilson. 1980. *Striped Maple in Massachusetts.* EC 61 (3), 490–496.

Hillman, William S. and Dudley D. Culley, Jr. 1978. *The Uses of Duckweed.* American Scientist 66, 4: pp. 442–451. July-Aug., 1978.

Hirt, Ray R. and Savel B. Silverborg. 1963. *Cedar Apple Rust.* Leaflet No. 18 of Trees Pest Information Service, SUNY E.S.F.-Syracuse.

Hitchcock, Albert Spear. 1950. *Manual of the Grasses of the United States.* U.S. Dept. of Agriculture Miscellaneous Publication #200. First edition, 1935. Revised by Agnes Chase in February 1951, 1051 pp., as 2nd edition.

Hogg, Edward H. and Ross W. Wien. 1988. *The Contribution of Typha Components to Floating Mat Bouyancy.* Ecology 69 (4): pp. 1025–1031. (On Cattails.)

Holway, J.G., Jon T. Scott et al. 1969. *VegetationEnvironment Relations at Whiteface Mountain in the Adirondacks.* Atmospheric Sciences Research Center of the State University of New York at Albany. Report #92, 236 pp.

Holweg, Arthur W. 1964. *Some Shrubs and Vines for Wildlife Food & Cover.* NYSC Oct.-Nov., 1964, pp. 21–27.

Holweg, Arthur W. 1971. *Our Wildflowers.* NYSC Aug.-Sept., 1971, pp. 14–16. (On protection.)

Horsley, Stephen B. 1984. *Ferns: Shapers of Tomorrow's Northern Hardwood Forests?* ADK, Oct.-Nov., 1984, pp. 20–23.

Hotchkiss, Neil. 1967, 1970. *Common Marsh, Underwater, and Floating-leaved Plants of the United States and Canada.* Dover Publications, Inc., New York, pp. 99 & 124

Houghton, Frederick. 1948. *Native Cold-Hardy Plants.* Bulletin to the Schools. Buffalo Museum of Science, pp. 191–194.

Huber, Ed H. 1956. *The Sugar Maple—Our New State Tree.* NYSC April-May, 1956, p. 32.

Huber, Ed H. 1964. *Our State Tree—The Sugar Maple.* NYSC April-May, 1964, p. 16.

Hudler, George W. 1984. *Diseases of Maples in Eastern North America.* Cornell Tree Pest Leaflet A-13. Cornell Cooperative Extension Publication.

Hugo, Nancy Ross. 1989. *A Fly In Their Soup: Amateur Botanist Phil Sheridan Feeds His Love of Carnivorous Plants.* Harrowsmith (magazine) #21, May-June 1989, pp. 58–67.

I

Isachsen, Yngvar W. and Donald W. Fisher, compilers and editors. 1970. *Geologic Map of New York.* New York State Museum and Science Service, Map and Chart Series number 15. (New York State is divided into five sheets. Northern New York is included in the Adirondack and Hudson-Mohawk sheets.)

Isherwood, Justin. 1986. *The King's Broad Arrow Tree.* NYSC Jan.-Feb., 1986, pp. 13–16. (Eastern White pine.)

Ives, Laurie. 1975. *Milkweed for Asparagus—Lamb's-Quarters for Spinach.* NYSC, June-July, 1975, pp. 27–28. (Edible wild plants include also Dandelion and Winter cress (*Barbarea*).)

J

Jackson, Stephen T. 1989. *Postglacial Vegetational Changes Along an Elevational Gradient in the Adirondack Mountains (New York): A Study of Plant Macrofossils.* NY State Museum Bulletin 465. 29 pp.

Jaynes, Richard A. 1968. *Interspecific Crosses in Kalmia.* AJB 55 (9): 1120–1125. (Bog laurel and Sheep laurel.)

Johnson, Charles W. 1985. *Bogs of the Northeast.* University Press of New England, Hanover, N.H. 269 pp.

Johnson, W. Carter and Curtis S. Adkisson. 1986. *Airlifting the Oaks.* NH, October, 1986, pp. 41–46, 108. (Blue jays distributing acorns.)

K

Keddy, Paul A. 1976. *Lakes as Islands: The Distributional Ecology of Two Aquatic Plants, Lemna minor and Lemna triscula.* EC 57 (2): 353–359. (Duckweeds in Ontario.)

Ketchledge, Edwin H.
  In *New York State The Conservationist*:
    1962, April-May, pp. 30–31: *Maples of New York.*
    1962, June-July, pp. 26: *Popples and Poplars of New York.*
    1962, Aug.-Sept., pp. 31: *Poison Ivy.*
    1962, Oct.-Nov., pp. 20–21: *Birches of New York.*
    1963, Feb.-Mar., pp. 14–15, 36: *Oaks of New York.**
    1963, Aug.-Sept., pp. 19–21: *Walnuts and Hickories of New York.*
    1964, April-May, p. 12: *Elms of New York.*
    1964, June-July, pp. 23–27: *Ecology of a Bog.*
    1964, Aug.-Sept., p. 10: *Ashes of New York.*
    1964, Oct.-Nov., pp. 32–33: *Native Trees With Lobed Leaves.* Includes Sycamore.)
    1965, Feb.-March, pp. 29–31, 34: *Changes in the Forests of New York.*
    1969, Oct.-Nov., pp. 20–27, 38: *Recognizing Woody Plants in Winter.*
    *Note that in the Oaks article, Figure 7, acorn part, photos are inverted on page 15!
  In *Adirondac,* the Magazine of the Adirondack Mountain Club, in a series of articles entitled "Adirondack Insights":
    # 1 July, 1982, pp. 22, 23, *The Soil Beneath.*
    # 2 August, 1982, pp. 20, 21, *The Sensitive Summits.*
    # 3 September, 1982, pp. 18, 19, *Dwarf Willows and Birches.*
    # 4 October, 1982, pp. 18, 19, *The Original Forest.*
    # 5 November, 1982, pp. 26, 27, *Rare Ferns.*
    # 6 December, 1982, pp. 22, 23, *Summit Stability* (alpine plants).
    # 7 February, 1983, pp. 18, 19, *Clubmosses.*
    # 8 March, 1983, pp. 20, 21, *The Nighttime Sky* (not on plants).
    # 9 April, 1983, pp. 24, 25, *The Age of Things* (tree rings).
    #10 May, 1983, pp. 20, 21, *The Willows.*
    #11 July, 1983, pp. 26, 30, 31, *The Spruces.*
    #12 October-November 1983, pp. 22, 23, *Insectivorous Plants.*
    #13 February-March 1984, pp. 26, 27, 36, *The Peat Mosses.*
    #14 September, 1984, pp. 18, 19, *Timberline.*
    #15 June, 1984, pp. 17–20, *Alpine Flora*
    #16 December, 1987, pp. 8, 9, *Forest Decline in the Adirondacks.*
    #17 February-March 1988, pp. 14, 15, 16, *Red Spruce Decline in the Adirondack High Country*
    #18 April, 1988, pp. 16, 17, *April Wonder* (the woods in early spring)
    #19 June, 1988, pp. 19–21, *Fir Waves.*
    #20 April, 1989, pp. 16–17, *Heart Lake History.*
    #21 June, 1989, pp. 12, 13, *Geography of Adirondack Tree Species..*
    #22 Sept., 1989, pp. 6, 7, *Timberline Versus Type-Line.*
  Additional publications:
    1967. *Trees of the Adirondack High Peak Region: A Hiker's Guide.* Adirondack

Mountain Club, Gabriels, NY. 101 pp.

1970. *Trees and Forests.* A chapter in *The Adirondack High Peaks and the Forty-Sixers,* pp. 197-213. The Peters Print, Albany.

1972. *Projections from a Crystal Ball.* (Adirondack Forest Preserve in Year 2050.) NYSC Feb.-March, 1972, pp. 4-7, 46.

1979. *A Guide to the Natural History of Mount Jo.* Adirondack Mountain Club, Glens Falls, NY, May, 1979. 24 pp.

King, Irwin H. 1965. *Sweet Springtime Adventure.* NYSC Feb.-March, 1965, pp. 20-21. (Maple Syrup production.)

Kingsbury, John M. 1971. *Common Poisonous Plants.* NY State College of Agriculture Extension Bulletin 538. Cornell University.

Kingsbury, John M. 1971. *Poison Ivy, Poison Sumac, and Other Rash-Producing Plants.* NY State College of Agriculture Extension Bulletin 1154. Cornell University.

Klaehn, Friedrich U. 1960. *Forest Tree Seeds and Fruits.* NYSC Oct.-Nov., 1960, pp. 20-22.

Klaehn, Friedrich U. 1960. *Forest Tree Flowers.* NYSC April-May, 1960, pp. 20-22.

Knaus, Fred and Robert A. Lubek. *Adirondack Orchids.* Adirondack Life, July-August, 1982, pp. 32-33.

Knight, Frank. 1987. *Rare and Protected Plants That Need Help.* NYSC May-June, 1987, pp. 3-9.

Kondo, Katsuhiko. 1972. *A Comparison of Variability in Utricularia cornuta and Utricularia juncea.* AJB 59 (1): 23-37. (Horned bladderwort.)

Kudish, Michael. 1975. *Paul Smith's Flora: A Preliminary Vascular Flora of the Paul Smith-Saranac Lake Area, the Adirondacks, New York, with Notes on the Climate, Geology, and Soils.* Paul Smith's College. 136 pp.

Kudish, Michael. 1981. *Paul Smiths Flora II: Additional Vascular Plants; Bryophytes (Mosses and Liverworts); Soils and Vegetation; Local Forest History.* Paul Smith's College. 162 pp.

Kudish, Michael. 1985. *Where did the Tracks Go: Following Railroad Grades in the Adirondacks.* The Chauncy Press. 148 pp.

L

LaBastille, Anne. 1974. *Maple Syrup the Modern Way.* Adirondack Life, Spring, 1974, pp. 4-9.

Lanier, Gerald N., 1975. *Blight Prospects for Control of Dutch Elm Disease.* New York Times, March 16, 1975.

Little, Elbert Luther, Jr., 1971. *Atlas of United States Trees,* Volume 1: Conifers and Important Hardwoods. U.S. Department of Agriculture, Forest Service, Miscellaneous Publication #1146, March, 1971. (Superbly-detailed range maps down to the county level.)

Lindberg, Olga. 1959. *The Wandering Wildflower.* NYSC April-May, 1959, p. 16. (On naturalized species.)

Littlefield, E.W. 1950. *To Pick or Not To Pick—Fact and Fancy About the New York State*

Wildflower Law. NYSC April-May, 1950, p. 24.

Littlefield, E.W. 1960. *Jack Pine-Poor Relation or Pioneer?* NYSC, April-May, 1960, pp. 6 & 7.

Livingston, Robert B. 1972. *Influence of Birds, Stones, and Soil on the Establishment of Pasture Juniper and Red Cedar in New England Pastures.* EC 53 (6): 1141–1147.

Luer, Carlyle A. 1975. *The Native Orchids of the United States and Canada (Excluding Florida).* New York Botanical Garden.

Lull, Howard W. 1968. *A Forest Atlas of the Northeast.* Northeast Forest Experiment Station of the U.S. Forest Service, U.S. Department of Agriculture, Upper Darby, PA. 46 pp.

Lutz, Richard W. and Richard D. Sjolund. 1973. *Monotropa uniflora: Ultra-structural Details of Its Mycorrhizal Habit.* AJB 60 (4): 339–345. (Indian pipe-fungus relations.)

Lynch, John A. 1975. *Fern Fronds Are Unfurling.* New York Times, April 27, 1975, page D 39.

M

MacArthur, Mrs. Robert. 1958. *Staghorn Sumac.* NYSC Oct.Nov. 1958, p. 27

MacDaniels, L.H. 1974–1975. *Nut Trees of the Northeast.* NYSC Dec.-Jan., 1974–1975, pp. 22–24. (Chestnut, Beech, Walnuts.)

Macior, Lazarus Walter. 1970. *The Pollination Ecology of Dicentra cucullaria.* AJB 57 (1): 6–11. (Dutchman's breeches.)

Marks, Peter L. 1974. *The Role of Pin Cherry in the Maintenance of Stability in Northern Hardwood Ecosystems.* EM 44 (1): 73–88.

Marie-Victorin, Frère. 1964. *Flore Laurentienne.* Presses de l'Université de Montréal. (This is a flora of the Laurentians, written in French and published by the University of Montréal Press. It contains some superb maps showing the northern limits of various tree spp. in Quebec: pp. 145, 160, 397, 72.)

McGrath, Anne. 1981. *Wildflowers In The Adirondacks.* North Country Books, Sylvan Beach, NY. 109 pp.

McMartin, Barbara: three articles in "Adirondac":
1982, Sept., pp. 8, 17, *Acid, Roots, and Forests.*
1983, Aug., pp. 24–26: *Acid Rain and Forests.*
1984, June, pp. 12–13, *Do You Really Want To Bushwhack?* (On alpine plants.)

McMartin, Barbara. 1983. *With The Ferns.* ADK, Dec. 1983, pp. 18–19.

McMartin, Barbara. 1985. *Citizen's Guide to Adirondack Wetlands.* Adirondack Park Agency, Ray Brook. 28 pp.

Mickel, John T. 1979. *How To Know The Ferns and Fern Allies.* Pictured Key Nature Series. Wm. C. Brown Company, Dubuque, Iowa. 229 pp.

Miller, Gertrude N. 1955. *The Genus Fraxinus, the Ashes, in North America, North of Mexico.* Memoir 335, Cornell University Agricultural Expt. Station. Feb. 1955, 64 pp.

Miller, Howard C. New York State Tree Pest Leaflets published by the State University of New York College of Environmental Science and Forestry (the College of Forestry before 1971), Syracuse:
    With co-author Savel B. Silverborg:
        1967. *Maple Tree Problems.* Leaflet F-12.
    With co-author Jack L. Krall:
        1965. *Spruce Gall Aphids.* Leaflet F-4.
        1969. *White Pine Weevils.* Leaflet F-2.
        1970. *Birch Leaf Miner.* Leaflet F-6.
        1970. *Larch Sawfly.* Leaflet F-25.
        1970. *Pine Sawfly.* Leaflet F-14.
    With co-author Douglas Allen:
        1972. *Maple Gall Mites.* Leaflet F-11.
        1971. *Pine Needle Scale.* Leaflet #13.
        1972. *Elm Leaf Beetle.* Leaflet #1.

Mitchell, Richard S. 1979. *Preliminary Lists of Rare, Endangered, and Threatened Plant Species in New York State.* New York State Museum Leaflet #21, 18 pp.

Mitchell, Richard S.:
    1983. *Berberidaceae through Fumariaceae of New York State.* New York State Museum Bulletin 451. 66 pp. (Includes Papaveraceae.)
    1984. *Atlas of New York State Ferns.* NY State Museum Bulletin 456. 28 pp.
    1986. *A Checklist of New York State Plants.* New York State Museum Bulletin No. 458, 272 pp.
    1988. *Platanaceae through Myricaceae of New York State.* NY State Museum Bulletin 464, 98 pp.

Mitchell, Richard S. and Ernest O. Beal. 1979. *Magnoliaceae through Ceratophylaceae of New York State.* NY State Museum Bulletin 435. (Includes Water lilies and Hornworts.)

Mitchell, Richard S. and J. Kenneth Dean:
    1978. *Polygonaceae (Buckwheat Family) of New York State.* NY State Museum Bulletin #431. 81 pp.
    1982. *Ranunculaceae (Crowfoot) Family of New York State.* NY State Museum Bulletin #336. (Buttercup Family.) 100 pp.

Mitchell, Richard S., Charles J. Sheviak, and J. Kenneth Dean. 1980. *Rare and Endangered Vascular Plant Species in New York State.* Published by the State Botanists' Office, NY State Museum, in cooperation with the U.S. Fish & Wildlife Service. 38 pp.

Mitchell, Richard S. and Charles J. Sheviak. 1981. *Rare Plants of New York State.* NY State Museum Bulletin #445, 96 pp.

Moore, Alma Chesnut. 1965. *Progress in Elm Disease Research.* New York Times, March 14, 1965.

Morgenstern, E.K. and J.L. Farrar. 1964. *Introgressive Hybridization in Red and Black Spruce.* Dept. of Forestry of Canada, Forest Research Branch Contribution #608.

Muenscher, W.C. 1930. *Aquatic Vegetation of the Lake Champlain Watershed.* In A Biological Survey of the Champlain Watershed. New York State Department of Conservation. Section VIII: pp. 164-185. J.B. Lyon Company, Albany.

Muenscher, W.C. 1931. *Aquatic Vegetation of the St. Lawrence Watershed.* In A Biological Survey of the St. Lawrence Watershed. Section V, pp. 121-144. New York State Conservation Department. J.B. Lyon Company, Albany.

Muller, Robert N. 1978. *The Phenology, Growth, and Ecosystem Dynamics of Erythronium in the Northern Hardwood Forest.* EM 48 (1): 1-20. (Trout lily in Hubbard Brook, New Hampshire.)

Muller, Robert W. and F. Herbert Bormann. 1976. *Role of Erythronium americanum in Energy Flow and Nutrient Dynamics of a Northern Hardwood Forest Ecosystem.* Science vol. 193, #4258, September 17, 1976, pp. 1126-1128. (Trout lily in Hubbard Brook Forest, New Hampshire.)

Musselman, Robert C. et al. 1975. *Localized Ecotypes of Thuja occidentalis in Wisconsin.* EC 56 (3): 647-655. (Northern white cedar growing in swamps vs. dry limestone soils.)

N

Nellis, John. 1973. *Giants in the Earth (Largest New York Trees: National Champions and Challengers).* NYSC Aug.-Sept. 1973, pp. 18-20.

New York State *The Conservationist* Staff (articles unsigned):

New York State Forest Practice Board. 1982. *New York State Big Tree Register.*
  The Arums. NYSC April-May, 1965, pp. 38, 39.
  Bracken, Maidenhair, and Walking Ferns. NYSC Oct.-Nov. 1963, pp. 38, 39.
  Find Big Trees. NYSC Feb.-Mar. 1972, p. 38.

Nicholson, Stuart A., Jon T. Scott, and Alvin R. Breisch. 1979. *Structure and Succession in Tree Stratum at Lake George, New York.* EC 60 (6): 1240-1254.

Northeastern Forest Experiment Station. 1970. *An Ally To Protect Elms.* NYSC Feb.-March 1970, p. 21. (An insect.) (The Station is run by the U.S. Forest Service at Upper Darby, Pennsylvania.)

O

Ogden, Eugene C. 1953. *Key To The North American Species of Potamogeton.* NY State Museum Circular 31. February, 1953. 11 pp. (Pondweeds.)

Ogden, Eugene C. 1974. *Potamogeton in New York.* NY State Museum Bulletin #423, 20 pp.

Ogden, Eugene C. et al. 1976. *Field Guide to the Aquatic Plants of Lake George, New York.* NY State Museum Bulletin # 426, 65 pp.

Ogden, Eugene C. 1981. *Field Guide to Northeastern Ferns.* NY State Museum Bulletin #444, 122 pp.

P

Pack, Charles Lathrop. 1935. *Another American Tragedy: Elm Trees Threatened With Death by Dutch Elm Disease.* American Tree Association, Washington, D.C. and White

Plains, NY. 8 pp.

Palmer, E. Laurence. 1962. *Some New York Ferns.* NYSC Aug.-Sept. 1962, pp. 21–26.

Palmer, E. Laurence. 1967. *Some Common Edible Wild Plants of NY State.* NYSC Feb.-March, 1967, pp. 23–27.

Pangburn, Jack. A series of articles in *Adirondac,* the Magazine of the Adirondack Mountain Club, entitled "Trailside," preceding those by Ruth Schottman:
  Aug., 1980, Cover photo of Indian Pipes.
  June, 1981, p. 27, Purple (or Red) Trillium.
  Aug., 1981, p. 23, Black-eyed Susan.
  Dec., 1981, p. 29, White pine.

Paterson, Tom. 1981. *Growing Hardy Native Ferns and Wildflowers.* ADK, August, 1981, pp. 10–11.

Peck, Charles Horton. 1899. *Plants of North Elba.* New York State Museum, Volume 6, No. 28, 266 pp.

Peck Charles Horton. 1911. *Cranberry and Averyville Marshes.* NY State Museum Bulletin #150, pp. 69–73.

Pellettieri, George. 1968. *Woodsman, Burn That Elm!* NYSC Oct.-Nov., 1968, pp. 10–11.

Peterson, Roger Tory and Margaret McKenny, 1968. *A Field Guide to Wildflowers.* The Peterson Field Guide Series, Houghton Mifflin Company, Boston, MA, 420 pp.

Petrides, George A. 1958, 1972. *A Field Guide to Trees and Shrubs.* The Peterson Field Guide Series. Houghton Mifflin Company, Boston, MA, 428 pp.

Phelps, Orra. 1970. *Mountaintop Flora.* Chapter in *The Adirondack High Peaks and the Forty-Sixers.* Adirondack Forty-Sixers, Adirondack, NY. pp. 277–290.

Phelps, Orra. 1976. *Slippers of the Queen.* NYSC May-June, 1976, p. 37. (Pink ladyslipper and other *Cypripedium.*)

Phelps, Orra. 1964. *Adirondack Mountain Flora: A Preliminary Check List of the Vascular Plants Growing on the Open Summits in the High Peak Region of the Adirondack Mountains.* Adirondack Mountain Club, Gabriels, NY, 14 pp.

Pohl, Richard W. 1954. *How To Know The Grasses.* Pictured Key Nature Series by William C. Brown Company, 192 pp.

R

Randorf, Gary, Richard Beamish, and Jocelyn Jerry. 1987. *Beside The Stilled Waters.* The Adirondack Council. 24 pp. (On acid precipitation.)

Register, Thomas E. and William R. West. 1976. *Horsetail Life Cycle.* Carolina Tips, Vol. XXXIX, No. 5, May 1, 1976. Carolina Biological Supply Company, Burlington, N.C.

Richards, Norman A., R.R. Morrow, and Earl L. Stone. 1962. *Influence of Soil and Site on Red Pine Plantations in New York.* Cornell University Agriculture Expt. Station Bulletin 977, Oct. 1962, Part I.

Riebesell, John. 1981. *Photosynthetic Adaptations in Bog and Alpine Populations of Ledum groenlandicum.* EC 62 (3): 579–586. (Studies done in the Adirondacks on Labrador tea.)

Risely, John H. and Arthur R. Hastings. 1961. *White Pine Weevil.* NYSC April-May, 1961, pp. 12–13.

Risely, John H. 1964. *Your Tree in Sickness and in Health.* NYSC June-July, 1964, pp. 21–22.

Roman, John Ross. 1980. *Vegetation-Environment Relationships in Virgin, Middle-Elevation Forests in the Adirondack Mountains, New York.* Ph.D. Dissertation, State University of New York College of Environmental Science and Forestry, Syracuse.

Runyon, Linda. 1985. *A Survival Acre: 50 Northeastern Wild Foods & Medicines.* The Chauncy Press. 43 pp.

Russell, Helen Ross. 1971. *Nature's Plastic Surgeon.* NYSC June-July, 1971, p. 17. (Fireweed.)

Russell, Norman H. 1957. *The Violets of Minnesota.* Proceedings of the Minnesota Academy of Science, Vol. XXV, 1958, pp. 126–191.

Ryan, Bill. 1986. *Witch Hazel Tradition Alive in Essex.* The Hartford (Connecticut) Courant: December 28, 1986, pp. C1 and C2.

S

Sayles, John M. 1959. *Big Pines.* (white pines near High Falls on the Oswegatchie River.) NYSC Aug.-Sept. 1959, p. 44.

Schaedle, Michail and Knowlton C. Foote. 1971. *Seasonal Changes in the Photosynthetic Capacity of Trembling Aspen Bark.* Forest Science 17: 308–313.

Schaefer, Paul. 1945. *Adirondack Forests in Peril.* National Parks Magazine. July-Sept., 1945, pp. 19–23. (On the Forest Preserve.)

Schemske, Douglas W. 1978. *Evolution of Reproductive Characteristics in Impatiens capensis (Balsaminaceae): The Significance of Cleistogamy and Chasmogamy.* EC 59 (2): 596–613. (Touch-me-not or Jewelweed.)

Schemske, Douglas W. et al. 1978. *Flowering Ecology of Some Spring Woodland Herbs.* EC 59 (2): 351–366. (Includes Dutchman's breeches and Squirrel corn in Illinois.)

Schopmeyer, C.S. 1974. *Seeds of Woody Plants in the United States.* Forest Service, US Department of Agriculture Handbook No. 450. Washington, D.C. 883 pp.

Schottman, Ruth. "Trailside" articles in *Adirondac,* the Magazine of the Adirondack Mountain Club:

1982, May, p. 17, Lady's Slippers.

1982, Aug., p. 23, Parasitic Flowering Plants in the Adirondacks (includes Beechdrops and Indian pipes.)

1983, May, p. 26, The Fascinating Arums II: Jack-in-the-Pulpit.

1983, Aug., p. 30, Wetland Plants with Large Leaves and Beautiful Names (includes Pickerel Weed and Arrowhead.)

1984, June, p. 21, Lilies Along Adirondack Trails: Trout Lily, Clintonia, and Canada mayflower.

Witchhobble (*Viburnum lantanoides*) Witchhobble invented the division of labor. The large, showy, outer, sterile flowers function only to attract insects. The inner, inconspicuous, fertile flowers produce pollen and seeds. © Jim Kraus 1992.

1985, July, p. 18, Goldthread.
1985, Aug., p. 18, Evening primrose.
1985, Sept., p. 11, Mullein.
1986, June, p. 11, Violets.
1986, Aug., p. 25, Jewelweed.
1987, May, p. 5, Wintergreens (includes *Mitchella, Polygala,* and *Gaultheria.*)
1987, June, p. 12, Three Trilliums.
1987, Aug., pp. 26-27, Stinging Nettles.
1988, Sept.,pp. 8-9, Milkweed.
1989, May, p. 13, Blue Cohosh.
1989, July, p. 10, Campions.
1990, May, p. 9, *Dicentra cucullaria* and *Dicentra canadensis.*
1990, June, p. 6, Wild Sarsaparilla: *Aralia nudicaulis.*
1990, July-Aug., p. 9, Baneberry.
1990, Sept.-Oct., pp. 11, 26, Cattail.

Sculley, Francis X. 1987. *Bittersweet: Candidate for Conservation.* ADK, Oct.-Nov., 1987, p. 15. (*Celastrus scandens,* not *Solanum dulcamara.*)

Sears, Paul B. 1942. *Post Glacial Migration of Five Forest Genera.* AJB 29 (8): 684–691. (Involves oaks, hemlock, hickories, beech, and basswood.)

Sears, Paul B. 1948. *Forest Sequence and Climatic Change in Northeastern North America Since Early Wisconsin Time.* EC 29 (3): 326-333. July, 1948.

Selkow, Paula. 1976. *Insect-Eating Plants Will Thrive in Terrariums.* New York Times, December 5, 1976, p. 38D.

Siccama, Thomas G., F.H. Bormann, and G.E. Likens. 1970. *The Hubbard Brook Ecosystem Study: Productivity, Nutrients, and Phytosociology of the Herbaceous Layer. Ecological Monographs* Vol. 40, No. 4, pp. 389-402, Autumn, 1970. (Hubbard Brook is in the White Mountains of New Hampshire.)

Sievers, Fred U. 1955-1956. *Blister Rust.* NYSC Dec.-Jan., 1955-1956, pp. 28-30. (Involves white pine, currants, and gooseberries.)

Silverborg, Savel B. 1969. *Spruce Canker.* New York State Tree Pest Leaflet F-17. State University of NY College of Environmental Science and Forestry, Syracuse.

Silverborg, Savel B., J.H. Risley, and E.W. Ross. 1963. *Ash Dieback Spreads.* NYSC Feb.-March, 1963, pp. 28-29.

Simmons, Fred C. 1971. *A Woodland Fantasy.* New York Times, December, 1971. (On beech and hemlock in the Adirondacks.)

Skilling, Darroll D. and James T. O'Brien. 1973. *How To Identify Sclerroderris Canker and Red Pine Shoot Blight.* U.S. Dept. of Agriculture, North Central Forest Expt. Station, St. Paul, Minn.

Slate, George L. 1961. *Hazelnuts.* NYSC Oct.-Nov., 1961, pp. 20-21.

Small, John Kunkel. 1935 and 1975. *Ferns of the Vicinity of New York.* Dover Publications, New York, 1975. 285 pp.

Smith, Brent H., Paul D. Forman, and Amy E. Boyd. 1989. *Spatial Patterns of Seed Dispersal and Predation of Two Myrmecochorous Forest Herbs.* Ecology 70 (6): pp.

1649–1656. (On Hepatica and ants.)

Smith, Morgan P. 1960. *Nut Trees of New York.* NYSC, Oct.-Nov., 1960, pp. 23–27.

Smith, Ralph H. 1965. *Some Marsh and Aquatic Waterfowl Food Plants.* NYSC, Aug.-Sept., 1965, pp. 23–26.

Smith, Ralph H. 1965. *Some Plants To Leave Alone.* NYSC, June-July, 1965, pp. 16–17. (Includes Poison ivy, Moccasin flower, Cypress spurge, Nettles, Buttercups, Iris.)

Smith, R.T. and J.A. Taylor, editors. 1986. *Bracken: Ecology, Land Use and Control Technology.* The Parthenon Publishing Group, Inc., Park Ridge, New Jersey, 464 pp.

Smith, Stanley J. 1962. *Purple Loosestrife—Weed or Beauty?* NYSC, Oct.-Nov., 1962, p. 32.

Smith, Stanley J. 1965. *Checklist of the Grasses of New York State.* NY State Museum Bulletin No. 403. 44 pp.

Smith, Stanley J. 1967. *How Plants Get Around.* NYSC, Oct.-Nov., 1967, pp. 23–26.

Solbrig, Otto T. 1971. *The Population Biology of Dandelions.* AS 59 (6): 686–694. November-December, 1971.

Soper, James H. and Margaret L. Heimburger. 1982, 1985. *Shrubs of Ontario.* Royal Ontario Museum, Toronto, A Life Science Miscellaneous Publication. 495 pp. (Detailed range distribution maps.)

Sprugel, Douglas G. 1984. *Density, Biomass, and Nutrient-Cycling Changes During Development in Wave-regenerated Balsam Fir Forests.* Ecol. Monographs 54 (2): pp. 165–186. (On Whiteface Mountain.)

Stauffer, R. Eliot. 1972. *Nature's Switch—Plants That Eat Bugs.* NYSC, Aug.-Sept., 1972, pp. 30–32, 47.

Stock, John. 1983. *Cryptococcan Economics.* ADK, June, 1983, pp. 8–10. (On the Beech bark disease.)

Stock, John W. 1986. *The 1986 Adirondack Park Big Tree Register.* Adirondack Park Big Tree Program, Adirondack Park Agency.

Storey, Michael. 1977. *Heartland: A Natural History of Onondaga County, NY.* Onondaga Audubon Society, Inc., 102 pp.

Storey, Michael. ca. 1979. *A Natural History of the Adirondack Park, New York.* Adirondack Park Agency. 52 pp.

Stuckey, Ronald L. 1978. *The Decline of Lake Plants.* NH 87 (7): Aug.-Sept., 1978, pp. 67–69.

Sullivan, Walter. 1968. *The Vanishing American Elm.* New York Times, on or about August 20, 1968.

Sullivan, Walter. 1975. *Scientists Test Odors To Trap Bark Beetles That Carry Elm-killing Fungus.* New York Times. October 21, 1975.

Swan, Frederick R. Jr. and Lawrence S. Hamilton. 1966. *Fire Shaped Our Forests.* NYSC Oct.-Nov., 1966, pp. 28–30.

Symonds, George W.D. Publications by William Morrow & Company, New York:

The Tree Identification Book. 1958. 272 pp. and *The Shrub Identification Book*. 1963. 200 "Master" pages and many others.

T

Tappeiner, John C. II and Hugo H. John. 1973. *Biomass and Nutrient Content of Hazel Undergrowth*. EC 54 (6): 1342-1348. (On Beaked hazelnut, Jack and Red pines in Minnesota.

Thomson, James D., Mary A. McKenna, and Mitchell B. Cruzan. 1989. *Temporal Patterns of Nectar and Pollen Production in Aralia hispida: Implications for Reproductive Success*. Ecology 70 (4), pp. 1061-1068.

Tirrito, Louis. 1989. *Beech Bark Disease*. CFA News, Winter 1989, Volume VI, No. 2, pp. 5, 6, 10. Catskill Forest Association, Arkville, New York.

Totemeier, Carl. 1979. *The Apple's Evolution: From Eden to Now*. New York Times, November 4, 1979.

V

Vander Kloet, S.P. 1988. *The Genus Vaccinium in North America*. Research Branch, Agriculture Canada, Publication 1828. 201 pp.

Van Diver, Bradford B. 1976. *Rocks and Routes of the North Country, New York*. W. F. Humphrey Press, Inc., Geneva N.Y. 205 pp.

Van Diver, Bradford B. 1980. *Upstate New York*. A K/H Geology Field Guide Series book. Kendall/Hunt Publishing Company, Dubuque, Iowa. 276 pp.

VanLeeuwen, Donald G. 1977. Master's thesis on the Roakdale Bog east of Onchiota. Available at the State University College at Plattsburgh.

W

Webster, Bayard. 1975. *The Beech Trees of North America Are Threatened by a Westward-Spreading Fungal Disease From Europe*. New York Times, September 21, 1975, page L35.

Whitehead, Donald R., Donald F. Charles, Stephen T. Jackson et al. 1986. *Late-Glacial and Holocene Acidity Changes in Adirondack (NY) Lakes*. Chapter 18 in Smol, J.P. et al. "Diatoms and Lake Acidity," pp. 251-274. Dr. W. Junk, Publishers, Dordrecht. (Includes macrofossils of vascular plants as well as diatoms at Heart Lake, Wallface Ponds, and Lake Arnold.)

Wiedmann, Carl P. 1987. *New York's Famous and Historic Trees*. NYSC, March-April, 1987, pp. 32-36.

Wilhelm, Stephen. 1974. *The Garden Strawberry: A Study of its Origin*. AS 62 (3): 264-271.

Williams, Timothy. 1962. *Some Exotic Plants of New York State*. NYSC June-July, 1962, pp. 5-7, 36. (Includes Dandelion, Chickweed, Norway Spruce, Scotch Pine, Queen Anne's Lace, Daisy, Foxtail grass.)

Wilson, Kenneth A. 1965. *Biology of Reproduction in Ferns*. NH June-July, 1965, pp. 53-60.

Winch, Fred E. Jr. 1959-1960. *The Conifers of New York*. NYSC Dec.-Jan., 1959-1960, pp. 22–27.

Woodin, Howard W. 1950. *Establishment of a Permanent Vegetational Transect Above Timberline on Mount Marcy, New York.* EC 40 (2): 320–322.

Wyckoff, Jerome. 1967. *The Adirondack Landscape: Its Geology and Landforms, A Hiker's Guide.* Adirondack Mountain Club, Glens Falls, NY. 72 pp.

Y

Yops, Chester and William E. Smith. 1964. *Red Pine Scale-Will It Spread?* NYSC Oct.-Nov., 1964, p. 17.

Z

Zabel, Robert, D.K. Collins, and Chester J. Yops. 1963. *Research on Ash Dieback.* NYSC Feb.-March, 1963, pp. 29, 30, 38.

Zahl, Paul A. 1961. *Plants That Eat Insects.* National Geographic, pp. 643–659.

Zimmerman, Craig A. 1976. Growth Characteristics of Weediness in *Portulaca oleracea* L. EC 57 (5): 964–974. (Purslane.)

Zon, R. 1914. *Balsam Fir.* U.S. Department of Agriculture Bulletin No. 55. 68 pp.

Additional Bibliography: Publications in Series:

National Oceanic and Atmospheric Administration (formerly the National Weather Service) of the U.S. Department of Commerce publishes both monthly and annual summaries of New York State climatic data. (These can be obtained from the Environmental Data and Information Service, National Climatic Center, Federal Building, Asheville, N.C.)

New York State Museum and Science Service in Albany publishes a series of New York State Museum Bulletins on the geology of various U.S. Geological Survey topographic quadrangles.

Soil Conservation Service of the U.S. Department of Agriculture with the New York State College of Agriculture publishes soil surveys of various New York State counties. (These are available from the County Extension Offices.)

The New York State *Conservationist* published a series of bibliographic listings on various topics, entitled "Conservation Library." The parts relevant to this Flora are:

Part 1. On Animals.
Part 2. Woody Plants. April-May, 1956, p. 30.
Part 3. Books on Herbs and Grasses. June-July, 1956, p. 13.
Part 4. Fungi and Aquatic Plants. Aug.-Sept., 1956, p. 30.
Part 5. Lichens, Mosses, and Ferns. Dec.-Jan., 1956–57, p. 30.
Part 6. Soils, Rocks, and Minerals. Feb.-March, 1957, p. 37.

Bibliographic Listings Arranged by Topics (*Denotes Tree Diseases)

Alpine Plants
    Adams, et al. (1920)
    Cate (1979)
    DiNunzio (1972)
    Houghton (1948)
    Ketchledge (Aug, 1982), (Sept, 1982), (Dec, 1982), (Sept, 1984), (June, 1984), (Sept, 1989)
    McMartin (1984)
    Phelps (1964)
    Phelps (1970)
    Riebesell (1981)
    Woodin (1950)

Aquatic Plants (†pertains to bogs)
    †Ashley (1987)
    Brumstead (1956)
    deVlaming & Proctor (1968)
    Fassett (1940, 1975)
    †Heinrich (1976)
    Hillman & Culley (1978)
    Hotchkiss (1967, 1970)
    †Johnson (1985)
    Keddy (1976)
    †Ketchledge (June/July, 1964), (Feb/Mar, 1984)
    Muenscher (1930), (1931)
    McMartin (1985)
    Ogden (1953), (1974)
    Ogden, et al. (1976)
    Peck (1911)
    †Riebesell (1981)
    Smith, R.H. (1956)
    Stuckey (1978)
    Schottman (Aug, 1983)
    Whitehead, et al. (1986)

Carnivorous and Insectivorous Plants
    Argo (1964)
    Ashley & Gennaro (1971)
    Buttrick (1980)
    Doeffinger (1978)
    Eisner (1967)
    Heslop-Harrison (1978)
    Hugo (1989)
    Kondo (1972)
    Ketchledge (Oct/Nov, 1983)
    Selkow (1976)
    Stauffer (1972)
    Zahl (1961)

Climate:
    National Oceanic and Atmospheric Administration (formerly National Weather Service)
    Lull (1968)
    Dethier (1966)
    Dethier & Vittum (1967)
    Frederick, et al. (1959)

Dicot Herbs
    Abrahamson (1975)
    Anderson & Loucks (1973)
    Bazzaz (1974)
    Beattie & Lyons (1975)
    Bierzychudek (1982)
    Brown, et al. (1985)
    Cohn & Kucera (1969)
    Cook, R.E. (1983)
    Davis (1973)
    Devlin (1988)
    Egler (1971)
    Gross & Werner (1983)
    Hartnett & Abrahamson (1979)
    Hass (1970)
    Ives (1975)
    Kondo (1972)
    Lutz & Sjolund (1973)
    Macior (1970)
    Mitchell (1983) (1988)
    Mitchell & Beal (1979)
    Mitchell & Dean (1978), (1982)
    Pangburn (1980), (Aug. 1981)
    Russell (1957)
    Schemske (1978)
    Schemske et al. (1978)

Schottman (Aug. 1982), (July, Aug. & Sept. 1985), (June & Aug. 1986), (May & Aug. 1987), (Sept. 1988), (May & July 1989), (May, June, & July-Aug. 1990)
Smith, R. (1965)
Smith, S.J. (1962)
Solbrig (1971)
Thomson, et al. (1989)
Wilhelm (1974)
Williams (1962)
Zimmerman (1976)

Edible, Toxic, & Medicinal Plants: e-edible; t-toxic
Brody (1979) e
Crooks & Kephart (1945, 1951) t
Headstrom (1957–58) t
Ives (1975) e
Ketchledge (Aug/Sept 1962) t
Kingsbury (1971) t
Palmer (1967) e
Runyon (1985) e
Smith, R.H. (1965) e

Ferns, Clubmosses and Horsetails
Baum (1984)
Cobb (1956)
Cook, R.E. (1983)
Cook, D.B. (1959)
Horsley (1984)
Ketchledge (Feb & Nov 1982)
Lynch (1975)
McMartin (1983)
Mickel (1979)
Mitchell (1984)
Ogden (1981)
NYS Conservationist staff (1963)
Paterson (1981)
Palmer (1962)
Register & West (1976)
Smith & Taylor (1986)
Small (1935, 1975)
Wilson (1965)

Floras (List of plants for specific areas)
Ashley (1987)
Cutler & Murray (1976)
Fernald (1950)
Gleason (1953)
Heady (1940)
Kudish (1975), (1981)
Marie-Victorin (1964)
Mitchell (1986)
Mitchell & Dean (1982, 1978)
Peck (1899), (1911)

Geology
Fisher, et al. (1979)
Isachsen & Fisher (1970)
Van Diver (1976), (1980)
Wyckoff (1967)
New York State Museum Bulletins

Monocots
Bierzychudek (1982)
Brody (1979)
Brown (1979)
Brown, et al. (1985)
Clemants (1990)
Doust & Cavers (1982)
Handel (1976)
Hillman & Culley (1978)
Keddy (1976)
Knaus & Lubek (1982)
Luer (1975)
Muller (1978)
Muller & Bormann (1976)
Ogden (1953), (1974)
NYS Conservationist Staff (1965)
Pangburn (1981)
Phelps (1976)
Pohl (1954)
Smith, S.J. (1965)
Schottman (May, 1982), (April, 1983), (August, 1983), (June,

1984), (June, 1987), (May, 1983), (Sept.-Oct., 1990)

Postglacial Vegetation and current changes (includes pollen studies)
    Bray (1915)
    Cox (1959)
    Ford (1990)
    Jackson (1989)
    Johnson & Adkisson (1986)
    Ketchledge (Sept. 1989)
    Sears (1942), (1948)
    Storey (1977)
    VanLeeuwen (1977)
    Whitehead, et al. (1986)

Protected, Rare, Threatened and Endangered Plants
    Aldrich (1975), (1974)
    Birmingham (1988 & 1990)
    Clemants (1986 & 1989)
    Holweg (1971)
    Knight (1987)
    Ketchledge (Nov. 1982)
    Littlefield (1950)
    Mitchell (1979 & 1986)
    Mitchell, et al. (1980)
    Mitchell & Sheviak (1981)

Shrubs & Lianas (Woody Vines) including Subshrubs
    Crooks & Kephart (1945)
    Drahos (1953)
    Flynn (1967-68)
    Headstrom (1957-58)
    Jaynes (1968)
    Holweg (1964)
    Kingsbury (1971)
    Ketchledge (Aug/Sept, 1962), (Oct/Nov, 1969), (Sept, 1982), (May, 1983)
    Livingston (1972)
    MacArthur (1958)
    Petrides (1958, 1972)
    Ryan (1986)
    Sculley (1987)
    Slate (1961)

    Soper & Heimburger (1982, 1985)
    Sievers (1955-56)
    Schottman (May, 1987)
    Tappeiner & John (1973)
    Vander Kloet (1988)
    Symonds (1963)

Soils
    Cline & Marshall (1977)
    U.S. Dept. Agriculture & N.Y. State College of Agriculture County Soil Surveys

Trees: Adirondack Forest Decline
    Follos (1986)
    Ford (1990)
    Gaffney (1984)
    Hibben (1961)
    Ketchledge: (Dec. 1987) (Feb./Mar. 1988)
    McMartin (1982), (1983)
    Schaefer (1945)
    Randorf, et al. (1987)

Trees: Conifers
    Bakuzis & Hansen (1965)
    Buck (1985)
    Cook, Smith & Stone (1952), (1973)
    Cook & Stout (1959-60)
    DeMent & Stone (1968)
    Drahos (1958-59)
    *Faber (1980), (1985), (1978)
    Greason (1986)
    *Hirt (1963)
    Isherwood (1986)
    Ketchledge (July, 1983), (June, 1988)
    Littlefield (1960)
    Livingston (1972)
    *Miller & Allen (1971)
    *Miller & Krall (1969), (1970), (1970), (1965)
    Morgenstern (1964)
    Musselman (1975)
    Pangburn (1981)
    Richards et al. (1962)

Risely (1961)
*Silverborg (1969)
Simmons (1971)
*Skilling & O'Brien (1973)
Sprugel (1984)
Sayles (1974)
*Sievers (1955–56)
Winch (1959–60)
*Yops & Smith (1964)
Zon (1914)

Trees: General
Dady (1974)
Fowells (1965)
Klaehn (1960)
Klaehn (1960)
Ketchledge (1967), (1970)
Nicholson, et al. (1979)
Petrides (1958, 1972)
*Risely (1964)
Schopmeyer (1974)
Symonds (1958)

Trees: Hardwoods
*Bishop (1970)
Borland (1976)
*Buzzard (1970)
Drahos (1955)
*Faust (1977), (1984)
Flint (1972)
Foote & Schaedle (1975)
Gadomoski (1987)
Greason (1986)
Headstrom (1958)
*Hibben (1961)
Hibbs (1980)
*Hirt (1963)
Huber (1956)
*Hudler (1984)
Johnson & Adkisson (1986)
King (1965)
Ketchledge (Apr/May, 1962), (June/July, 1962), (Oct/Nov, 1962), (Feb/Mar, 1963), (Aug/Sept, 1963), (Apr/May, 1964), (Aug/Sept, 1964),
(Oct/Nov, 1964)
LaBastille (1974)
*Lanier (1975)
MacArthur (1958)
MacDaniels (1974–75)
Marks (1974)
*Miller & Allen (1972), (1972)
*Miller & Krall (1970)
*Miller & Silverborg (1967)
*Moore (1965)
*Northeast Forest Experiment Station (1970)
*Pack (1935)
*Pellettieri (1968)
Schaedle & Foote (1971)
Simons (1971)
Smith, M.P. (1960)
*Stock (1983)
*Sullivan (1968), (1975)
*Silverborg et al. (1963)
Totemeier (1979)
*Webster (1975)
*Zabel, et al. (1963)

Trees: Largest Specimens
Baar (1962)
Nellis (1973)
Sayles (1959)
Stock (1986)
ESF (1987)
NYS Forest Practice Board (1982)
Wiedemann (1987)
NY State Conservationist (1972)

Vegetation (Ecology in addition to the floristic listing)
Adams et al. (1920)
Bray (1915)
Cook, J.C. (1974)
Curran (1974)
DiNunzio (1984)
Heimburger (1934)
Holway & Scott (1969)
Ketchledge (Feb/Mar, 1965),

(Feb/Mar, 1972), (Oct., 1982)
Kudish (1975), (1981)
Marie-Victorin (1964)
Nicholson, Scott & Breisch (1979)
Roman (1980)
Smith, S.J. (1967)
Sprugel (1984)
Storey (1977)

Swan (1966)
Wildflowers (in general)
    Ellison & Kingsbury (1965)
    Eldblom (1988)
    Lindberg (1959)
    McGrath (1981)
    Paterson (1981)
    Peterson (1968)
    Schemske (1978)

Index

A dagger (†) signifies a plant beyond the area of this Flora.

Abies balsamea 88
Acer
    negundo 168
    pensylvanicum 168
    platanoides 169
    rubrum 36, 169,
                photo front cover
    saccharinum ..... 170, map 171
    saccharum ..... 170, photo 239
    spicatum 173
Aceraceae 168
Achillea millefolium 18, 197
acid deposition in rain
    and snow 37, 40
acidity of soil 49, 55–59
Actaea
    alba 107
    pachypoda 107
    rubra 107
    spicata 107
Adder's Tongue Family 81
adiabatic lapse rate 30
Adiantaceae 82
Adiantum pedatum 60, 82
Adirondack Upland   3, 18, 32, 53
advanced guard populations 74
aestival flowering 42
Agalinis purpurea † 13
ages of trees 248
agriculture 17
Agrimonia 150
agrimony 150
Agropyron repens 221
Agrostis
    alba 222
    borealis 222
    capillaris 222

    hyemalis 222
    mertensii 222
    perennans 222
    scabra 222
    tenuis 18, 222
Alaska 7
Albany 12
alder 16, 43
    black (holly) 165
    green ... 15, 76, 122, map 123
    mountain 15, 76, 122,
                      map 123
    speckled 54, 121
    tag 121
alexanders 178
alexanders, golden † 14
alfalfa 159
Alisma plantago-aquatica † 13
Alismataceae 211
Allium tricoccum 228
Alnus
    crispa ...... 15, 122, map 123
    incana 121
    rugosa 121
    viridis ..... 15, 122, map 123
alpine zone 3, 7–10
aluminum 54, 55
Amaranthaceae 127
Amaranthus retroflexus 127
Ambrosia artemisiifolia 197
Amelanchier
    arborea 150
    bartramiana 151
    laevis 150
    stolonifera 151
Amerindians 12, 17
ammonium 54, 55

Anacardiaceae 173
Anacharis canadensis 211
Anaphalis margaritacea 197
Andromeda
    glaucophylla 141
    polifolia 141
Anemone
    quinquefolia 107
    riparia † 13
anemone
    five-leaved 107
    wood 107
Angelica atropurpurea 178
animals
    pollinating 7
anorthosite 50
Antennaria
    neglecta 198
    neodioica 198
Anthoxanthum odoratum 222
Apiaceae 178
Apocynaceae 179
Apocynum
    androsaemifolium 179
    cannabinum † 13
Appalachian Mts. 9
apple 153
Aquifoliaceae 165
Aquilegia canadensis 108
Araceae 213
Aralia
    hispida 176
    nudicaulis 177
    racemosa 177
Araliaceae 176
arbor vitae 104
Arctic tundra 3, 7–10
Arctium lappa 198
Arctostaphylos uva-ursi 142
Arethusa bulbosa 233, 244
Arisaema triphyllum 60, 213
Aronia melanocarpa 151
arrowhead 211
arrowwood 196

Artemisia vulgaris 198
Arum Family 213
Asarum canadense † 13
Asclepiadaceae 179
Asclepias
    amplexicaulis † 13
    incarnata 180
    syriaca .... 179, photo 243
ash 245
    black 54, 184
    green † 14
    mountain 7, 158
    prickly † 14
    red † 14
    white 51–64, 183
aspect 44, 46
aspen 7, 42, 60
    bigtooth 41, 42, 136, 248
    largetooth 136, 244
    quaking 136, 244
    trembling 41, 136, 248
Aspleniaceae 83
Aster
    acuminatus 198
    cordifolius ... 14, map 199, 200
    divaricatus 14, 200
    lanceolatus 200
    macrophyllus 200
    novae-angliae 201
    prenanthoides 201
    puniceus 201
    simplex 200
    umbellatus 201
aster
    crooked-stemmed 201
    flat-topped white 201
    heart-leaved   14, 74, map 199, 200
    large-leaved 200
    New England 74, 201
    panicled 200
    rough-stemmed 201
    sharp-leaved 38, 198
    tall white 200

white woodland   14, 74, 200
    whorled wood 198
Aster Family 197
Asteraceae 197
Athyrium
    asplenioides 83
    filix-femina 83
    pycnocarpon 83
    thelypteroides 60, 84
augite 50
Ausable Forks 3, 14
Ausable Valley 3, 12, 14, 17
autotrophs 16
autumn, arrival of 34, 43
autumnal equinox 27
avens
    purple 153
    water 153
    white 153

B horizons of soil 51, 54
balm-of-Gilead 136
balsam fir ... 7, 11, 16, 49, 51, 54, 61,
        71, 73, 88, map 89
balsam poplar 136
Balsaminaceae 176
baneberry 73
    red 107
    white 107
Barbarea vulgaris 140
Barberry Family 110
bark, green 42
basil 181
basswood 51–64, 131
bastard-toadflax 165
beans 17
beak-rush 220
bearberry 142
bedrock 53, 61
bedstraw
    fragrant 189
    marsh 189
    rough 189
    small 189

    swamp 189
    sweet-scented 60, 189
    yellow 190
beech
    American .... photo xii, 7–16,
        41, 42, 51–64, 118, 248
    blue 125
    water 125
beech drops 16, 186
Beech Family 43, 118
beggars ticks 202
    swamp 202
bellflower 187
Bellis perennis 201
bellwort 14
    sessile-leaved 232
Berberidaceae 110
Berteroa incana 140
Betula
    alleghaniensis 122
    cordata 121
    glandulosa† 10, 242, 244
    lenta† 13
    lutea 122
    papyrifera 124
    populifolia 125
    pumila 124, 244, 246
Betulaceae 121
Bidens
    beckii 201
    cernua 202
    connata 202
    frondosa 202
    tripartita 202
bilberry, alpine 10, 145, 244
bindweed 180
biological clocks 36
biotite mica 50
Birch Family 121, 249
birch
    black† 7, 12, 13
    canoe 124
    dwarf† 10, 238
    gray 40, 41, 125

paper 7–11, 40, 41, 42, 49, 124
silver 122
swamp 124, 244, 246
white 124
yellow 7–15, 16, 41, 43, 51, 122, 248
birdsfoot trefoil 159
bittercress
    Pennsylvania 141
bittersweet
    Cesastrus 165, 242, 245
    Solanum 73, 180
black alder 165, 238, 241
blackberry 42, 73, 157
black-eyed susan 207
black gum † 7
black locust 160
black medick 159
Black River Lowland 11–15
Bladderwort Family 187
bladderwort 54
    common 187
    eastern 187
    great 187
    horned 187
    humped 187
    intermediate 187
bloodroot † 13, 242
blowdowns 41
bluebell 188
Bluebell Family 187
blueberry 12, 17, 41, 42
    alpine 10, 145, 244
    highbush † 12
    late sweet 144
    low sweet ..... photo 66, 144
    high mountain † 10, 244
    northern † 10
    sourtop 145
    velvetleaf 145
bluecurls † 14
blue-eyed grass 233
blue flag 233

blueweed 181
bluets .... 190, photo 250
    long-leaved 190
    pale 190
blue vervain 181
bogbean 180
bog candle 234
bog rosemary 40, 141
bogs 7, 11, 54
boneset † 13
Boquet River 12, 14, 17
Borage Family 181
Boraginaceae 181
boreal forest 8–11
Botrychium
    dissectum 81
    virginianum 81
bouncing bet 128
box elder 168
Brachyelytrum erectum 60, 222
bracken 42, 83, 238, 241
Brasenia
    peltata 106
    purpurea 106
    schreberi 106
Brassica kaber 141
Brassicaceae 140
breakage of tree limbs 40, 63
Bromus
    ciliatus 222
    inermis 223
brooklime, American 186
Broomrape Family 186
buckbean 180
Buckbean Family 180
buckwheat, climbing ... 38, 58, 129
Buckwheat Family 129
buds 43, 44
buffaloberry 13
buffering 49
bugleweed 182
Bulbostylis capillaris 215
bulrush
    hardstem 220

northern	221
swaying	221
tufted †	244
water	221
bunchberry	4, 16, 41, photo 240
burdock	198
bur marigold, nodding	202
bur reed	228
butter and eggs	185
Buttercup Family	109
buttercup	
common	109
hooked	109
kidneyleaf	60, 109
tall meadow	18, 109
butternut	15, 74, 116, map 117
buttonbush	188
C horizon of soil	51
Calamagrostis canadensis	223
calcite	49
calcium	49, 54, 55
Calluna vulgaris	142
Calla palustris	213
Callitrichaceae	183
Callitriche palustris	183
Calopogon	
pulchellus	233
tuberosus	233
Caltha palustris	photo viii, 108
Calystegia sepium	180
Cambombaceae	106
Campanula	
aparinoides	187
rotundifolia	187, 242, 245
Campanulaceae	187
campion	
bladder	18, 128
white	128
Canada	7–15, 49, 71
canopy, forest	38
Cape Cod	7, 12
Caprifoliaceae	191
Capsella bursa-pastoris	140
carbon 14 dating	8
Cardamine	
diphylla	140
pensylvanica	141
pratensis	141
cardinal flower	188, 242, 245, photo 247
Carex	
arctata	215
argyrantha	216
bigelowii †	10, 239
brunnescens	216
canescens	216
communis	216
crinita	216
debilis	216
exilis	217
filiformis	217
flava	217
flexuosa	216
houghtonii †	244, 246
intumescens	217
lasiocarpa	217
lurida	217
magellanica	217
oligosperma	217
pallescens	218
pauciflora	218
paupercula	218
pedicellata	218
pedunculata	218
plantaginea	60, 218
rostrata	218
scabrata	219
scoparia	219
stricta	219
trisperma	219
vulpinoidea	219
Carnation Family	127
carnivorous plants	275
carpenter's square	185
carpetweed	127
Carpetweed Family	127
Carpinus caroliniana	15, 125

carrion flower 233
Carrot Family 178
Carya cordiformis † 13
Carya ovata † 13
Caryophyllaceae 127
Cashew Family 173
Castanea dentata † 13
cation exchange capacity 53
cat's ear 206
Catskills 9, 14, 71
cattail 16
    narrow-leaved † 14
    wide-leaved 54, 228
Caulophyllum thalictroides . . . 60, 110
Ceanothus americanus † 13
cedar
    eastern red . . . 7, 15, 49, 74, 104,
        map 105
    ground 78
    northern white . . . 54, 104, 248
Cedar Family 103
celandine 111
Celastraceae 165
Celastrus scandens 165, 242, 245
Celtis occidentalis † 13
Centaurea maculosa 202
Cephalanthus occidentalis 188
Ceratophyllaceae 107
Ceratophyllum demersum 107
Chamaedaphne calyculata 142
Chamaesyce maculata 166
Champlain Lowland 3, 11–15,
        30, 73
chance and vegetation 74
Chapel Pond 14
charlock mustard 141
charnockitic gneiss 50
Chateaugay River 17
Chazy River 17
checkerberry 143
Chelidonium majus 111
Chelone glabra 184
Chenopodiaceae 126
Chenopodium album 18, 126

cherry
    bird 154
    black 11, 42, 51, 60,
        photo 75, 156
    choke 42, 53, 156
    fire 154
    pin 7, 154
    sand 156, 244
    red 40, 41, 42, 154, 246
chestnut, American † 13
chickweed 18, 128
chicory 18, 202
Chimaphila umbellata 146, 242, 245
Chiogenes hispidula 142
chokeberry, black 23, 151
Chrysanthemum leucanthemum 206
Chrysosplenium americanum 149
Cichorium intybus 18, 202
Cicuta bulbifera 178
Cimicifuga racemosa † 13
Cinna latifolia 223
cinquefoil
    alpine . . . . 154, map 155
    common 154
    marsh 154
    Norway 154
    old field 154
    rough-fruited 18, 154
    shrubby 153
    silvery 18, 153
    three-toothed 10, 76, 154,
        map 155
Circaea
    alpina 60, 162
    lutetiana 15, 162
    quadrisulcata 162
Cirsium
    muticum 202
    vulgare 203
Cistaceae 133
clay 51–53
Claytonia caroliniana photo 4, 127
clearstem 115
clearweed 115

Clematis virginiana 108	coontail 107
cliffbrake 82	coördinates,
climate 21, 31–46, 275	latitudinal and longitudinal
Clinopodium vulgare 181	252–256
Clintonia borealis 228	*Coptis*
clover 18	*groenlandica* 108
alsike 160	*trifolia* 108
rabbit foot 160	*Corallorhiza maculata* 233
red 160	coralroot, spotted 16, 233
white 160	Cornaceae 163
yellow hop 160	*Cornus*
cloudiness 43	*alternifolia* 163
clubmoss 77, 78, 242	*amomum* 15, 163
alpine † 10	*canadensis* .... 164, photo 240
bog 77	*florida* † 242
bristly 58, 77	*foemina* 164
running 77	*racemosa* 164
shining 42, 78, 249	*rugosa* 15, 164
staghorn 77	*sericea* 165
stiff 77	*stolonifera* 165
swamp 77	*Corydalis sempervirens* 111
tree 78	corydalis
wolf's foot 77	pale 111
clubrush 221	pink 111
Clusiaceae 130	*Corylus cornuta* 125
cohosh	cottongrass sedge 220
black † 13	cottonwood † 13, 36
blue 60, 110	cow parsnip 178
coldest day 22, 23	cowslip 108
coltsfoot 210	cow wheat 41, 185
columbine 108	cranberry
Comandra umbellata 165	large-fruited 144
common names 71	small-fruited 145
Compositae 197	cranberry bush 40, 197
Composite Family 197	cranesbill † 13
Comptonia peregrina 15, 116	*Crataegus* 152
conifers 42, 44, 53, 60, 61, 88, 105	creeping snowberry 42, 142
	Crepis tectorum 20, 203
Connecticut River Lowland 7, 15	crowberry † 10, 244
Conopholis americana † 13	Crowfoot Family 107
Convolvulaceae 180	crown of trees 39
Convolvulus Family 180	Cruciferae 140
Convolvulus sepium 180	*Cryptogramma stelleri* 82
Conyza canadensis 203	cucumber tree

magnolia † 14
cuckoo flower 141
Cucurbitaceae 135
Cucumber Family 135
cudweed, low 20, 204
Cupressaceae 103
curly dock 130
currant 73
    bristly black 60, 149
    skunk 149
    wild black † 13
    wild red 149
Currant Family 148
Cyperaceae 215–222
Cyperus
    bipartitus 219
    esculentus 219
    filiculmis 220
    rivularis 219
Cypress Family 103
Cypripedium
    acaule 233, 242
Cystopteris
    bulbifera 84
    fragilis 84

Dactylis glomerata 18, 223
daisy . . . . 18, 206, photo 313
    English 201
    oxeye 206
    white 206
Dalibarda repens 152
dandelion 18, 42, 210
Danthonia spicata 223
Daucus carota 18, 178
day
    coldest 22, 23
    length of 22, 23
    longest 26
    shortest 28
    warmest 26
Daylight Saving Time 23
deciduous hardwoods 40
Decodon verticillatus 161

deer's hair † 10
dehydration 22, 40, 41
Dennstaedtia punctilobula 83,
                              238, 245
Dennstaedtiaceae 83
Dentaria diphylla 140
depth of soil 53, 61–64
dependence of one species
    upon another 60
Deschampsia flexuosa 223
desert plants 55
devil's paintbrush 18, 205
dewberry 157
diameters of trees 248
Dianthus deltoides 127
Diapensia lapponica † 10, 244
Dicentra
    canadensis 60, 111
    cucullaria 60, 111
Dicots 106, 275
Diervilla lonicera 191
Digitaria ischaemum 224
dimorphic flowers 249
dimorphic fronds 249
dioecism 249
diopside 50
Dirca palustris 161
disturbance of forest 3, 61
ditch moss 211
dockmackie 194
dogbane 179
dogberry 148
dogwood
    alternate-leaved 163
    dwarf 164
    flowering † 7, 12, 242
    gray 74, 164
    pagoda 163
    panicled 74, 164
    red osier 165
    red-stemmed 165
    rough-leaved 74, 164
    round-leaved 15, 164
    silky 15, 163

doll's eyes 107
dolomitic limestone and dolostone
                                     49
dormancy break 43
Draba arabisans † 13
dragon's mouth 233
Drosera 242
      intermedia 132
      rotundifolia . . . . . photo 45, 132
Droseraceae 132
drought 36, 38
Dryopteris
      campyloptera 84
      cristata 84
      disjuncta 86
      goldiana 60, 85
      intermedia 85, 242
      marginalis 60, 85
      noveboracensis 87
      phegopteris 86
      spinulosa var. *americana* 84
      spinulosa var. *intermedia* 85
      thelypteris 87
dry season 40
duck potato 211
duckweed 214
Dulichium arundinaceum 220
Dutchman's breeches 42, 60
dwarf cornel 164
Eagle Bay 3
Earth Day 241
Echinochloa
      crusgalli 224
      muricata † 13, 224
      pungens 224
Echinocystis lobata 135
Echium vulgare 181
elderberry
      black 192
      red 53, 192
Eleocharis
      acicularis 220
      robbinsii 220
      smallii 220

elevation
      and mean annual temperature
                                30, 31, 32
      and growing degree days
                                30, 31, 32
      and frost free season 33, 34,
                                         35
      and phenology 34, 35
      limits, lower and upper,
          of species 71
      limits of this Flora 2, 3
elm,
      American 51–64, 114, 248
      cork † 14
      slippery † 14
      white 114
Elodea canadensis 211
Elymus
      canadensis † 13
      hystrix 224
Empetrum nigrum † 10, 244
enchanter's nightshade 15, 162
      small 60, 161
endangered plants 10, 244–246
Epifagus virginiana 186
Epigaea repens 142, 242, 245
Epilobium
      adenocaulon 162
      angustifolium 162
      ciliatum 162
      glandulosum 162
      leptophyllum 162
      lineare 162
Epipactis helleborine 19, 234
equinoxes 23, 27
Equisetaceae 79
Equisetum
      arvense 79
      fluviatile 80
      hyemale 80
      limosum 80
      scirpoides 80
      sylvaticum 80
      variegatum 80

Eragrostis
   minor 224
   poaeoides 224
   spectabilis 224
Ericaceae 141
Erigeron
   annuus 203
   canadensis 203
   philadelphicus 203
   strigosus 203
Eriocaulaceae 214
Eriocaulon
   aquaticum 214
   septangulare 214
Eriophorum
   spissum 220
   vaginatum 220
   virginicum 221
Erythronium americanum 229
escapes, plant 18
Esopus Valley 14
Euphrasia
   condensata 184
   stricta 184
Eupatorium
   maculatum 204
   perfoliatum † 13
   rugosum 15, 204
   urticaefolium 204
Euphorbia
   cyparissias 166
   supina 166
Euphorbiaceae 166
European plants 18–20
Euthamia graminifolia 204
evening prim-rose 163
Evening Primrose Family 161
evening catchfly 128
evening lychnis 18, 128
evergreen, adaptation 32, 42
everlasting
   clammy 205
   pearly 197
exchange acidity 54, 55

exploitably vulnerable plants
                10, 242–246
exposure to wind 40, 41
extirpated plants 246
eyebright 184

Fabaceae 159
Fagaceae 118
Fagus grandifolia .... photo xii, 118
fall, arrival of 34, 35
false hellebore .... 232, photo 316
false medic 227
families of plants 67
farms 17
fens 11
fern 238, 241
   beech 86
   Boott's 84, 85
   bracken 83, 242, 245
   brake 83
   Braun's holly 42, 87
   bulblet 84, 245
   cinnamon ... photo iv, 81, 245
   Christmas 42, 60, 87
   cliff brake 82
   crested shield 42, 84
   crested wood 84
   eagle 83
   fancy 85
   fiddlehead 86
   fragile 84
   fragrant woodsia 87
   giantwood 85
   glade 83
   Goldie's 60, 85
   grape 81
   Hart's tongue † 238
   hay-scented 83, 242, 245
   interrupted 81
   lady 83
   maidenhair 60, 82
   marginal shield 42, 60, 85
   marsh 16, 87
   mountain wood 84

New York 87
oak 86
ostrich 86
polypody 38, 42, 82
rattlesnake 81
rock brake 82
royal 82
rusty cliff 87
rusty woodsia 87
sensitive   16, 86, 242, 245, 249
silvery spleenwort 60, 84
spinulose wood 85
sweet 15, 116
wood 84, 85, 238
Festuca
   elatior 224
   nutans 224
   obtusa 224
   ovina 225
field notes ix
fields, abandoned 18
figwort 20, 185
Figwort Family 184
fir, balsam ... 7–11, 16, 34, 39, 41, 42,
       49, 51, 61, 88, map 89
fir waves 40
fires, forest 44
fireweed 162
first growth forest 17
flagged trees 40
fleabane 203
   common 203
   daisy 203
floating mats of vegetation   15, 16
floods 36
floras 276
flowering, dates of 35, 36, 42
"Flowering" Fern Family 81
foamflower 60, 150
food plants 18
forest
   boreal 8, 9–14
   burned 41
   fires 12, 44

first growth 17
logged 17
northern hardwood 11
old growth 17
original 17
Preserve, NY State 17
southern hardwood 11
spruce-fir 8
types 16, 17
virgin 17
forestry 17
forget-me-nots 181
fossils
   macro- 7
   pollen 7
Four Brothers Islands 19
Fragaria virginiana 152
Fraxinus
   americana 183
   nigra 184
   pennsylvanica †
freezing of lakes
   frequency................ 72
fringed polygala 167
Frog's Bit Family 211
frost, etc. 30, 33, 35, 43
frozen ground 40
fruiting, dates of 22, 44
full-leaf 33
Fumariaceae 111
Fumitory Family 111

gabbro 49, 50
Galeopsis tetrahit 181
Galinsoga 204
Galium
   asprellum 189
   circaezans 189
   palustre 189
   trifidum 189
   triflorum 60, 189
   verum 190
Garden heliotrope 197
gardens 18

garnet 50
Gaspé 9
Gaultheria
    hispidula 142
    procumbens 143
Gaylussacia baccata 143
gay wings 167
genera 67
gentian 245
    bottle 179
    closed 179, 242
    narrow-leaved 179
Gentiana 245
    andrewsii † 242
    crinita † 242
    linearis 179
Gentianaceae 179
geology 64
Geraniaceae 174
Geranium
    maculatum † 13
    robertianum 174
germander, American † 14
Gerardia † 13
Geum
    canadense 152
    rivale 153
gill-over-the-ground 181
Ginseng Family 176
ginseng, dwarf 177
glaciers 8, 49, 50
glasswort 126
Glechoma hederacea 181
Glyceria
    canadensis 225
    grandis 225
    maxima 225
    melicaria 225
    striata 225
Gnaphalium
    decurrens 205
    macounii 205
    uliginosum 19, 204
    viscosum 205

gneisses 49, 50, 71
goat's beard, yellow 210
Goodyera repens 234
goldenrod
    blue-stemmed 60, 207
    bog 209
    Canada 42, 208
    Cutler's † 10
    downy 209
    early 208
    large-leaved 208
    grass-leaved 204
    gray 209
    puberulent 209
    rough-leaved 42, 209
    wide-leaved 60, 208
goldthread 34, 42, 108
gooseberry, prickly 73, 148
Goosefoot Family 126
goosefoot, white 18, 126
Gourd Family 135
Gramineae 221
granitic gneiss 49, 50
grape
    frost 167
    riverbank 167
grass
    autumn bent 222
    Brachyelytrum 223
    barnyard 13, 224
    blue-eyed 232
    blue joint 223
    bottlebrush 224
    Canada blue 227
    colonial bent 222
    common hair 223
    cotton 220
    drooping reed 224
    eel 211
    fly-away 223
    fringed brome 223
    fowl manna 226
    foxtail 227
    hair 219, 224

herd	226
June	224
Kentucky blue	227
love	225
meadow fescue	225
millet	226
nodding fescue	225
northern bent	222
nut	219
orchard	18, 224
pigeon	227
poverty	224
purple love	225
quack	222
rattlesnake	225
red top	18, 222
reed canary	227
reed meadow	225
Rhode Island bent	222
rice cut	226
sheep fescue	225
slender manna	226
smooth brome	223
smooth crab	224
spreading rice	226
sweet vernal	223
tape	211
tickle	223
timothy	18, 227
tumble	225
upland bent	222
whitlow †	13
wild oat	224
wild rye	13
wirestem muhly	226
witch	226
wood blue	227
wool	221
yellow-eyed	214
Grass Family	221
grass pink	233
Grass River	17
gravel	51
Great Lakes Lowland	14
Green Mountains	9
Greenbrier Family	233
Grossulariaceae	148
growing degree days	30, 31, 33, 34, 35
growing season	3, 30, 31, 34, 36
Guttiferae	130
Gymnocarpium dryopteris	86
Gymnosperms	88
Habenaria (see *Platanthera*)	
hackberry †	13
hackmatack	136
hail	36
Haloragaceae	161
Hamamelidaceae	112
Hamamelis virginiana	15, 112, map 113
hardhack	
Ostrya	126
Spiraea	159
harebell	188, 242, 245
hare's tail	220
hat pins	214
hawk's beard	20, 203
hawkweed	19
Canadian	205
orange	18, 205
panicled †	13
rough	206
yellow	18, 205
hawthorn	73, 152
hazelnut, beaked	53, 125
he balsam	96
heal all	18, 182
heat islands	31
Heath Family	141
heather	142
Hedyotis	
caerulea	190, photo 250
longifolia	190
Helianthus tuberosus	205
heliotrope	197
hellebore	

false ..... 73, 232, photo 316
    white 232
helleborine 234
hemlock
    bulblet-bearing water   73, 178
    eastern 8, 11, 39, 41, 42, 44,
            49, 51, 54, 102, 248
    ground 88
hemp nettle 181
Hepatica
    acutiloba 109
    nobilis 109
Heracleum
    lanatum 178
    maximum 178
herb Robert 174
herbarium specimens ix
hickory† 7, 11–16, 245
    bitternut† 13, 49
    shagbark† 13, 49
Hieracium
    aurantiacum 18, 205
    caespitosum 18, 205
    canadense 205
    kalmii 205
    lachenalii 19, 205
    paniculatum† 13
    pratense 205
    scabrum 206
    vulgatum 205
highbush cranberry 197
High Peaks 7, 10, 17
hoary alyssum 140
hobblebush 53, 194, 245,
               photo 274
Holly Family 165, 242, 245
    black alder 165
    mountain 41, 42, 166
    winterberry 54, 165
Honeysuckle Family 191
    bush 191
    Canada 191
    fly 191
    hairy 191

mountain 191
mountain fly 192
northern 192
Tartarian 192
horehound, cutleaf
    water 182
hornbeam
    American 15, 74, 125
    hop 42, 51–64, 126
hornblende 50
hornwort 107
horsetail
    common 79
    dwarf 80
    field 79
    rough 80
    swamp 80
    variegated 80
    water 80
    wood 80
horseweed 203
Houstonia
    caerulea 190
    longifolia 190
huckleberry 41, 54, 143
Hudson Lowland 11–15, 17
human invasions 17
humidity 30
humus, pH of 55–59
Hydrocharitaceae 211
Hydrocotyle americana 178
hydrogen 55
Hydrophyllaceae 181
Hydrophyllum virginianum ... 15, 181
Hypericaceae 130
Hypericum
    canadense 130
    ellipticum 130
    mutilum 131
    perforatum 18, 131
    virginianum 131
hypersthene 50
Hypochoeris radicata 206
hypsithermal period 8

Hystrix patula 224

ice-out, Lower Saint Regis Lake,
    mean date 23
ice rime 39
identification
    manuals for plants x
Ilex verticillata 165, 242, 245
ilmenite 50
Impatiens
    biflora 176
    capensis 60, 176
    pallida 176
Indian
    chickweed 127
    cucumberroot 230
    hemp † 13
    pipe 16, 147
    poke 232
    tobacco 73, 188
    turnip 213
independence of species 16
industry 17
insect pollination 7, 245
insectivorous plants 275
invasions
    human 17
    plant 18
Iridaceae 233
Iris versicolor 233
Iris Family 233
iron 54, 55
ironwood
    Carpinus 125
    Ostrya 126
islands and red pine 41
Isoetaceae 79
Isoetes 79

Jacob's ladder 233
Jack-in-the-pulpit 23, 60, 213
Jerusalem artichoke 205
jewelweed
    pale 176

    spotted 176
joe-pye weed 204
jointweed 129
Jones Pond 15
Juglandaceae 116
Juglans cinerea ... 15, 116, map 117
Juncaceae 214
Juncus
    articulatus 214
    canadensis 215
    effusus 215
    nodosus 215
    pelocarpus 215
    tenuis 215
    trifidus † 10, 244
Juneberry 150
    Bartram's 151
    bush 151
juniper
    common 103
    pasture 73, 103
Juniperus
    communis 103
    virginiana ... 15, 104, map 105

Kalmia 242, 245
    angustifolia 143
    latifolia † 242
    polifolia .... photo 48, 144
Keene Valley 3, 14
king devil 18, 205
knapweed 202
knotgrass 18, 129
knotweed 18
    Japanese 130

Labiatae 181
Labrador 49
Labrador tea 40, 41, 54, 144,
photo back cover
Lactuca canadensis 206
Lady's slipper 233, 242
Lady's sorrel 18, 174
Lady's thumb 18, 130

299

Ladies' tresses 235
Lake Champlain 3, 11–15, 73
Lake Erie Plain 14
Lake George 3
Lake Ontario Plain 14
lakes, effects of
    on growing season 31, 41
lamb kill 143
lamb's quarters 18, 126
Lamiaceae 181
Lapland rosebay † 10, 244
Laportea canadensis 60, 115
Lapsana communis 206
larch 7–11, 16, 41, 42, 90,
                               map 91
Larix laricina 73, 90,
                        map 91, 248
latitude and phenology 35
laurel 242, 245
    bog . . . . 40, photo 48, 144
    mountain † 242
    pale 144
    sheep 41, 54, 73, 143
Laurentians 12
leaf color 33, 36, 43
leaf fall 22, 33, 36, 40, 43
leafing 22, 23, 33, 42, 43, 44
leatherleaf 40, 54, 142
leatherwood 161
leaves, needle vs. broad 40
lecture notes ix
Lechea maritima 133
Ledum groenlandicum 144,
                   photo back cover
leek, wild 42, 228
Leersia oryzoides 225
Legume Family 159
Leguminosae 159
Lemna minor 214
Lemnaceae 214
Lentibulariaceae 187
Lepidium
    apetalum 141
    densiflorum 141

Leucanthemum vulgare 18, 206,
                             photo 313
life-of-man 177
light intensity 22
lightning 44
Liliaceae 228
Lilium philadelphicum † 13, 242
Lily Family 228
lily 242
    bluebead 228
    bullhead 106
    calla 16, 213
    Clinton's 228
    corn 228
    fragrant water 106
    trout 42, 229
    water 106
    white water 106
    wood † 13, 228
    yellow pond 106
lily-of-the-valley
    false 230
    wild 230
lime 49
limestone 49, 71
limits of species' ranges:
    elevation
        lower 3
        upper 3
Linaria vulgaris 185
linden, American 131
Linden Family 131
Linnaea borealis 191
liverberry 230
liverleaf 109
Lobelia
    cardinalis 188, 242, 245,
                           photo 247
    dortmanna 188
    inflata 188
lobeliaceae 187
logging 17, 18, 19
logging roads 18, 19
long-day plants 42

longest day 22, 26
Lonicera
    caerulea 192
    canadensis 191
    dioica 191
    hirsuta 191
    tatarica 192
    villosa 192
longitude 252–257
loosestrife
    fringed 147
    purple 161
    spiked 148
    swamp 148, 161
    tufted 148
    whorled 147
lopseed † 13
Lotus corniculatus 159
lousewort 16, 185
Lychnis alba 128
Lycopodiaceae 77
Lycopodium 242
    annotinum 77
    clavatum 77
    complanatum 77
    digitatum 77
    inundatum 77
    lucidulum 78
    obscurum 78
    selago † 10
    tristachyum 78
Lycopus
    americanus 182
    virginicus 182
Lysimachia
    ciliata 147
    quadrifolia 147
    terrestris 148
    thyrsiflora 148
Lythraceae 161
Lythrum salicaria 161

macrofossils 7, 64
Madder Family 189

mafic minerals 50
magnesium 54, 55
magnetite 50
Maianthemum canadense 230
Maidenhair Family 82
maize 17
Mallow Family 132
Malus pumila 153
Malva moschata 132
Malvaceae 132
mandarin
    rose 231
    white 231
Mangosteen Family 130
maple 249
    ash-leaved 168
    box elder 36, 168
    goosefoot 168
    hard 170
    leaved viburnum . . . 15, 74, 194,
        map 195
    moosewood 168
    mountain 53, 60, 61, 173
    Norway 36, 169
    red . . . photo front cover, 11,
        16, 36, 41, 44, 51–54, 169
    rock 170
    silver . . . 36, 76, 170, map 171
    soft 169
    striped 38, 42, 168, 248
    sugar . . . 7–16, 41, 42, 43, 51–64,
        170, photo 240, 248
    swamp 169, 170
    whistlewood 168
marble, dolomitic 49
marigold
    marsh . . . . photo viii, 108
    nodding bur 202
    water 201
marshes 11
Matricaria matricarioides 206
mats, floating vegetation . . . 15, 16
Matteucia struthiopteris 86
Mayflower 142

Canada 42, 229
meadowsweet 53, 158
Medeola virginiana 230
Medicago
    lupulina 159
    sativa 159
medicinal plants 276
Megalodonta beckii 201
Melampyrum lineare 185
melilot
    white 159
    yellow 159
Melilotus
    alba 159
    officinalis 160
Mentha arvensis 182
Menyanthes trifoliata 180
Menyanthaceae 180
Mezereum Family 161
mica 50
Midwest 14
migration routes . . . . . 5–20, map 6
milfoil 18, 197
    green 161
    water 161
    whorled 161
Milium effusum 60, 225
milkweed
    common . . . . . 180, photo 243
    curlyleaf † 13
    swamp 73, 179
Milkweed Family 179
Milkwort Family 167
mineral nutrients 49
mining 17
mint 182
Mint Family 181
Minuartia groenlandica † 10, 244
Mitchella repens 190
Mitchell's Checklist 67
moccasin flower 233
Mohawk Lowland 11–15
Molluginaceae 127
Mollugo verticillata 127

Moneses uniflora † 146
Monocots 211
monoecism 249
Monotropa
    hypopithys 147
    uniflora 147
Monotropaceae 147
Montana 8
Montréal 12
mooseberry 194
moosewood 168
Moraine Ground 50
Morning Glory Family 180
mountain ash,
    American 7, 158
mountain bride † 10, 244
mountain sandwort † 10, 244
mugwort, common 198
Muhlenbergia frondosa 226
mullein, great 18, 19, 185
musclewood 15, 125
musk mallow 132
Mustard Family 140
Myosotis sylvatica 181
Myrica
    asplenifolia 116
    gale 116
Myricaceae 116
Myriophyllum verticillatum 161

naiad 213
Najadaceae 213
Najas flexilis 213
names of plants
    common 71
    scientific 71
nannyberry 15, 196
Nasturtium officinale 141
National Weather Service 36, 43,
    46
native species 67, 77–235
Naturalized species . . . 36, 67–71, 77,
    236
Naumbergia thyrsiflora 148

needles, conifer 40
Nelumbo lutea † 242
Nemopanthus mucronatus 166
nettle 60, 73, 115
    hemp 73, 181
Nettle Family 115
Newfoundland 8, 49, 71
New England 7–15
New Hampshire 9, 12
New Jersey 7
New Jersey tea † 13
New York State,
    south-eastern 7–15
nightshade
    bittersweet 73
    enchanter's 15, 60, 162
Nightshade Family 180
Nineteenth Century 17
nipplewort 206
nitrate 54
nitrogen 54
non-green flowering plants 16
northern hardwoods 7–15
northern limits of plants 12
Nuphar
    advena 106
    luteum 106
    microphyllum 106
nutrients, mineral to plants 54, 55
Nymphaea odorata 106
Nymphaeaceae 106

oak 7, 11–16, 245
    black † 13, 49
    bur † 14
    chestnut † 13, 49
    mossycup † 14
    northern red 44, 49, 61, 76,
        118, map 119
    scrub † 13
    swamp white † 13
    white † 13, 15, 49
Oakesia sessilifolia 232
Oenothera

    biennis 163
    perennis 163
old growth forest 17
Oleaceae 183
Olive Family 183
Onagraceae 162
Onoclea sensibilis 86, 238, 241
Ophioglossaceae 81
orchid 242, 245
    bog green woodland 235
    large roundleaf 237
    leafy white 235
    ragged fringed 235
    white bog 235
    white fringed 235
    weed 19, 234
Orchid Family 233
Orchidaceae 233, 242, 245
original forest 17
ornamentals 18, 36
Orobanchaceae 186
Orthilia secunda 146
Oryzopsis asperifolia 226
Osmorhiza claytonii 60, 178
Osmunda
    cinnamomea . . . . . photo iv, 81
    claytoniana 81
    regalis 82
Osmundaceae 81
Ostrya virginiana 126
Oswegatchie River 17
outwash, glacial 50, 51
Oxalidaceae 174
Oxalis
    acetosella 174
    corniculata 18, 174
    montana 174

Pacific Northwest 7
Panax trifolius 177
Panicum capillare 226
Papaveraceae 111
parasitic plants 16
Parsley Family 178

parsnip,
    cow 178
    water 179
Parthenocissus quinquefolia 166
partridgeberry 42, 190
pastureland 17
Pea Family 159
peat 7
Pedicularis canadensis 15, 185
peninsulas and red
    pine 41
Pennsylvania 7
people,
    Amerindian 12, 17
    European ancestry 18
peppergrass 141
pepperwort 140
petty morel 177
pH of humus 55–61
Phalaris arundinacea 226
Phaseolus polystachios † 13
Phegopteris connectilis 86
phenology 31, 32, 34, 43, 71
Phleum pratense 18, 226
phosphorus 54, 55
Phryma leptostachya † 13
Picea
    glauca .... 92, map 93
    mariana .... 94, map 95
    rubens .... 96, map 97
pickerel weed 228
pigweed 127
Pilea pumila 115
pimbina 194
Pinaceae 88
pine,
    eastern white 8, 11, 16,
        39, 40, 41, 42, 44, 49,
        100, 248
    ground 78
    jack 96, 244
    northern ground 78
    Norway 98
    pitch 7, 12, 49, 74, 100,
        map 101
    prince's 78, 146
    red 11, 39, 41, 42, 54, 76, 98,
        map 99, 248
    running 77
    Scotch 102
    Scots 102
pine sap 16, 147
pineappleweed 206
Pink Family 127
pink,
    grass 233
    maiden 127
    swamp 233, 244
pinkweed † 13
Pinus
    banksiana 96, 244
    resinosa .... 98, map 99
    rigida .... 100, map 101
    strobus 100
    sylvestris † 102
pinweed 133
pipes 80
pipewort 214
pipsissewa ... 41, 42, 146, 242, 245
pitcher plant ... photo 24, 54, 132,
        242, 245
Plantaginaceae 183
Plantago
    lanceolata 18, 183
    major 18, 183
    rugelii 183
plantain, ... 19
    broad-leaved 183
    common 18, 183
    narrow-leaved 18, 183
    rattlesnake 234
    water † 13
Plantain Family 183
Platanthera
    blephariglottis 234
    clavellata 234
    dilatata 234
    lacera 235

orbiculata	235	large-leaved	211

orbiculata 235
Platanus occidentalis † 13
plum, †
    wild 246
Poa
    compressa 226
    nemoralis 226
    pratensis 227
Poaceae 222
poison ivy ...... 73, 174, map 175
poison sumac † 12
poisonous plants 73
Pogonia ophioglossoides 235
pollen
    fossil record 7, 64
pollination 249
Polygala paucifolia 167
Polygalaceae 167
Polygonaceae 129
Polygonatum
    pubsecens 231
Polygonella articulata 129
Polygonum
    amphibium 129
    aviculare 18, 129
    cilinode 129
    cuspidatum 130
    lapathifolium † 13
    pensylvanicum † 13
    persicaria 18, 130
    sagittatum 130
Polypodiaceae 82
Polypodium
    virginianum 82
    vulgare 82
Polypody Family 82
Polystichum
    acrostichoides 60, 87
    braunii 87
ponds 15, 16
Pondweed Family 211
pondweed, ... 20, 211, 212
    clasping-leaf 212
    floating 212

large-leaved 211
red-head 212
Pontederia cordata 228
Pontederiaceae 228
poplar 41, 245
    balsam ...... 7, 42, 76, 136, map 137
    silver 135
    white 135
popples 136
Poppy Family 111
Populus
    alba 135
    balsamifera ..... 136, map 137
    deltoides † 13
    grandidentata 136
    tremuloides 136
postglacial vegetation ... 5–20, 277
Portulaca oleracea 127
Portulacaceae 127
Potamogeton
    amplifolius 212
    bupleuroides 212
    epihydrus 20, 212
    gramineus 212
    natans 212
    perfoliatus 212
    pusillus 212
    richardsonii 212
    robbinsii 212
    spirillus 212
Potamogetonaceae 212
potassium 54, 55
Potentilla
    anserina 153
    argentea 18, 153
    fruticosa 153
    norvegica 154
    palustris 154
    recta 18, 154
    simplex 154
    tridentata ... 10, 154, map 155
precipitation 22, 36, 37, 38, 43
(see also rain, snow, and sleet.)

Prenanthes altissima 206
Primrose Family 147
Primulaceae 147
protected species 10, 244–246
Prunella vulgaris 18, 182
Prunus
    nigra † 246
    pensylvanica 154
    pumila 156, 244
    serotina ...... photo 75, 156
    virginiana 156
Pteretis pensylvanica 86
Pteridium aquilinum ... 83, 238, 241
Pulse Family 159
purslane 127
    milk 166
Purslane Family 127
pussy toes 198
Pyrola
    elliptica 146
    secunda 146
Pyrolaceae 146
Pyrus
    americana 158
    malus 153
    melanocarpa 151

Quaker ladies 190
quarries, abandoned 18
quartzite 49
Québec 12, 13, 71
Queen Anne's lace 18, 178
Quercus
    alba † 13
    bicolor † 13
    ilicifolia † 13
    macrocarpa † 14
    prinus † 13
    rubra .... 118, map 119
    velutina † 13
quillwort 79

ragweed 197
railroads and

railroad grades 17
rainfall,
    acid 22, 30, 36
ramp 228
randomness of vegetation 49
Ranunculaceae 107
Ranunculus
    abortivus 60, 109
    acris 18, 109
    aquatilis 110
    recurvatus 109
    trichophyllus 110
Raphanus raphanistrum 141
Raquette River 3, 17
rare species 10, 72, 242–246
rarity codes and ranks 10, 239
raspberry, ... 42
    purple-flowering 158
    red 73, 157
rattlesnakeroot 206
references,
    bibliographic 72, 255–279
remnants 63, 73
reproduction of plants 249
    asexual 249
    of trees 249
    strategies 249
    sexual 249
    vegetative 249
residences, of people 17
Rhinanthus cristagalli 185
rhododendron 242
rhododendron, rosebay 12
Rhododendron lapponicum † 10,
                          242, 244
Rhus
    radicans .... 174, map 175
    typhina 173
Rhyncospora
    alba 221
    capitellata 221
Ribes
    americanum † 13
    cynosbati 148

    glandulosum 149
    lacustre 60, 149
    triste 149
rich site species 50, 61
rime, ice 39
Robbins' ragwort 207
Robinia pseudoacacia 160
rock brake 82
Rockrose Family 133
roots 52
Rosa palustris 157
Rosaceae 150
rose,
    pogonia 235
    swamp 157
Rose Family 150
Rubiaceae 188
Rubus
    alleghaniensis 157
    hispidus 157
    idaeus 157
    odoratus 158
    strigosus 157
Rudbeckia
    hirta 207
    serotina 206
Rumex
    acetosella 18, 130
    crispus 130
Rush Family 214
rush, ... 214, 215
    Arctic† 10, 244
    beak 220
    bul- 220, 221
    candle 215
    club 221
    jointed 214
    knot 215
    marsh 215
    path 215
    sand 215
    scouring
        common 80
        dwarf 80

    soft 215
    spike 219, 220
    yard 215
Sagittaria
    graminea 211
    grass-leaved 211
    latifolia 211
St. Johnswort Family 130
St. Johnswort
    Canadian 130
    common 18, 131
    elliptical 130
    marsh 16, 131
St. Lawrence Lowland 3,
                              11–15, 73
St. Regis Lake 17
St. Regis River 17
Salicaceae 135
Salicornia europea 126
Salix
    alba var. *tristis*† 138
    bebbiana 138
    discolor 139
    fragilis 139
    lucida 139
    pedicillaris 139
    pentandra† 138
    pyrifolia 139
    rigida 139
    sericea 140
    uva-ursi† 10, 244
Salmon River 17
Sambucus
    canadensis 193
    pubens 193
    racemosa 193
sand 51
sand spurrey 128
Sandalwood Family 165
sandstone 49
Sanguinaria canadensis† 13, 242
Santalaceae 165
Saponaria officinalis 128

Saranac River 3
Sarracenia purpurea .... photo 24, 132, 242, 245
Sarraceniaceae 132
sassafras † 12
sarsaparilla,
    bristly 176
    wild 177
Satureja vulgaris 181
savin 104
Saxifraga virginiensis 150
Saxifragaceae 149
Saxifrage Family 149
saxifrage,
    early 150
    golden 149
scent bottle 234
schist 49
Schizachne purpurascens 227
Schroon River 12, 14
scientific names 71
Scirpus
    acutus 221
    atrocinctus 221
    atrovirens 221
    cespitosus † 10, 244
    microcarpus 222
    occidentalis 221
    rubrotinctus 221
    subterminalis 222
Scolopendrium † 242
Scrophularia marilandica 20, 185
Scrophulariaceae 184
Scutellaria
    epilobiifolia 182
    galericulata 182
    lateriflora 182
Sedge Family 215
sedge, ... 215–222
    Bigelow's † 10, 244
    fox 219
    Houghton's † 244, 246
    hummock 218
    plantain-leaved 60, 218
    three-way 219
    tussock 218
seeds, dispersal of 16, 18, 249
self-heal 18, 182
Senecio
    robbinsii 207
    schweinitzianus 207
serviceberry 150
    Bartram's 151
Setaria glauca 227
settlement, by people 17
seventeenth century 17
severity of winter 45
shad 150
shadblow 150, 151
shadbush 41, 150, 151
shade tolerance and intolerance ... 61–64
shagbark hickory † 13, 49
shale 49
shallow soils 11, 53, 61–64
sheep kill 143
Shepherdia canadensis † 13
shepherd's purse 140
shinleaf 42, 146
    one-sided 146
short-day plants 42
shortest day 22
shrubs 277
Silene
    cucubalus 18, 128
    latifolia 18, 128
    vulgaris 18, 128
silt 51–53, 61–64
silverrod 207
silverweed 153
Sinapsis arvensis 141
Sisyrinchium angustifolium 233
site 71
Sium suave 179
skid roads 18
skullcap 182
    mad-dog 182
skunk cabbage † 13

sleet 36
slopes 40, 44
smartweed,
    pale † 13
    water 129
Smilacaceae 233
Smilacina
    racemosa 231
    stellata † 13
    trifolia 231
Smilax herbacea 233
snake mouth 235
snakeroot,
    white 15, 73, 74, 204
snapweed 176
snow 30, 36, 38, 39, 43
snowpack 38, 44
soapwort 128
soils 47–64
    acidity 49, 55–59
    clayey 53, 61–64
    humus pH 55–61
    limestone 71
    minerals 49
    nutrients 54
    outwash 50
    pits . . . . ix, 62
    poorly-drained 54
    sandy 11, 51
    shale 49
    shallow 11, 61–64
    silty . . . 50, 51, 53, 57, 61, 62
    texture 50
    till 50, 61–64
    water-logged 54
    water table 54
    wet 11
Solanaceae 180
Solanum dulcamara 180
Solidago
    bicolor 207
    caesia 60, 207
    canadensis 208
    cutleri † 10

    flexicaulis 208
    graminifolia 204
    juncea 208
    latifolia 60, 208
    macrophylla 208
    nemoralis 209
    puberula 209
    rugosa 209
    uliginosa 209
Solomon's plumes 23, 231
Solomon's seals,
    false 231
    starry false † 13
    three-leaved false 231
solstices
Sonchus oleraceus 210
Sorbus americana 158
sorrel,
    lady's 18, 174
    sheep 18, 130
    wood 174
southern hardwoods 7, 8
southern limits of
    natural ranges 10
sow thistle 209
Sparganiaceae 228
Sparganium
    americanum 228
    angustifolium 228
spatterdock 106
specimens, herbarium ix
speedwell,
    American 186
    common 186
    marsh 186
    thyme-leaved 186
Spergula arvensis 128
Spergularia rubra 128
spikenard 177
spikerush,
    needle 219
    slender 219
    triangle 219
spiny wild cucumber 135

Spiranthes cernua 235
Spiraea
    latifolia 158
    tomentosa 159
Spleenwort Family 83
spleenwort 83, 84
spores, dispersal of 16
spring, arrival of ... 23, 34, 35, 42
spring beauty .... photo 4, 31, 42, 58, 127
spruce, ... 7–11, 15, 44
    black ..... 16, 41, 49, 74, 94, map 95, 248
    bog 94
    cat 92
    red ... 16, 40, 41, 42, 51, 54, 61, 73, 96, map 97, 248
    white .... 49, 73, 92, map 93, 248
spruce decline 39
spruce-fir forest 7–11
Spurge Family 166
spurge,
    cypress 166
    spotted 166
spurrey 128
squash 17
squashberry 194, 244
squawroot † 13
squirrel corn 42, 60
Stachys palustris † 13
Staff Tree Family 165
starflower 148
star thistle 202
stations, for species 72
steeplebush 159
Steironema ciliatum 147
Stellaria
    graminea 128
    media 18, 128
stick tights 202
stinking benjamin 231
stitchwort 128
stomates 40

strategies of plants 245
strawberry 152
streams and pH 60
Streptopus
    amplexifolius 231
    roseus 231
stunting of trees 40
subtle pioneers 61–64
sumac,
    poison † 12
    staghorn 173, 249
summer solstice 26
sundew .... photo 45, 54, 132, 242
    round-leaved 132
    spatulate-leaved 132
sundrops 163
sunlight 41, 42, 44
sunrise 26
Susquehanna River
    Lowland 12
swamp,
    candles 148
    pink 233
    rose 157
swamps 11
sweet
    cicely 60, 178
    clover, ... 159
        white 159
        yellow 160
    fern 16, 115
    gale 16, 116, 249
    jarvil 178
sycamore † 13
Symplocarpus foetidus † 13
synonymy, of scientific names 71

tacamahac 136
tall meadow rue 60, 110
tally of species, genera, families;
    native and naturalized 67–70
tamarack .... 90, map 91
Tanacetum vulgare 210
tansy 210

tape grass 211
Taraxacum officinale 19, 210
Taxaceae 88
Taxus canadensis 88
teaberry 143
tearthumb 130
temperature 22, 23, 30, 33,
 34, 35, 43
Teucrium canadense † 14
Thalictrum
 polygamum 110
 pubescens 60, 110
Thelypteris
 noveboracensis 87
 palustris 87
thirty-mile radius 3
thistle,
 bull 203
 sow 210
 star 202
 swamp 202
thornapple 152
threatened species 10,
 244– 246, 277
Thuja occidentalis 104
Thymelaeaceae 161
Thymus
 pulegoides 183
 serpyllum 183
Tiarella cordifolia 60, 150
Tilia americana 131
Tiliaceae 131
till, glacial 50, 61–64
tolerance to shade 61, 64
toothwort 140
touch-me-not, ... 16, 54
 pale 176
 spotted 60, 176
toxicity 73
Toxicodendron radicans 174,
 map 175
Tragopogon pratensis 210
trailing arbutus 41, 42, 142,
 242, 245

Triadenum virgincium 131
Trichostema dichotomum † 14
Trientalis
 americana 148
 borealis 148
Trifolium 19
 arvense 160
 aureum 160
 agrarium 160
 pratense 160
 repens 160
Trillium 242, 245
 erectum 232
 grandiflorum † 14
 undulatum 232
trillium 242, 245
 painted 232
 purple 232, 249
 red 232
 stinking benjamin 232
 wake robin 232
 white † 14
Tsuga canadensis 102
tule 220
tulip tree † 7, 12
tundra 7–10
turtlehead 184
Tussilago farfara 210
twinflower 42, 191
twisted stalk,
 large-leaved 231
 rosy 231
Typha
 angustifolia † 14
 latifolia 228
Typhaceae 228

Ulmaceae 114
Ulmus
 americana 114
 rubra † 14
 thomasii 14
Umbelliferae 178
Urtica dioica 115

Urticaceae 115
Utricularia
   cornuta 187
   gibba 187
   intermedia 187
   vulgaris 187
Uvularia
   perfoliata † 14
   sessilifolia 60, 232

Vaccinium
   angustifolium ... photo 66, 144
   boreale † 10, 244
   canadense 145
   macrocarpon 144
   myrtilloides 145
   oxycoccus 145
   pallidum † 13
   pensylvanicum 144
   uliginosum 10, 145, 244
   vacillars 13
Valerian Family 197
Valeriana officinalis 197
Valerianaceae 197
valleys (see individual rivers and lowlands)
Vallisneria americana 211
vascular plants ix
vegetation mats, floating ... 15, 16
vegetative reproduction 249
Veratrum viride .... 233, photo 316
Verbascum thapsus 19, 185
Verbena hastata 181
Verbenaceae 181
Vermont 9, 12
vernal equinox 23
vernal flowering herbs 42, 43
Veronica
   americana 186
   officinalis 186
   scutellata 186
   serpyllifolia 186
vetch 18, 160
vervain 181

Viburnum
   acerifolium .... 15, 194, map 195
   alnifolium 194
   cassinoides .... 194, photo 254
   dentatum 196
   edule 194, 244
   lantanoides .... 194, photo 274
   lentago 15, 196
   opulus 197
   recognitum 196
   trilobum 197
Vicia cracca 19, 160
Vine Family 166, 277
Viola
   blanda 134
   canadensis 60, 133
   conspersa 133
   incognita 133
   macloskeyi 133
   papilionacea 134
   pubescens 134
   rotundifolia 134
   sororia 134
Violaceae 133
Violet,
   American dog 133
   Canada 60, 133
   common blue 134
   downy yellow 134
   false 42, 152
   round-leaved 134
   sweet white 133
   yellow dog-tooth 229
viper's bugloss 181
Virginia creeper 166
virgin's bower 108
Vitaceae 166
Vitis riparia 167

wake robin 231
walnut, white 116, 249
wapato 210
warmest day 22, 26

Daisy (*Leucanthemum vulgare*)
Plants on the Move. A European species marching over North American fields and colonizing them. © Jim Kraus 1992.

Washington State 8
water
    arum 213
    celery 211
    cress 141
    hemlock, bulblet-bearing . . . .
        73, 178
    horehound 182
    leaf 15, 181
    lily
        white 106
        yellow 106
    lobelia 188
    marigold 201
    milfoil 160
    parsnip 179
    pennywort 178
    plantain † 13
    shield 106
    smartweed 129
    starwort 183
    table 54
    weed 211
    willow 161
    naiad 212
Water Milfoil Family 160
Water Plantain Family 210
Waxmyrtle Family 116
weather stations 30, 34–39
weed orchid 19, 234
weeds 18–20
whistlewood 168
white buttons 214
White Mountains 9
white water crowfoot 110
white goosefoot 18, 126
whitlow grass 13
wild
    bean † 13
    calla 213
    carrot 18, 178
    celery 211
    ginger † 13
    iris 233

leek 228
lettuce 206
licorice 189
morning glory 180
mustard 141
oats 60, 232
radish 141
raisin 41, 53, 54, 194,
    photo 254
rye 13
thyme 183
Willow Family 43, 135
willow 16, 249
    balsam 139
    bearberry † 10, 244
    Bebb's 138
    bog 139
    crack 36, 139
    pussy 139
    shining 139
    silky 140
    water 161
    weeping † 138
willowherb,
    narrow-leaved 162
    northern 162
wind
    and floating vegetation mats
        15, 16
    and pollen 249
    exposure of plants 63
    speed 40
windflower 107
winterberry holly 54, 165
winter severity 43
winter solstice 20
Wintergreen Family 146
wintergreen 41, 42, 143
    one-flowered 146
    flowering 167
witchhobble 42, 53, 194, 249,
    photo 274
witch hazel 15, 74, 112,
    map 113

Witch Hazel Family	112
withe rod	194
wood betony	15, 16, 185
wood-sorrel	42, 174
Wood sorrel Family	174
Woodsia ilvensis	87
woundwort †	13
Xyris montana	214
Xyridaceae	214
yarrow	18, 197
yellow	
adder's tongue	229
dogtooth violet	229
eyed-grass	214
rattle	185
rocket	140
yew	73, 88, 245
Zanthoxylum americanum †	14
Zizia aurea †	14
Zosteraceae (see Potamogetonaceae)	

False hellebore (*Veratrum viride*) Parallel-veined leaves indicate a monocot. Monocots also have three petals, three sepals, three or six stamens, a three-carpelled ovary, a single embryonic leaf in the seed, and scattered vascular bundles in the stem. © Jim Kraus 1992.

FIELD NOTES

FIELD NOTES

FIELD NOTES

FIELD NOTES